Frank Vollertsen · Sybille Friedrich ·
Bernd Kuhfuß · Peter Maaß ·
Claus Thomy · Hans-Werner Zoch
Editors

Cold Micro Metal Forming

Research Report of the Collaborative Research Center "Micro Cold Forming" (SFB 747), Bremen, Germany

Final report of the DFG Collaborative Research Center 747

Editors
Frank Vollertsen
BIAS—Bremer Institut für angewandte Strahltechnik GmbH
Bremen, Germany

Sybille Friedrich
BIAS—Bremer Institut für angewandte Strahltechnik GmbH
Bremen, Germany

Bernd Kuhfuß
Fachbereich Produktionstechnik
Universität Bremen
Bremen, Germany

Peter Maaß
Zentrum für Technomathematik
Universität Bremen
Bremen, Germany

Claus Thomy
BIAS—Bremer Institut für angewandte Strahltechnik GmbH
Bremen, Germany

Hans-Werner Zoch
Stiftung Institut für Werkstofftechnik IWT
Bremen, Germany

https://doi.org/10.1007/978-3-030-11280-6

Preface

This book gives an overview of the main research results, which are valuable to enable a stable and understood mass production with lot sizes of more than 1,000,000 metal parts by micro forming. The results have been gained in the years 2007 to 2018 within the framework of the German Collaborative Research Center (CRC) "Micro Cold Forming—Processes, Characterization, Optimization" (SFB 747 "Mikrokaltumformen—Prozesse, Charakterisierung, Optimierung") of the University of Bremen.

Chapter 1 of this book covers the motivation for the research on a reliable production as well as information about the structure and the partners within the collaboration and gives an overview of the main results, which were gained for mastering effects occurring in mass production. Besides processes and methods for improving the forming processes itself, also the mastering of the complete production line and the aspect of high flexibility in the design of processes and production systems are addressed. Within this overarching description, references for more detailed information in the following chapters are given.

Each of Chaps. 2–6 focuses on a field of competence covered by the CRC. Fields of competence are micro forming, process design, tooling, quality control and characterization as well as materials, especially designed for micro forming. Within this frame, all subprojects of the third funding period present their results in Sects 2.1–6.4 in more detail, and each subproject is giving answers to a special aspect, which allows mastering mass production of micro parts. The interdisciplinary cooperation between the researchers from production engineering, mathematics and physics, who are working in several institutes, is an excellent base for research in the demanding field of micro metal forming.

All the editors of this contributed book are the members of the executive board of the CRC. Head of the board and of the CRC is Frank Vollertsen. All the authors, who contributed to this book, work in the relevant fields of the CRC. They are directors or staff members of the collaborating research institutes. This book extends the knowledge presented in the more fundamental book "Micro Metal Forming" (Ed. F. Vollertsen, published by Springer).

A prerequisite for successful cooperation is the funding of manpower and equipment, which was granted by Deutsche Forschungsgemeinschaft (DFG) and the University of Bremen. We gratefully acknowledge this support. The editors appreciate the powerful collaborations of the researchers within the CRC, which have been the key for the successful work.

On behalf of the editors

Bremen, Germany
November 2018

Frank Vollertsen
Head of the CRC (SFB 747 Mikrokaltumformen)

Acknowledgements The editors and authors of this book like to thank the Deutsche Forschungsgemeinschaft (DFG), German Research Foundation, for the financial support of the SFB 747 "Mikrokaltumformen—Prozesse, Charakerisierung, Optimierung" (Collaborative Research Center "Micro Cold Forming—Processes, Characterization, Optimization"). We also like to thank our members and project partners of the industrial working group as well as our international research partners, for their successful cooperation.

Contents

Contributors

Mostafa Agour BIAS—Bremer Institut für angewandte Strahltechnik GmbH, Bremen, Germany;
Faculty of Science, Department of Physics, Aswan University, Aswan, Egypt

Ralf B. Bergmann BIAS—Bremer Institut für angewandte Strahltechnik GmbH, University of Bremen, Bremen, Germany;
Physics and Electrical Engineering and MAPEX Center for Materials and Processes, University of Bremen, Bremen, Germany

Florian Böhmermann Laboratory for Precision Machining—LFM, Leibniz Institute for Materials Engineering—IWT, Bremen, Germany;
Leibniz-Institut für Werkstofforientierte Technologien—IWT, University of Bremen, Bremen, Germany

Ekkard Brinksmeier LFM Laboratory for Precision Machining, Leibniz-Institut fur Werkstofforientierte Technologien—IWT, Bremen, Germany

Brigitte Clausen Leibniz-Institut für Werkstofforientierte Technologien, Bremen, Germany

Chengsong Cui Leibniz Institute for Materials Engineering—IWT, University of Bremen, Bremen, Germany

Frederik Elsner-Dörge Labor für Mikrozerspanung—LFM, Leibniz Institute for Materials Engineering—IWT, Bremen, Germany

Claas Falldorf BIAS—Bremer Institut für angewandte Strahltechnik GmbH, Bremen, Germany

Andreas Fischer BIMAQ—Bremen Institute for Metrology, Automation and Quality Science, University of Bremen, Bremen, Germany

Michael Freitag Faculty of Production Engineering, University of Bremen, Bremen, Germany;
BIBA—Bremer Institut für Produktion und Logistik GmbH, Bremen, Germany

Sybille Friedrich BIAS—Bremer Institut für angewandte Strahltechnik GmbH, University of Bremen, Bremen, Germany

Gert Goch BIMAQ—Bremen Institute for Metrology, Automation and Quality Science, University of Bremen, Bremen, Germany;
The University of North Carolina at Charlotte, Charlotte, USA

Phil Gralla Center for Industrial Mathematics, University of Bremen, Bremen, Germany

Lukas Heinrich BIAS—Bremer Institut für angewandte Strahltechnik GmbH, Bremen, Germany

Marius Herrmann Bremen Institute for Mechanical Engineering (bime), Bremen, Germany

Svetlana Ishkina Bremen Institute for Mechanical Engineering (bime), Bremen, Germany

Mischa Jahn Zentrum für Technomathematik, Bremen, Germany

Bernd Köhler Leibniz-Institut für Werkstofforientierte Technologien, Bremen, Germany

Julien Kovac Leibniz Institute for Materials Engineering—IWT, University of Bremen, Bremen, Germany

Bernd Kuhfuss Bremen Institute for Mechanical Engineering (bime), University of Bremen, Bremen, Germany

Michael Lütjen BIBA—Bremer Institut für Produktion und Logistik GmbH, University of Bremen, Bremen, Germany

Peter Maaß Zentrum für Technomathematik, Bremen, Germany;
Center for Industrial Mathematics, University of Bremen, Bremen, Germany

Andreas Mehner Leibniz Institute for Materials Engineering—IWT, University of Bremen, Bremen, Germany

Salar Mehrafsun BIAS—Bremer Institut für angewandte Strahltechnik GmbH, Bremen, Germany

Axel Meier Labor für Mikrozerspanung—LFM, Leibniz Institute for Materials Engineering—IWT, Bremen, Germany

Hamza Messaoudi BIAS—Bremer Institut für angewandte Strahltechnik GmbH, Bremen, Germany

Merlin Mikulewitsch BIMAQ—Bremen Institute for Metrology, Automation and Quality Science, University of Bremen, Bremen, Germany

Eric Moumi Bremen Institute for Mechanical Engineering (bime), Bremen, Germany

Ann-Kathrin Onken Bremen Institute for Mechanical Engineering (bime), University of Bremen, Bremen, Germany

Iwona Piotrowska-Kurczewski Zentrum für Technomathematik, Bremen, Germany;
Center for Industrial Mathematics, University of Bremen, Bremen, Germany

Lewin Rathmann BIAS—Bremer Institut für angewandte Strahltechnik GmbH, Bremen, Germany

Oltmann Riemer Leibniz-Institut für Werkstofforientierte Technologien—IWT, University of Bremen, Bremen, Germany;
Laboratory for Precision Machining—LFM, Leibniz Institute for Materials Engineering—IWT, Bremen, Germany

Daniel Rippel BIBA—Bremer Institut für Produktion und Logistik GmbH, University of Bremen, Bremen, Germany

Christian Robert LFM Laboratory for Precision Machining, Leibniz-Institut für Werkstofforientierte Technologien—IWT, University of Bremen, Bremen, Germany

Christine Schattmann BIAS—Bremer Institut für angewandte Strahltechnik GmbH, Bremen, Germany

Christian Schenck Bremen Institute for Mechanical Engineering (bime), University of Bremen, Bremen, Germany

Alfred Schmidt Zentrum für Technomathematik, Bremen, Germany

Alwin Schulz Leibniz Institute for Materials Engineering—IWT, University of Bremen, Bremen, Germany

Joseph Seven BIAS—Bremer Institut für angewandte Strahltechnik GmbH, Bremen, Germany

Aleksandar Simic BIAS—Bremer Institut für angewandte Strahltechnik GmbH, Bremen, Germany

Benjamin Staar BIBA—Bremer Institut für Produktion und Logistik GmbH, Bremen, Germany

Claus Thomy BIAS—Bremer Institut für Angewandte Strahltechnik GmbH, University of Bremen, Bremen, Germany

Anastasiya Toenjes Leibniz Institute for Materials Engineering—IWT, University of Bremen, Bremen, Germany

Kirsten Tracht Bremen Institute for Mechanical Engineering (bime), University of Bremen, Bremen, Germany

Jost Vehmeyer Center for Industrial Mathematics, University of Bremen, Bremen, Germany

Frank Vollertsen Faculty of Production Engineering-Production Engineering GmbH, University of Bremen, Bremen, Germany;
BIAS—Bremer Institut für angewandte Strahltechnik GmbH, University of Bremen, Bremen, Germany

Axel von Freyberg BIMAQ—Bremen Institute for Metrology, Automation and Quality Science, University of Bremen, Bremen, Germany

Axel von Hehl Leibniz Institute for Materials Engineering—IWT, University of Bremen, Bremen, Germany

Philipp Wilhelmi Bremen Institute for Mechanical Engineering (bime), University of Bremen, Bremen, Germany

Melanie Willert Leibniz-Institut für Werkstofforientierte Technologien—IWT, University of Bremen, Bremen, Germany

Igor Zahn Bremer Goldschlaegerei (BEGO), Bremen, Germany

Hans-Werner Zoch Leibniz Institute for Materials Engineering—IWT, University of Bremen, Bremen, Germany

Chapter 1
Introduction to Collaborative Research Center Micro Cold Forming (SFB 747)

Frank Vollertsen, Sybille Friedrich, Claus Thomy, Ann-Kathrin Onken, Kirsten Tracht, Florian Böhmermann, Oltmann Riemer, Andreas Fischer and Ralf B. Bergmann

1.1 Motivation

Frank Vollertsen*, Sybille Friedrich and Claus Thomy

Micro systems technology is one of the key enabling technologies of the 21st century [Hes03], with increasing relevance due to a general trend towards miniaturisation in many industries. The main boosters for this trend are currently the consumer and communication electronics market and—to a lesser extent—medical technology (especially microfluidic devices, which had a market volume of approx. $2.5B in 2017 [Cle17]). As an example, for the companies organized in the industry association

F. Vollertsen (✉) · S. Friedrich · C. Thomy · R. B. Bergmann (✉)
BIAS—Bremer Institut für angewandte Strahltechnik GmbH, University of Bremen, Bremen, Germany
e-mail: info-mbs@bias.de

R. B. Bergmann
e-mail: Bergmann@bias.de

A.-K. Onken · K. Tracht (✉)
University of Bremen, Bremen, Germany
e-mail: tracht@bime.de

F. Böhmermann (✉) · O. Riemer (✉)
Laboratory for Precision Machining—LFM, Leibniz Institute for Materials Engineering—IWT, Bremen, Germany
e-mail: boehmermann@iwt.uni-bremen.de

O. Riemer
e-mail: riemer@iwt.uni-bremen.de

A. Fischer (✉)
BIMAQ—Bremen Institute for Metrology, Automation and Quality Science, University of Bremen, Bremen, Germany
e-mail: andreas.fischer@bimaq.de

F. Vollertsen et al. (eds.), *Cold Micro Metal Forming*, Lecture Notes in Production Engineering, https://doi.org/10.1007/978-3-030-11280-6_1

IVAM Microtechnology Network, medical technology is by far the most important market [NN14]. Nevertheless, in the context of electromobility, autonomous driving, and industry 4.0, there should also be a significant increase in the demand for existing MEMS (micro electromechanical systems) including connectors, as well as a need for further improvement, miniaturization and functional integration.

A typical example of the benefits and challenges of miniaturization is the ABS (anti-blocking system) in modern cars. Whilst the weight could be decreased to approx. 10% of the weight of the first system, the part complexity has significantly increased. This is indicated by the decrease in the number of parts by approx. 90% in current systems, compared to the first versions [Nos14], even though additional functions are integrated. The trend towards an increase in miniaturization can also be illustrated using the example of HF (high frequency) connectors, where nowadays the minimum pin diameters commercially available off the shelf are in the range of 0.7 mm, with allowable tolerances for functional features of several µm.

As in most of the applications discussed above, significant quantities of parts ranging from several thousands to literally billions (e.g. for resistor end caps) have to be produced, micro cold forming is among the dominant production processes. Examples of components (containing) micro parts produced by cold forming are (listed by their application areas):

Medical technology/chemical technology: Hearing aid devices, cardiac pacemakers, micro pumps and micro pump couplings, microfluidic reactor devices;
Automotive technology: ABS and other advanced assistance systems based on MEMS, contacts and connectors, fuel pumps, injection nozzles;
Electronics: Battery caps, displays, diodes, electrodes, connectors, resistors, nozzles, contactors;
General industry: sensors (e.g. for pressure), hydraulic and pneumatic connectors, micro pumps, micro valves;
Consumer electronics: smartphone speakers, electric blades, compact cameras, microphones.

As a first conclusion, we find three main trends: an increase in miniaturization, an increase in functional integration, and an increase in total batch sizes. Moreover, common to many of these applications is the need for zero-defect quality. This is often due not only to cost considerations, but also to the safety criticality of many of the components (ABS, medical devices).

Consequently, and as most micro forming parts are produced under extreme cost pressure, the complete process chain has to be optimized to enable further cost-efficient miniaturization. This means that not only aspects relating to the micro cold forming process as such (e.g. non-systematic scatter in material properties), but also preceding and succeeding process steps (e.g. heat treatment) as well as materials handling have to be considered and optimized along the process chain. Moreover, in order to increase process stability and part quality, novel tool materials and tool production processes have to be investigated to minimize tolerances and improve wear behavior. Finally, and this is among the most urgent industrial needs in view of zero-defect requirements, methods and systems for 100% inline measurement and inspection at production rates of from 60 to 300/min and more are required (depending, of course, on the part complexity). This is especially true for systems allowing the optical inspection and measurement of complex features, which are often inside longer cavities.

Research on micro metal forming in Germany was developed by Engel and Geiger, starting in the 1990s in Erlangen. Discussion of the scientific advances, achieved also in other countries like Japan and the USA, was held (not limited to but also) in The International Academy for Production Engineering (CIRP), documenting the milestones in numerous papers and 2 keynote papers. These keynote papers 'Microforming', [Gei01] and 'Size effects in manufacturing of metallic components' [Vol09] are key documents about the development of micro metal forming. The relevance of size effects is due to the fact that these effects are the reason why knowledge from (macro) metal forming cannot be transferred easily to the micro range. These effects have been the topic of a Priority Program (SPP 1138) funded by the Deutsche Forschungsgemeinschaft (DFG) in the period from 2002 to 2008.

Three categories of size effects are specified, taking their names from the feature that is kept constant: density, shape and structure. Fundamental knowledge concerning size effects is documented in the book "Micro Metal Forming" [Vol13].

Specific features are:

1. The size or at least one of the dimensions of the produced parts is comparable with the grain size of the material used, resulting in hard to control material behavior.
2. The very small volume of the parts changes the failure behavior due to different probabilities of the occurrence of a defect in a particular workpiece, if homogeneous defects with low density exist in the raw material.
3. The very low weight (typically between 100 μg and 10 mg) of the (raw) parts makes handling difficult due to, for example, adhesion effects. Therefore parts should integrate multiple functions to reduce the number of components in an assembly and to minimize the number of handling and joining operations. On the other hand, the small weight might allow the use of more expensive materials or enable new processes.
4. Quality assurance becomes more difficult compared to macro parts, as many methods usually employed cannot be used for the measurement of micro part dimensions. Also the (scaled down) tolerances interfere greatly with the precision of the metrology, making the use of methods like statistical process control (SPC) impossible.

Research in Collaborative Research Centers (CRC) started in 1998 by several CRC, each addressing a special aspect of micro technology. The aim of SFB 440 was the Assembly of Hybrid Microsystems, SFB 499 focused on the development, production and quality assurance of molded micro components of metallic and ceramic materials and SFB 516 addressed the design and production of active microsystems. From 2010–16 a research group "Small Machine Tools for Small Workpieces" made new approaches for machine tools. In the period 2007–2018 the Collaborative Research Center "Micro Cold Forming" (SFB 747) with about 60 scientists worked on topics relevant for the further development of mass micro metal forming.

1.2 Aim of the SFB 747

Frank Vollertsen* and Sybille Friedrich

The central concern of the Collaborative Research Center (CRC) "Micro Cold Forming" is the provision of processes and methods for the production of metal micro parts by forming, whereby all essential aspects for the forming process, from material development through component testing to process design, are included. With the resulting knowledge of the mechanisms and correlations, a purposeful process design is made possible for the process-reliable production of metallic components with a size of less than 1 mm and the necessary tools. Batch sizes over 1 million parts are in focus. As a basis for this, the CRC serves industrial requirements with a manufacturing frequency of typically 300 parts/minute. By definition [Gei01], micro forming deals with parts having dimensions less than 1 mm in at least two directions. For further limitation of the research program, sheet thicknesses of 10–200 μm and wire diameters of 200–1,000 μm are specified.

The Collaborative Research Center focuses on micro components that are produced in unit quantities or batch sizes over 1 million parts. The increase in the number of variants results in the need for the reconfigurability of the production lines with the aim of making the production of micro parts more flexible. There is demand for individual processes that are easy to handle and flexible in use and thus support a modular production. Here, the planning of the processes, the definition of the interfaces and the monitoring strategies are of particular importance in order to be able to quickly realize economically the start-up of the production of different micro parts. The materials used in the central process chains are steel (1.4301), aluminum (99.5) and copper (E-Cu58), as well as their alloys. The microstructure in terms of homogeneity, grain size and isotropy plays an important role in the formability. These factors are of particular importance for the alloys of the metals listed above, since the production-related variables determine them. In addition, other materials and combinations of materials are used, which can exploit the opportunities offered with regard to component and workpiece design. The goal is the production of the already mentioned micro technical components. Microsystem technology (MST) and micro electromechanical systems (MEMS) are explicitly not the subject of the CRC research.

1.3 Structure and Partners

Frank Vollertsen* **and Sybille Friedrich**

The research program gives the Collaborative Research Center a broad basis, from the development of materials through the processes and their optimization to the planning aspects of micro forming technology production. In order to visualize the internal collaboration, three perspectives on the structure of the CRC were defined, by means of which the CRC is presented as a whole (see Fig. 1.1). As a superordinate element, a demonstrator was realized, on which all subprojects describe their research progress in terms of the hardware, concept or virtual contribution. Research work was coordinated using two further structural elements—the project structure and the content relation.

The CRC offers a comprehensive overview of all aspects of micro forming technology for sheet metal and massive forming with regard to the safe production of micro components. Based on this objective, the structure of the Collaborative Research Center with the project areas processes, characterization and optimization results. Table 1.1 shows the project structure of the Collaborative Research Center.

A—Processes

Project area A of the CRC "Micro Cold Forming" deals with fundamental questions of the single processes. The forming processes themselves are examined, as well as

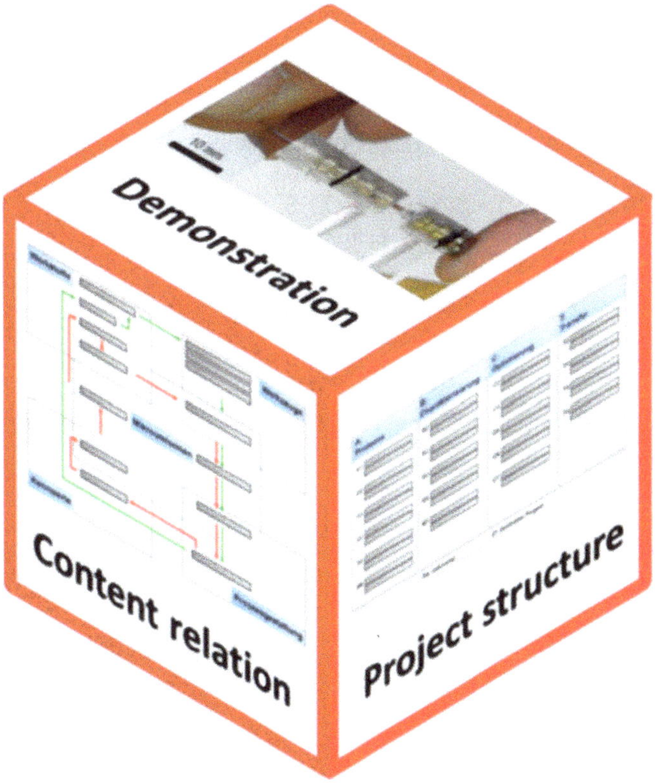

Fig. 1.1 Perspectives of the structure of the CRC allowing optimal internal interaction and collaboration

Table 1.1 Project Structure of the Collaborative Research Center (SFB 747). Line 1: DFG Short number and title, Line 2: Running Head(s) of the project, Line 3: Duration of the project (start and end), Line 4: Link for further details

A: Processes	B: Characterization	C: Optimization	T: Transfer
A1 PVD-sheets Zoch, Mehner 2007–2018 Section 6.3	B1 Deformation behavior Vollertsen 2007–2010 [Hu10]	C1 Tool Materials Partes 2007–2013 [Feu13]	T2 Refining Kuhfuß 2015–2016 Section 2.4
A2 Heat treatment Zoch 2007–2018 Section 6.4	B2 Distribution based simulation Brannath, Schmidt, Hunkel 2007–2014 [Lüt14]	C2 Surface optimization Riemer, Maaß 2007–2018 Section 4.5	T3 Micro cavity Bergmann, Lütjen 2015–2016 Section 5.3
A3 Material accumulation Vollertsen, Schmidt 2007–2018 Section 2.2	B3 Tool duration Vollertsen, Bergmann 2007–2018 Section 4.2	C4 Simultaneous Engineering Lütjen 2007–2018 Section 3.3	T4 Micro milled dental products Riemer, Maaß 2015–2017/1. Hj. Section 4.6
A4 Material displacement Kuhfuß 2007–2018 Section 2.3	B4 Material strength Zoch, Clausen 2007–2018 Section 5.5	C5 Linked parts Tracht, Kuhfuß 2011–2018 Section 3.2	T5 TEC-Pro Vollertsen 2015–2016 Section 4.4
A5 Laser contour Goch, Vollertsen 2007–2018 Section 4.3	B5 Safe processes Bergmann, Goch, Lütjen 2007–2018 Section 5.2	C6 Spray-graded tool steels Schulz 2011–2017 Section 6.2	
A6 Friction polishing Brinksmeier 2007–2017 Section 4.6	B7 Process stability Vollertsen 2011–2018 Section 2.5		

further process steps before and after the forming step. At the beginning of the interacting processes is the production of semi-finished products for the micro forming production.

B—Characterization

With regard to new materials, tools and processes for micro forming, an exact knowledge concerning the material behavior of both the workpieces and the tools and their interaction is essential. This is subject of the project area B characterization of this Collaborative Research Center.

C—Optimization

In order to meet the precision and speed requirements of a reliable and cost-effective production process, this project area uses the results of process development and the characterization of basic material properties and product parameters to optimize the key production steps.

T—Transfer

Research enhancing the basic research results of the CRC is examined in transfer projects, each realized in cooperation with an industry partner.

The results of the complete funding period of all projects running in 2016 or later are described in Sects. 2.1–6.4. In addition, a special approach for in situ geometry measurement in fluids, using confocal fluorescence microscopy, is presented in Sect. 5.4.

Internal Cooperation

Collaborating Institutes

Eight institutes, located on the campus of the University of Bremen, collaborate with their special knowledge to achieve the collective aim. They are listed in alphabetical order together with the most important research areas covered within the CRC. The names of the actual responsible heads of the projects are given in brackets:

BIAS—Bremer Institut für angewandte Strahltechnik: Laser material processing, sheet and bulk metal micro forming (Prof. Dr.-Ing. Frank Vollertsen); optical metrology (Prof. Dr. rer. nat. Ralf Bergmann).

BIBA—Bremer Institut für Produktion und Logistik: Logistics and simultaneous engineering (Dr.-Ing. Michael Lütjen).

BIMAQ—Bremer Institut für Messtechnik, Automatisierung und Qualitätswissenschaft: Process control including metrology, quality assurance (Prof. Dr.-Ing. Andreas Fischer, Prof. Dr.-Ing. Gert Goch).

bime—Bremer Institut für Strukturmechanik und Produktionsanlagen: Bulk metal forming including machine development (Prof. Dr.-Ing. Bernd Kuhfuß); process chain layout and automatization (Prof. Dr.-Ing. Kirsten Tracht).

IfS—Institut für Statistik: Monte-Carlo simulation and statistics (Prof. Dr. Mag. rer. nat. Werner Brannath).

Leibniz—IWT Leibniz-Institut für Werkstofforientierte Technologien: Physical vapor deposition, heat treatment and mechanical testing (Prof. Dr.-Ing. Hans-Werner Zoch, Prof. Dr.-Ing. Brigitte Clausen, Dr.-Ing. Andreas Mehner, Dr.-Ing. Alwin Schulz, Dr.-Ing. Martin Hunkel).

Leibniz—IWT (LFM) Leibniz-Institut für Werkstofforientierte Technologien, Laboratory for Precision Machining: Cutting, machining and polishing (Prof. Dr.-Ing. Prof. h.c. Dr.-Ing. E.h. Ekkard Brinksmeier, Dr.-Ing. Oltmann Riemer).

ZeTeM—Zentrum für Technomathematik: Industrial mathematics (Prof. Dr. Dr. h.c. Peter Maaß), simulation systems (Prof. Dr. Alfred Schmidt).

Fields of Competence

The global approach, which spans a bridge from the base materials to the finished component in terms of micro forming technology, is a specialty of this Collaborative Research Center. Two exemplary cycles are shown in Fig. 1.2. Within the fields of competence, working meetings focused on material aspects, on process design including the demonstrator and on the aspects of component characterization as well as on tool-specific issues.

In Chaps. 2–6, more can be learned about the knowledge gained in the fields of competence micro forming, tooling, quality control and characterization, as well as concerning new materials for thin sheets and adapted tools, which were especially developed to fulfill the needs of micro forming aspects.

Communication with the International Scientific Community

An essential device is the discussion of research results at relevant conferences and congresses, as well as publications in international journals and books. In addition, there are contacts with renowned scientific networks, such as acatech, AGU, AHMT, AWT, CIRP, euspen, ICFG, IDDRG, I^{2}FG, GCFG, GQW, MHI, SME, WAW, WGP und WLT by the principal investigators and participating institute directors.

The book “Micro Metal Forming”, written by the scientists of SFB 747 and edited by F. Vollertsen, was published by Springer in the book series “Lecture Notes in Production Engineering” in 2013 [Vol13].

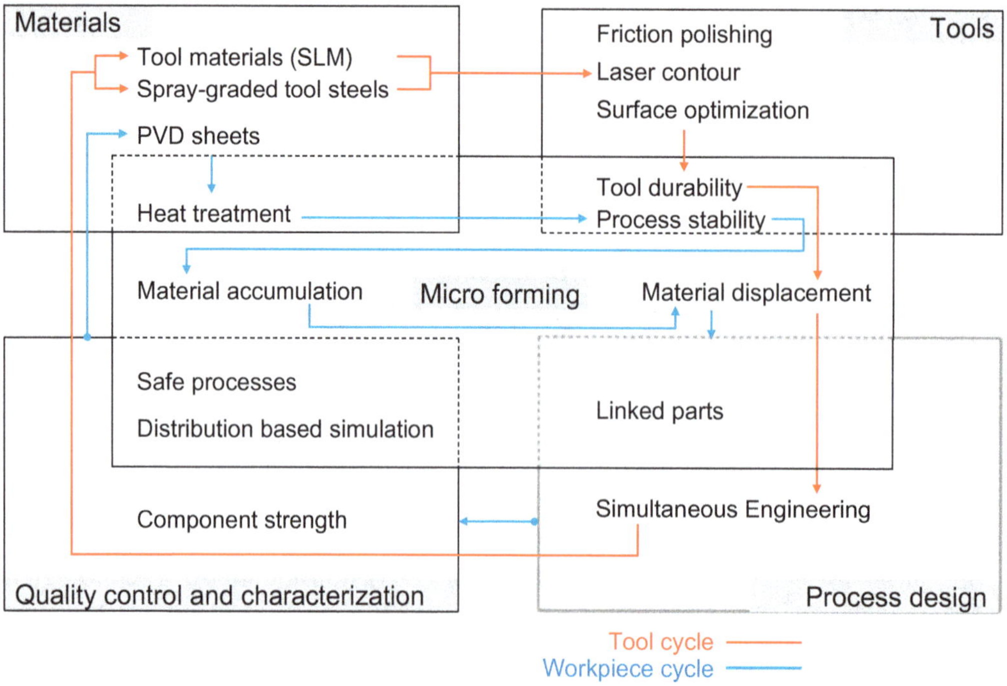

Fig. 1.2 Exemplary display of one workpiece and one tool cycle within the fields of competence of CRC (for space reasons without transfer projects)

Further, the research of the CRC has been sustainably supported by numerous impulses gained by close collaboration with international partners in relevant research fields, such as:

Micro forming (Prof. Ming Yang, Tokyo Metropolitan University, Japan)

Laser material processing and micro milling (Prof. Frank E. Pfefferkorn, University of Wisconsin-Madison, USA)

Microstructuring of polycrystalline diamond (Dr. Yiqing Chen and Prof. Liagchi Zhang of the University of New South Wales in Australia)

Wear (Prof. Hans Norgaard Hansen and Dr. Guido Tosello of the Technical University of Denmark (DTU))

Mathematical methods (Prof. Pham Quy Muoi, The University of Danang, Vietnam; Dr. Jonathan Montalvo-Urquizo, CIMAT Centro de Investigación en Matemáticas, Guanajuato, Mexico; Prof. Eberhard Bänsch, University Erlangen).

The CRC (co)organized the following scientific conferences:

The national Colloquium Micro Production has taken place biannually since 2003. The joint organization of the collaborative research groups of micro technology and their follow-up projects have the overall objective of micro technical questions. The conference was hosted in Bremen in 2009 and 2017. The event was also used intensively by the industrial partners for the exchange of information. Since 2014 the CRC has established a session "micro forming technologies" at the International Conference on Nanomanufacturing (nanoMan), which is biannually organized by the International Society for Nanomanufacturing. The International Conference on New Forming Technology (ICNFT) is an inspiring forum for researchers and professional practitioners to discuss aspects of leading-edge novel forming technologies. The steering committee of the conference comprises members from BIAS, Germany, Harbin Institute of Technology, China, the University of Birmingham, UK and the University of Strathclyde, UK. The CIRP and DFG sponsored 5th ICNFT 2018 took place in Bremen, Germany. In 2018, special sessions focusing on micro cold forming were included. Within this scope, the CRC held its final Colloquium in exchange with international specialists.

Additionally, the CRC presented itself at fair trade stands, for example at the International Conference on Technology of Plasticity (ICTP) in 2011 and 2017. Around 700 international, mainly scientific, participants were informed about the collaborative focus and research results.

Cooperation with Industry

In addition to the scientific national and international networks, an industrial working group was established in the first funding period, with the aim of transferring the research results to industry and orienting the scientific research to current needs. Members of the industrial workgroup are:

BEGO Medical GmbH, Robert Bosch GmbH, Uni Bremen Campus GmbH, Felss GmbH, Harting Applied Technologies GmbH & Co. KG, Hella Fahrzeugkomponenten GmbH, Huber und Suhner AG, IFUTEC GmbH, Keyence GmbH, Philips Consumer Lifestyle B.V., SITEC Industrietechnologie GmbH, SLM

Fig. 1.3 A selection of national and international conferences and exhibition stands on trade fairs for research and industry, (co)organized by SFB 747

Solutions GmbH, Stüken GmbH & Co. KG, Tyco Electronics GmbH & Co. KG, Wafios Umformtechnik GmbH, Werth Tool MT GmbH, Weidmüller Interface GmbH & Co. KG.

The consortium has been regularly provided with information from the CRC and an annual industry colloquium was held to identify co-operation issues. From this consortium also the transfer themes emerged (Fig. 1.3).

Definition of Demands on Mastering Mass Production of Micro Parts

Research with the aim of providing new processes and methods especially adapted to the production of micro parts in high numbers took place in three funding periods (see Fig. 1.4).

From 2007 to 2010, the researchers developed new processes and methods that are especially designed for micro forming. The findings were improved in the second period (2011–14) to be more stable and to be used for the production of more complex products, while the processes were kept as easy as possible. In the last period, the knowledge was increased by upscaling the number of parts and the transfer to multistage processes.

The challenges arising from the goal of mastering the mass production of micro parts, which are to be solved with different weighting in the individual subprojects (see Table 1.1), consisted in the shortening of the process times, the increase in the forming speed, the modification of the tribology and the heat balance, as the higher clock rates result in an effectively higher power dissipation in the process zones. This is followed by the demand for faster transport, faster measurement and control processes, mastery of the tribology, thermal balance of the processes, and the dispersion of material properties (see Fig. 1.5). The main results of the

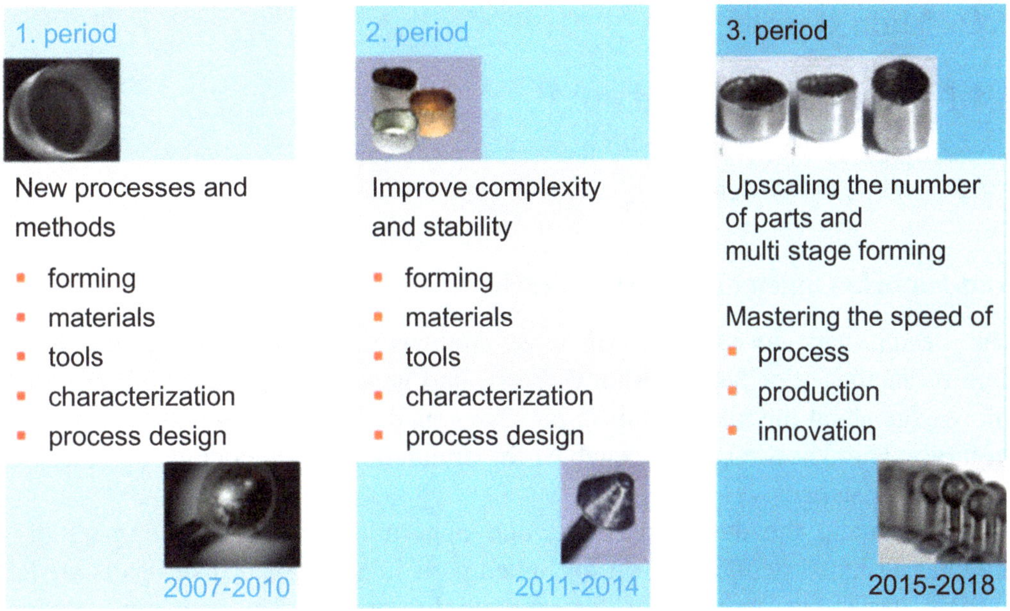

Fig. 1.4 Funding periods and their main research focus

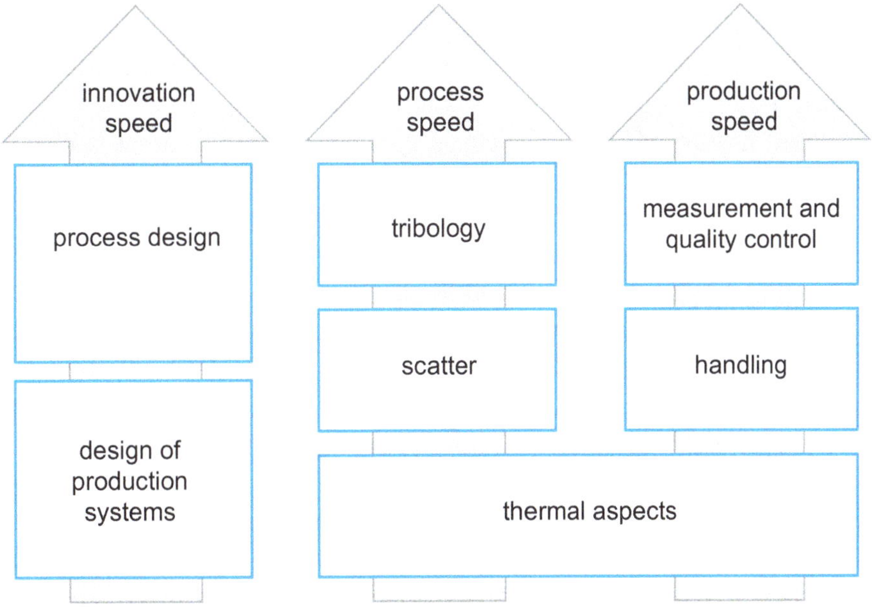

Fig. 1.5 Mastered aspects for a high quality and economic mass production of more than one million batches/micro parts

Collaboration Research Center "Micro Cold Forming" concerning mastering these aspects of innovation speed, process speed and production speed are summarized in Sect. 1.4.

1.4 Main Results

1.4.1 Innovation Speed

1.4.1.1 Process Design

Ann-Kathrin Onken and Kirsten Tracht*

The design and development of new products and production processes is time-consuming. For faster design of parts, and hence a higher innovation speed, knowledge about the manufacturing processes, parts and their interdependencies is required. This knowledge also enables the simulation and modeling of processes and process steps.

To speed up the design of parts, one opportunity is the development of a modular design system, as shown for linked parts in Sect. 3.2. Linked parts are left in the material they are made of, for example, foil or wire. The remaining material of the interconnection can be used for the implementation of functions that assist in manufacturing processes. Besides the orientation of the parts, the interconnections are appended with functional elements, which are relevant for the handling, such as for the positioning of parts as concerned in the modular design system.

Basic knowledge about the process parameters and their influence on the workpiece are major results from the CRC 747 and build an important basis for fast adaptation to other products, parts or manufacturing processes.

Laser-based free-form heading (Sect. 2.2) is an alternative process chain in the micro range for the production of conventional upsetting and metal forming. It can be applied on rod and on thin metal sheet. With this method, which only works in the micro range, thermal upsetting relations to 500 are possible. The aim of three hundred line-linked parts per minute are realized, which can be formed in a further step. A validated 3D simulation enables the parameter identification and process analysis for free form heading.

Rotary swaging, which is well implemented in the macro range, for example, for producing parts for the automotive industry, has been adapted to the micro range. The two main variations of the process, infeed and plunge rotary swaging, were investigated in the CRC (Sect. 2.3). In order to increase the productivity in the micro range, the process speed can be increased. However, this could lead to failures, particularly due to inappropriate material flow against the feed direction. The findings show that, with additional measures, both the radial and the axial material flow can be controlled, and in both cases the productivity in micro rotary swaging can be increased by up to a factor of four. Furthermore, adjusting the axial positioning of the workpiece during plunge rotary swaging allows a higher diversity to be achieved for the geometry of the swaged part.

On the basis of rotary swaging, the refining (Sect. 2.4) deals with the conditioning of wire semi-finished products for the subsequent forming processes. The results show how the formability and the geometrical diversity of semi-finished parts are prepared by process modifications in rotary swaging tools and kinematics. These results can be adapted to other forming processes like deep drawing and extrusion.

In the case of single and multistep deep drawing, the prediction of the influence of the process parameters and tools on the geometrical properties of parts is also an aim of the investigations about the process stability (Sect. 2.5). The objective is the definition of the interaction of the tool geometry and the stability of a multi-stage micro-drawing process. The results show how to determine allowable and achievable manufacturing tolerances, and wear of the tools' geometry to ensure a stable manufacturing process while producing very large quantities.

Within the competence field of tools, especially the progress in laser chemical machining (Sect. 4.3) enables precise quality control during the manufacturing of metallic micro forming tools. The combination of thermal modeling to define the temperature regime for a disturbance-free removal and closed-loop quality control to compensate the deviations of quality features opens up the possibility of dimensionally and geometrically flexible micro machining. In addition, a developed machining strategy consisting of roughing and finishing steps is used to improve both the microstructural and the topographical tool quality.

By developing the method of non-linear inverse problems, it was possible to allow optimized process parameters for the manufacturing process of tribologically optimized micro forming dies (Sect. 4.5).

The impact of the process parameters on the material properties is presented by the results of two investigated processes. The results of the PVD sheet production in Sect. 6.3 show how magnetron sputtering is established for the generation of thin foils with high strength and good forming behavior. The results clarify how the thickness of the coating is adjustable by changing the transport speed and the size of the exposure area.

In the micro range, heat treatment in a drop-down tube furnace, where the parts are heat-treated as they are falling, is possible because of the size effects (Sect. 6.4). The investigations make it possible to retain the accuracy of the shape as well as the specific setting of properties. Therefore, parameters like the falling time, temperature areas within the furnace, and quenching medium are considered.

Several manufacturing processes for forming tools with tailored material properties that lengthen the tool lifetime were established. One option is tool manufacturing by selective laser melting [Flo16]. Co-spray forming with selective heat treatment afterwards allows adapted tool steels and properties in the specific regions of the tool, with a gradual material transition in between to ensure good bonding (Sect. 6.2).

1.4.1.2 Design of Production Systems

Ann-Kathrin Onken and Kirsten Tracht*

The planning and configuration of process chains constitutes a major factor of success for the industrial production of metallic micromechanical components, due to the occurrence of size effects, the inherently small tolerances, and the small geometrical dimensions of the workpieces. For this reason, methodologies and recommendations for the design of production systems for micro mass production are required. Within this context, CRC 747 focuses on two different aspects with two different perspectives.

The first is the widening of the tolerance field (Sect. 3.2). This methodology follows a super-ordinated point of view on the process chain. Based on measured geometrical dimensions of linked parts, trends occurring that are, for example, caused by wear are used for matching, and design adjustments. Therefore, a reduction of surplus parts as well as the increased durability of tools is achieved. The production of linked parts is crucial for this methodology. Due to the retention of the production order, the parts are measured to identify trends concerning geometrical deviations, which are used for building homogeneous trend sections. These sections are used for matching linked parts for building assemblies. Hence a softening of the tolerances, similar to selective assembly, is achieved. Selective assembly is usually applied to scenarios where processes are hard to control. In micro mass production, this specific matching enables a widening of the tolerance field. The feedback loop to the design stage enables further increases in the outcome by a stepwise adjustment of the nominal value and new combinations of parts. With the adjusted nominal value, the identified sections, and matchings, further batches are produced and assembled.

The second perspective focuses on detailing the interdependencies between and within processes by considering the effects of single parameters on the production result. Small variations in single parameters can have significant influences along the process chain and finally interfere with the compliance with tolerances. For this reason, the μ-Process Planning and Analysis methodology (Sect. 3.3) covers all phases from the process and material flow planning to the configuration and evaluation of the processes and process chain models. The process configuration relies on so-called cause–effect networks, subsuming the relevant logistic and technical parameters of the corresponding processes and describing their relationships to each other. By using these networks, μ-Process Planning and Analysis enables a fast evaluation of different process configurations (e.g. the use of different materials or different production velocities) already during the planning stages. The networks enable an assessment of the impacts of different choices on the follow-up processes or the production system in general. Thereto, μ-Process Planning and Analysis can directly reflect changes to the configuration in the integrated material flow simulation and evaluate these configurations, e.g., regarding work-in-progress levels, lead times or the products' estimated qualities.

1.4.2 Micro Mass Forming

1.4.2.1 Tribology

Florian Böhmermann* and Oltmann Riemer*

Tribology is one major concern in the development of micro cold forming processes. Size effects lead to changes in the predominant friction mechanism towards adhesion, as well as a reduction of the effectiveness of lubrication. This is associated with an increase of friction and has an impact on the wear behavior. Furthermore, micro forming dies and work pieces show anisotropic behavior in wear or non-predictable malfunctions, as the geometrical features and micro structure are of about the same size. Thus, the development of robust micro cold forming processes presumes an understanding of the mechanisms of friction and wear in the micro regime. The main work within CRC 747 was first to determine the wear mechanisms in micro cold forming processes and to derive measures to avoid and reduce wear (Sect. 4.2). The second area of investigation was the development of dry micro deep drawing processes utilizing textured forming dies. The work here comprised the development of new and geometrically scaled down experimental setups for tribological investigations and the generation and application of textured surfaces with friction-reducing properties (Sect. 4.5).

Investigation and Avoidance of Wear

The investigation of forming die wear in micro deep drawing was carried out within a combined blanking and deep drawing process on a high dynamic forming press with two linear direct driven axes. Here, up to 300,000 micro cups were machined in a row. Replication techniques were applied to allow for an ex situ forming die wear investigation and the forming die to remain in the machine. The combined blanking and forming die suffered from failure of the cutting edge rather than abrasive wear of the drawing edge radius (see Sect. 4.2). The dies' malfunction was traced back to the comparatively coarse grain of the applied cold working steel. The application of forming dies made from fine-grained tool steel generated by powder metallurgy helped to overcome this issue.

Furthermore, a combined blanking and forming die from the alternative material Stellite was introduced. The die was manufactured by laser selective melting (SLM) and micro grinding for contour finalization. Comparatively high hardness and excellent toughness allowed for the successful forming of 231,000 micro cups in a row (see Sect. 4.5 and [Flo16]).

With regard to the investigation of adhesive wear in micro rotary swaging, dry machining of aluminum Al99.5 workpieces was carried out. Distinct amounts of cold-welded aluminum were found on the dies after the experiments (Sect. 2.3). With the aim of avoiding adhesion, a fracture-tough, tungsten-doped diamond-like carbon (DLC) hard coating system was developed suitable for the particular application on micro rotary swaging dies with small geometric features (Sect. 6.3).

Subsequent dry micro rotary swaging experiments against aluminum workpieces applying DLC hard coated forming dies showed a distinct reduction in adhesive wear. This allowed the feed velocity to be doubled, achieving higher degrees of deformation in a single process step and increasing the forming die life by a factor of three.

With the aim of providing wear-resistant tool steels with improved and locally adapted mechanical properties, a new material generation process was developed: co-spray forming. In co-spray forming, melts of different materials are sprayed onto a substrate, forming a deposit with the finest microstructure and gradient zones between different material layers. Subsequently, the generated material is hot rolled and heat treated. Micro rotary swaging dies were made from co-sprayed material composed of two different hot working steels. Subsequently, these dies were successfully applied in infeed rotary swaging experiments (Sect. 6.2).

The application of cemented carbides as material for micro forming dies is a promising measure to reduce the abrasive wear and plastic deformation of forming dies. This is due to cemented carbides' significantly higher hardness compared to hardened tool steels, together with their toughness against impact loads. However, machining of cemented carbides in small geometrical dimensions with features sizes down to 100 µm is challenging, e.g. due to strong geometrical limits to the application of grinding. Micro milling applying novel ball endmills with cutting edges from binderless polycrystalline diamond (PCD) is an alternative approach that provides both comparatively high material removal rates and a sufficient surface finish. Machining experiments with binderless PCD ball endmills were carried out on tungsten carbide samples of different compositions. It was found that especially fine-grained materials (grain sizes <1 µm) provide the best machining results. Crack-free machining was achieved and ductile inter-crystalline cutting was identified as the predominant material removal mechanism. Further work showed the distinct dependence of the machining result, i.e. the surface roughness, on the machining strategy: this is up- and down-milling. Down-milling provoked distinct chatter and reduced the machining quality. Furthermore, wear mechanisms of the applied micro endmills were identified in dependence on the process parameter feed per tooth. This allowed suitable machining parameters to be derived for the machining of micro forming dies from cemented carbide with a geometrically defined cutting edge (Sect. 4.5).

On top of these measures with respect to channeling and avoiding wear in micro forming, the application of diamond as a tool material has been advanced. The outstanding properties of diamond, with a low friction coefficient in tribological contact with metals, were applied beneficially, and additionally the diamond tool surfaces were micro-structured for friction control. Therefore, a friction polishing process was developed and the principal mechanisms governing the machining of diamond by using thermo-chemical effects were elucidated (Sect. 4.7).

Micro Tribometry

Micro deep drawing processes, obviously, are characterized by only small areas of contact between the tool—this is the die and the blank holder—and the workpiece.

Even though the nominal process forces are low, the reduced interface provokes significant loads on the microscopic level determining friction and wear. Tribological investigation in the micro regime is crucial, for example, for the development of dry deep drawing processes utilizing novel textured forming dies. This requires sensitive instruments and well-developed methodologies to guarantee meaningful results.

For the micro tribological investigation within CRC 747, a new methodology using a micro tribometer in ball-on-plate configuration with balls exhibiting facet areas was developed. Applying the micro tribometer allowed the successful determination of the impact of the micro geometry of textured surfaces on their friction coefficient under dry conditions (Sect. 4.5).

The in-depth tribological investigations of micro deep drawing processes were carried out on a specialized forming press equipped with piezo-electric force measurement equipment to precisely determine the blank holder and punch force in dependence on the punch's position. The aerostatic bearings of the machine, furthermore, helped to minimize the impact of frictional losses during machine movement. Tribological investigations were carried out to determine the impact of the tool geometry, i.e. the drawing edge radius geometrical inaccuracies as a result of the micro machining process, on process forces and the drawing ratio. Furthermore, they were used to confirm the results of the finite element method (FEM)-based development of multistage deep drawing processes (Sect. 2.5).

Micro Textured, Tribologically Active Surfaces

The micro geometry or roughness of interfacing surfaces is of greatest relevance for the development of dry micro deep drawing processes. The micro geometry determines the real area of contact, the predominant friction mechanism—adhesion or abrasion—and thus the friction and the wear behavior. Micro milling, as the most common process for the manufacture of micro forming dies made from hardened tool steels, due to its kinematic and size effects, allows textured surfaces to be generated. The design of these textures is most dependent on the hardness of the machined material, the feed per tooth, and the line pitch. The friction-reducing properties were shown in strip drawing tests and micro tribological investigations using a micro tribometer. It was found for dry strip drawing tests on stainless steel strips that textured surfaces allow the friction to be reduced by up to 20% compared to a polished reference sample. Such textures were also transferred to the surfaces of micro forming dies for the deep drawing of rectangular cups. However, an increase of, for example, the achievable drawing ratio has so far not been found (Sect. 4.5).

The derivation of the most suitable textured surfaces generated by micro milling still remains a research topic. To cover this demand, micro contact and friction modeling is applied. With the purpose of micro modeling, for the first time the combination of a stochastic micro contact model, i.e., the model of Greenwood and Williamson, and feature parameters according to the ISO 25178 standard was shown. This allows for the implantation of actual textured surfaces, e.g., generated by micro milling, into the model and thus the calculation of micro contact conditions in dependence on the normal load. For selected textured surfaces, the contact parameter real area of contact and average force acting on a single roughness peak were calculated. The results show excellent correlation with the frictional properties of the surfaces. For the smallest real area of contact, the lowest friction was measured in ball-on-plate tribological experiments. This new expanded Greenwood and Williamson model builds the foundation for the development of a comprehensive friction model for the derivation of the most promising textured surfaces design, allowing the reduction of friction in dry deep drawing. The friction modeling is the subject of ongoing research (Sect. 2.5).

1.4.2.2 Scatter

Andreas Fischer*

Dispersion (or scatter) indicates the distribution of the available data of a physical quantity. It can be characterized by different measures, e.g., variance, standard deviation, quantile or span. Dispersion is quantified and interpreted using stochastic methods (probability theory and statistics). It can be used, for example, to evaluate the stability of processes and process parameters, the uncertainty in the characterization of materials and components, as well as the efficiency of signal processing algorithms.

Due to the increasing influence of the grain structure of metals in micro production, dispersion plays an even more important role than in the production of macro parts. It has to be considered in all components of the micro process chain (see Fig. 1.6).

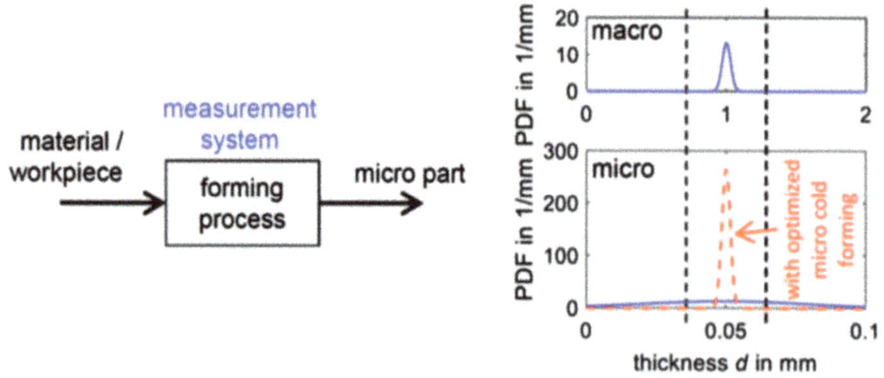

Fig. 1.6 Dispersion affects all components of a production process, the geometric or material properties of the workpiece, the process capability, the measuring process, and finally propagates to the quality features of the micro part. Considering the example of producing a foil with the thickness d, the probability density functions (PDF) shown visualize the scaling effect with respect to the dispersion. When scaling down the macro process (top) to smaller foil thicknesses (bottom), the (also scaled) tolerances (black dashed lines) are exceeded. Since the dispersion of the micro part feature results from the superposed dispersions of the material parameters, the production process and the measurement system, all three components have to be considered in order to minimize the resulting dispersion of the produced micro part and to fulfill given tolerances

Within the CRC, dispersions were considered

(a) to characterize material properties and processes,
(b) to optimize process chains, and
(c) to assess a measuring system for geometric micro features.

(a) The presentation of the characterization of material properties and processes is chronologically divided into the steps of a general process chain. The dispersion of two primary shaping processes and the material parameters, respectively, are discussed first. The dispersion attributed to the main forming process is described next, followed by the dispersion occurring in a subsequent process step such as heat post-treatment.

In the production of thin foils as semi-finished parts for micro deep drawing, it is of primary importance that the thickness is homogeneously distributed. Indeed, foils, like other components with dimensions of less than one millimeter, are subject to size effects. In particular, the strength and ductility and thus the drawability of thin foils significantly decrease when the thickness is reduced. Therefore, foils with a non-homogeneous thickness distribution may result in a higher number of waste parts in micro deep drawing due to locally weaker drawability. In order to reach a more homogeneous thickness distribution of thin foils, the magnetron sputtering process was optimized by an oscillating or rotating substrate holder (see Sect. 6.3). As a result, the standard deviation of the foil thickness was reduced by a factor of 9 from 4.7 to 0.5 µm.

Laser-based free form heading is a further original forming process, where dispersion effects were identified and reduced. The resulting preforms of the produced parts have a random eccentricity, i.e., the location of the intermediate shape relating to the shaft varies for each workpiece. The eccentricity of the preform depends on the applied laser intensity. For intensities where the absorption can be described by the model of Fresnel, the process design causes more material to melt on the laser beam affected side of the rod and results in a random variation of the eccentricity. When using higher intensities, the absorption over the entire rod cross-section is more uniform due to the keyhole formation and the dispersion of the eccentricity is reduced by about 40%. However, this requires a narrow process window, since the keyhole effect causes higher dynamics within the melt, which can also lead to an eccentric positioning of the solidifying melt (see Sect. 2.2). Independent of the laser process, the span of the eccentricity can also be reduced from a maximum of 100–30 µm by the subsequent forming step. Thus, the impact of the dispersion of the first process stage on the accuracy of the final part geometry is small as a result of the self-calibration during the second process stage.

The interaction between the dies and the workpiece during forming leads to material modifications, e.g., with respect to the microstructure and the mechanical properties. The dispersion of the Martens hardness of the part after micro rotary swaging reaches as much as $\sim 250\ \text{N/mm}^2$ compared to about $\sim 160\ \text{N/mm}^2$ before forming (see Sect. 2.3). The increased dispersion occurred because of the inhomogeneity of the microstructure (austenite soft, strain-induced martensite hard) as well as the distribution of the hardness across the diameter of the cylindrical workpiece. The observed distributions can be explained by the strain being inhomogeneously distributed over the diameter. The results of cyclic tension tests of micro rotary swaged parts show a dispersion over almost 4 decades (see Sect. 5.5). The effect can be explained by a size effect. In small volumes, the appearance of a critical failure is less probable than in a larger volume. The hypothesis is confirmed by the lower fatigue strength of macroscopic swaged parts. As a consequence, the fatigue strength of the micro rotary swaged parts should be defined as the value with a lower probability of failure (e.g. 10%) than the usual 50%.

Finally, it was observed that the heat treatment by means of a drop-down tube furnace produces high dispersions of the micro hardness of the processed micro cups. This finding is also a result of the size effect, as the relation between gravity and flow forces in the process is different for micro cups compared to macro parts. This leads to high variations in the falling time of the parts through the furnace, resulting directly in different material properties. In order to reduce this effect by stabilizing the heat treatment time in the furnace, an opposing gas flow was integrated into the process (Sect. 6.4).

(b) Concerning the optimization of process chains, the high dispersion in process parameters inherent to micro manufacturing leads to additional challenges during the planning of complete manufacturing process chains. For this reason, the method "Micro-Process Planning and Analysis" (μ-ProPlAn) was developed that enables process planners to plan and configure stable process chains (Sect. 3.3). μ-ProPlAn includes stochastic methods to characterize the expected value and the variance of the input process parameters based on an experimentally obtained dataset. The dispersion of productions can be improved by evaluating different configurations of the entire process chain and by using μ-ProPlAn's so-called cause–effect networks to estimate the required mean values of the input process parameters, as well as the resulting variances of the production output quantities.

The stochastic modeling approach in μ-ProPlAn is mainly based on a forward model of the process chain. In order to find suitable process parameters for a given production aim (e.g. the specified geometrical features of the product), however, the forward model has to be inverted. Since many forward models are ill-posed, meaning they cannot be inverted as they are non-injective or tend to amplify small deviations in the output, regularization methods such as Tikhonov regularization from the area of inverse problems are applied. In addition, existing, classical approaches have been altered to consider uncertainties when solving the inverse problem. As a result, process parameters can be identified that not only fulfill a given expected output but also stay within a given range (tolerance). Further information on this topic can be found in Sects. 4.5 and 4.6.

(c) Beside the manufacturing processes, dispersion also plays an important role in quality inspection. The dispersion of the measurement results is characterized by the measurement uncertainty. In micro production, the proportion of the geometric feature of the workpiece to be measured to its tolerance and the measurement uncertainty is different compared with the inspection of macro parts. In addition, high aspect ratios in small dimensions, steep surfaces and gradients, as well as a limited accessibility are challenging for the acquisition and evaluation of surface measurement data. To cope with these boundary conditions, a measurement system was developed within the CRC, which captures the part's three-dimensional surface with a digital holographic microscope setup and automatically evaluates its geometric features by a holistic approximation, as well as surface defects by means of a convolutional neural network. The details of this system are described in Sects. 5.2 and 5.3. Increased uncertainties in the evaluation of incomplete geometric features are reduced by acquiring as much as possible of the part's surface out of 4 directions in one shot. Another problem is the holographic reconstruction of optically rough surfaces, if the height variations exceed one quarter of the wavelength. Therefore, the two-wavelength contouring technique was applied. In addition, a method to reduce speckle decorrelation noise was implemented based on averaging the results obtained from multiple measurements with varying directions of illumination. Finally, by applying convolution-based low pass filtering, the standard deviation of the depth measurement could be reduced from 4.4 to 1.5 µm. Regarding the subsequent automatic evaluation of geometric parameters out of the surface point cloud of a micro cup, the holistic approximation approach was verified to yield measurement uncertainties below 0.5 µm.

1.4.3 Mastered Production

1.4.3.1 Measurement and Quality Control

Ralf B. Bergmann*

At the start of the CRC, there was no method available that enabled the measurement of the shape or deformation of micro parts with a complex geometry that is at the same time precise (low measurement uncertainty in the µm or sub-µm range), fast (typically less than a second) and robust (capable of operating outside a measurement lab). In the course of the CRC, a digital holographic measurement system was realized that uses four simultaneous directions of observation and enables optical measurement to be done within approx. 120 ms. The measurement and fully automated evaluation of 3D geometry and surface defects of a micro cup proceed within around 12 s. The dominant time fraction of 11 s is required for data transfer and hologram analysis to obtain the 3D point cloud which is used as input for geometry and surface defect analysis. Next, measurement algorithms are used for an automatic evaluation of dimensional deviations by combining a least squares approximation with an optimal decomposition of the point cloud into elementary geometries and for an automated recognition of surface defects. The system achieves an axial measurement uncertainty of 5 µm and a lateral resolution of 2 µm. In the framework of a transfer project, another holographic measurement technique was set up, that enables the detection of defects with a minimum lateral extent of 2 μm and a minimum depth of 5 μm within cavities of the micro formed parts investigated. For further details refer to Sects. 5.2 and 5.3.

The characterization of the laser chemical machining (LCM) quality as well as its comparison with competing processes were defined as key objectives toward widespread industrial acceptance. Using confocal microscopy in combination with the holistic approximation approach of the CRC to quantify geometric features, it was demonstrated that LCM is particularly suited for the manufacture of micro cavities with dimensions < 200 μm. In comparison with micro milling, laser chemical machining shows higher shape and dimensional accuracy due to sharp edge contours with mean edge radii of (11.2 ± 1.3) μm. However, the surface finish quality of micro milling (with Sa = 0.2 μm) could not be achieved due to the remaining surface waviness, which limits the LCM finishing to Sa > 0.7 μm. For further details refer to Sect. 4.3.

Tool wear is usually measured by using optical measurements directly on the tool. The challenge is, however, to determine the tool wear from a measurement of the forming product. In this case, experiments in lateral micro upsetting were carried out in which the tool wear is reproduced on the formed wire. The reproduced tool geometry on the wire was analyzed and compared to the original geometry of the tool. Both the tool and the wire were measured with a confocal microscope with the same measurement conditions. The development of this technique allows the measurement of the tool wear history on the wire reproductions with a forming accuracy of down to 1.5 μm. For further details refer to Sect. 4.2.

Prior to the CRC it was not possible to perform tensile tests or fatigue tests for micro swaged wires with enhanced strength, since the wires tend to break in the clamping device as a consequence of stress peaks. It was not possible to glue adequate tubes to the samples ends to enhance the diameter of the shafts, since for a sufficiently strong bond it is necessary to cure the adhesive at 180 °C, which leads to an unwanted heat treatment of the sample. Within the project a special clamping device for wires, following the approach for fibers, was constructed. The extreme strength of the wires demands a very high surface hardness of the device, which was implemented by a nitriding heat treatment. For further details refer to Sect. 5.5.

For the determination of a forming limit diagram, the ISO standard defines the dimensions of the experimental setup. The ISO standard is, however, created for material testing in the macro range and material thicknesses above 300 μm. Within the CRC, a test was developed to measure the forming limit diagram for thicknesses below 200 μm. For the detection of localized deformation, which occurs in the micro range, the test setup was scaled down to a diameter of 6 mm for the forming area. Thus the lateral resolution of the optical strain measurement is in the range of the grain size and localized deformation can be measured. The optical measurement method is based on a stereo camera system that uses digital image correlation for the strain measurement. Due to the new measurement method, the determination of the forming limit diagram in the micro range is simplified and the accuracy is increased. For further details refer to Sect. 2.5.

1.4.3.2 Handling

Ann-Kathrin Onken and Kirsten Tracht*

Micro mass production with conventional handling technologies like vibratory conveyors is limited by the sticking effects and small tolerances that occur. Basic functions for handling parts in bulk are functions like separation, orientation, positioning, and transfer. Hence, handling micro parts in bulk slows down the manufacturing and assembly of micro parts. For this reason, the preparation of micro parts has a major impact on the ability to produce micro parts in large quantities. The handling of micro parts is simplified and sped up by interconnecting the parts. Therefore, the linked parts are left in the material, such as foil or wire. While manufacturing linked parts, handling functions such as orientation, separation, and positioning is already covered by the application of the linked parts production method. The basic material of the parts offers the possibility of implementing assisting elements for subsequent production stages.

Technologies for the handling of linked micro parts are speeding up their manufacturing and assembly (Sect. 3.2). A fast transportation is enabled by specific conveyor technologies, which transport parts at rates of up to 500 parts/minute. In combination with referencing of the parts, a fast synchronization of pre-assemblies is implemented. The storage of micro parts is also simplified by the interconnection. Concepts ensure the safe buffering and long-term storage. For long-term storage, the well-known concept of coiling is combined with the usage of layers between the linked parts. This ensures fast and deformation-free storage. Buffering between the stations facilitates especially the required flexibility. Due to the retention of the orientation and position, simple concepts like extending the conveying distance using guiding rollers are applicable.

The production as linked parts offers several advantages, simplifies the handling, and significantly speeds up parts transport, but new challenges also result from the linking of the parts. This has been investigated in cooperation with the processes of laser melting (Sect. 2.2) and rotary swaging (Sect. 2.3). It was shown that the application of the feeder to the tasks of transport, positioning and eventual feed during processing results in different requirements, between which a compromise is needed. Further, the structural connection of the parts leads to a transfer of forces and the processes change the structure. The effects that occur are shown and measured to deal with the resulting challenges.

1.4.3.3 Thermal Aspects

Frank Vollertsen*

Within the current research on micro forming, questions of thermal aspects were mainly those where special effects occurred due to the upscaling of the production rate. Higher production rates very often lead to shorter processing time, while the total work (energy for a specific plastic deformation) per workpiece is held constant. This results in an increased process power, which has to be transferred by the tool. As a consequence of the dissipation of heat, which is the biggest share of energy consumption in metal forming processes, the input power into the tools is increased. As the tool dimensions and the tool material, i.e. its physical properties like thermal conductivity and other, are held constant, the conditions for cooling are unchanged. Only higher thermal gradients due to heating up to higher temperatures will enhance the apparent cooling power, but in spite of that, higher temperatures will result in tools and even in workpieces when scaling up the production rate. The increased temperature of the workpieces and tools can have many different effects. An increased tool temperature can have detrimental effects on tool wear, the workpiece material might have another plastic deformation behavior, and the increased temperature could affect lubricated and unlubricated friction. In order to master this, the temperature fields in tools and workpieces, the damage mechanisms and the development of the microstructure all have to be controlled. Therefore, research has been done in two main fields to control the thermal aspects. First, methods for the measurement, prediction and control of thermal fields were developed. Second, detailed questions, which were mainly induced by the higher production rates, were solved.

Within the first task, i.e., mastering thermal fields in micro metal forming, experimental, analytical and numerical methods were developed. The process of laser-induced chemical etching, which has already been used for two-dimensional processing like cutting and drilling, had to be extended to three-dimensional processing, i.e. to a process similar to free-form milling. As the basic chemical reaction of the process is very sensitive to the local temperature, the temperature fields should be known. Due to the difficulties of direct temperature measurements, the temperature field was calculated. An analytical model was developed (Sect. 4.3) that helps to keep the temperature above the threshold limit (typically in the range of 65 °C for the given materials) and below the limit above which some disturbances of the process occur, like boiling of the electrolyte leading to inferior work piece quality (slightly above 100 °C).

A method for the numerical simulation of a thermal process, e.g., thermal upsetting, was also developed. Thermal upsetting of wire and sheet material to get intermediate stages of the desired part was introduced within the research work. This process does not need any solid tools. It uses local melting of the raw part and the effect of surface tension, which is the dominant force component in the given range of workpiece sizes. A multiphase simulation was developed for the analysis of the thermal field during heating and cooling of the solid, liquid and gaseous material. Using an experimental determination of the solidification speed (Sect. 2.2), the anticipated transport mechanisms for the heat transfer were validated. In addition, the influence of the method for the absorption of the energy (heat conduction mode or keyhole mode) could be assessed concerning its influence on the overall cycle time (Sect. 2.2).

Within the second task, the thermal impacts due to higher production rates were analyzed with respect to the heat development, the impact on the microstructure of tool and workpiece material, and also wear and fatigue properties.

Measurements of the frictional heat due to a misalignment of tool components in deep drawing (Sect. 4.2) showed that the thermal impact can be measured, but is rather low (temperature increase below 5 K) and will not affect the process or the workpiece properties. Also, no critical effect on the tools was seen in rotary swaging. The tool temperature induced by the process heat in rotary swaging was as high as 300 °C (see Sect. 2.3), but an influence on the tool wear was excluded. This is because the annealing temperature of the tool materials used is above 500 °C (Sect. 6.2). For that reason, any impact on the tool properties, especially the wear properties, was excluded.

Significant influences of the time–temperature course were observed in the development of the microstructures of both the tool steel and workpiece materials. Tool steels were produced by spray forming. The substrate was initially a flat plate, which, in combination with spray forming of graded materials, led to excessive heating and coarsening of carbides. By changing the shape of the substrate to the form of a ring and thus increasing the surface during deposition, the cooling could be enhanced and finer carbides resulted (Sect. 6.2).

While the heating of the tools in rotary swaging did not affect the tool wear, the heating of the workpiece could have a significant influence on the workpiece properties, e.g. hardness and fatigue behavior. This was seen after measuring the material properties for samples which were processed by rotary swaging. Using a high feed-forward speed results in a lower hardness and endurance limit. The primary influences were strain hardening and strain-induced martensite formation and therewith the resulting martensite content, which was lower after forming at a higher speed (Sect. 5.5). Different mechanisms are discussed for this effect observed for high alloyed steel. One assumption is that a slow, stepwise deformation leads to higher martensite content than deformation in a few steps. The other assumption is that a higher feed-forward speed increases the mean temperature of the workpiece. This in turn leads to less strain-induced martensite (Sect. 2.3), which can be fully suppressed at slightly elevated temperatures.

References

[Cle17] Clerc, S., Roussel, B., Villien, M.: Status of the microfluidics industry. Market and technology report—May 2017, Lyon, France. https://www.imicronews.com/images/Flyers/MedTech/YDMT17017_Status_of_the_Microfluidics_Industry_Yole_April_2017_Report_Flyer.pdf. Accessed 16 May 2018

[Feu13] Feuerhahn, F., Schulz, A., Seefeld, T., Vollertsen, F.: Microstructure and properties of selective laser melted high hardness tool steel. Phys. Procedia 41, 843–848 (2013). https://doi.org/10.1016/j.phpro.2013.03.157. Lasers in Manufacturing Conference 2013 (LIM 2013)

[Flo16] Flosky, H., Feuerhahn, F., Böhmermann, F., Riemer, O., Vollertsen, F.: Performance of a micro deep drawing die manufactured by selective laser melting. In: Proceedings of the 5th International Conference on Nanomanufacturing (nanoMan), August 15–17, 2016, Macau, China (2016). (CD) http://www.sklumt.polyu.edu.hk/uploads/nanoman2016/nanoMan%201024.pdf. Accessed 10 Oct 2018

[Gei01] Geiger, M., Kleiner, M., Eckstein, R., Tiesler, N., Engel, U.: Microforming. CIRP Annals **50**(2), 445–462 (2001)

[Hes03] Hesselbach, J., Raatz, A., Wrege, J., Herrmann, H., Weule, H., Buchholz, C., Tritschler, H., Knoll, M., Elsner, J., Klocke, F., Weck, M., von Bodenhausen, J., von Klitzing, A: mikroPRO—Untersuchung zum internationalen Stand der Mikroproduktionstechnik. wt Werkstattstechnik online 93, Nr. 3, pp. 1436–4980 (2003)

[Hu10] Hu, Z., Wielage, H., Vollertsen, F.: Effect of strain rate on the forming limit diagram of thin aluminum foil. In: Dohda, K. (ed.) Proceeding of the International Forum on Micro Manufacturing (IFMM10). Nagoya Institute of Technology Nagoya, pp. 181–186(2010)

[Lüt14] Lütjens, J., Hunkel, M.: FE-sigma—Introducing micro-scale effects into FEM simulation of cold forming. In: Ferguson, B.L., Goldstein, R., Papp, R. (eds.) Proceedings of the 5th International Conference on Thermal Process Modeling and Computer Simulation, June 16–18, 2014, Orlando, USA. ASM International, pp. 20–25 (2014)

[NN14] NN: Pressemitteilung IVAM Fachverband für Mikrotechnik, Dortmund, März 2014. IVAM Survey 2014. Wirtschafts- und Stimmungslage deutscher Unternehmen der Mikrotechnik, Nanotechnik, neuen Materialien und optischen Technologien. http://www.ivam.de/news/journalists/pm_ivam_survey_2014. Accessed 20 May 2014

[Nos14] Nosper, T.J.: Unterlagen zur Vorlesung mechatronische Systemtechnik, HS Weingarten. http://www.hs-weingarten.de/~nosper/public/Download/Kapitel%205.1%20ABS%20Neues%20Layout.pdf. Accessed 16 May 2018

[Vol13] Vollertsen, F. (ed.): Micro Metal Forming. Springer, Berlin (2013). ISBN 978-3-642-30915-1

[Vol09] Vollertsen, F., Biermann, D., Hansen, H.N., Jawahir, I.S., Kuzman, K.: Size effects in manufacturing of metallic components. CIRP Ann. Manuf. Technol. **58**, 566–587 (2009)

Chapter 2
Micro Forming Processes

Bernd Kuhfuss, Christine Schattmann, Mischa Jahn, Alfred Schmidt, Frank Vollertsen, Eric Moumi, Christian Schenck, Marius Herrmann, Svetlana Ishkina, Lewin Rathmann and Lukas Heinrich

B. Kuhfuss (✉) · E. Moumi (✉) · C. Schenck · M. Herrmann · S. Ishkina (✉)
Bremer Institute for Mechanical Engineering (bime), Bremen, Germany
e-mail: kuhfuss@bime.de

E. Moumi
e-mail: moumi@bime.de

S. Ishkina
e-mail: ishkina@bime.de

C. Schattmann (✉) · F. Vollertsen · L. Heinrich
BIAS— Institut für angewandte Strahltechnik GmbH, Bremen, Germany
e-mail: schattmann@bias.de

M. Jahn · A. Schmidt
Zentrum für Technomathematik, Bremen, Germany

L. Rathmann (✉)
BIAS—Bremer Institut für angewandte Strahltechnik GmbH, Bremen, Germany
e-mail: rathmann@bias.de

F. Vollertsen et al. (eds.), *Cold Micro Metal Forming*, Lecture Notes in Production Engineering, https://doi.org/10.1007/978-3-030-11280-6_2

2.1 Introduction to Micro Forming Processes

Bernd Kuhfuss

The projects of this chapter describe micro forming processes that are studied as single processes but can also be combined as process chains. Proven examples are material accumulation and succeeding rotary swaging, or rotary swaging and extrusion.

Micro forming differs from macro forming due to scaling effects, which can mean both challenges and benefits. Problems may arise from the handling of fragile parts or adhesive forces between the micro parts or friction effects.

Benefits from scaling effects are made use of in the project "Generation of functional parts of a component by laser-based free-form heading". The aim is a material accumulation that is generated from short duration laser melting. This material accumulation gives the basis for succeeding cold forming operations. The first powerful application for the new technology was upsetting. In the macro range, upset ratios of about 2.3 are achievable, but this is reduced in the micro scale due to earlier buckling of the components. In the micro scale, where cohesive forces can exceed the gravitational force, the molten material forms a droplet that remains adhered to the rod. Thus, upset ratios of up to 500 were reached. The process development was accompanied by a mathematical model and allows for a deep insight into the thermodynamics of laser-induced material accumulation in the micro range.

The laser molten material accumulation could, for example, be further processed by micro rotary swaging. Though rotary swaging has been known in the macro range for a long time and is nowadays an intensively used process in the automotive industry to produce lightweight components from tubular blanks, there are only a few scientific works that have addressed material characteristics like the work hardening or residual stress that are linked to the process and machine parameters and the resulting material flow. Due to micro scale specifics that follow from the kinematics, i.e. relatively smaller stiffness against part buckling and wider tool gaps in the opened state, the feed rates achievable cannot compete with high throughput technologies that produce 500 parts per minute and more. One major aim of the project "Rotary swaging of micro parts" was to increase the productivity for the main process variants, namely infeed and plunge rotary swaging. This demanded also process modeling to understand how parameter variations like friction between tools and the work in the different zones of the swaging tools affect the process.

From the modeling and simulation, separate process variations were deduced and investigated. For infeed swaging, a special workpiece clamping allows for compensation of the pushback force, which results in a 10-fold increase of the maximum feed rate. For plunge rotary swaging, an approach was tested to close the tools gaps during opening using elastic intermediate elements that encapsule the workpiece against the forming dies and enable a 4-fold increase of the radial feed rate.

Micro rotary swaging could become a base technology for process chains. Besides the combination with material accumulation, where the final geometry is formed by swaging, swaging can also be a preliminary step for subsequent operations. This was in the focus of the project "Process combination micro rotary swaging/extrusion". In general, blanks for extrusion processes are produced by wire drawing, which provides high velocities, good surfaces and a diameter with close tolerances. Substitution of wire drawing by rotary swaging gives promising results in some applications. Whereas rotary swaging cannot compete with respect to process velocities, it can offer advantages when certain material properties are needed. In addition, the swaging process can be designed in such a way that the work surface shows micro lubrication pockets, which is also favorable in later extrusion processes.

One effect of the downscaling of the forming of micro parts is the relatively closer tolerances for production. In deep drawing processes, the geometries of the punch, center deviation of the punch, drawing gap and blank position with respect to the drawing die all influence the robustness of the process. These interdependencies are studied in the project "Influence of tool geometry on process stability in micro metal forming". In the micro range, friction between the work, punch and drawing radius plays an important role and leads to variations that limit the usable process windows. Numerical methods and experimental research allow the detection of the geometric influences and their individual contribution to the work quality and process stability.

2.2 Generation of Functional Parts of a Component by Laser-Based Free Form Heading

Christine Schattmann*, **Mischa Jahn, Alfred Schmidt and Frank Vollertsen**

Abstract To overcome the disadvantages like buckling in upsetting processes in micro range, an alternative two-stage shaping process has been developed. This two-stage process consists of a master forming and a subsequent forming step. During master forming, a laser heat source is applied to a workpiece melting the material. Because of size effects the melt pool, whose size depends on the radiation time and the radiation strategy, stays connected to the part. After switching off the laser, the solidifying melt preserves its shape to the so-called preform. Finally, a subsequent forming stage ensures gaining the desired geometry. Since the laser melting process is very stable, a constant part quality is reachable even in linked part manufacturing. Here, cycle rates up to 200 parts per minute enable an industrial application. The reached microstructure, which is defined by the solidification process, has good properties for the subsequent forming process. Thus, a high formability can be achieved.

Keywords Laser · Cold forming · Finite element method (FEM)

Miniaturization is becoming increasingly important for the generation of functional parts. Due to the decreasing the size and increasing the range of the functions of many components, new manufacturing processes are necessary to overcome difficulties that arise when processes such as bulk metal forming are transferred from the macro to micro range. Conventionally, a multi-stage upsetting process with an increased process complexity and number of cycles with high demands on handling is necessary to achieve large plastic forming. This results in a time consuming and, therefore, expensive process. Furthermore, the formability is reduced as the size of the specimen decreases [Eic10], which is also true for the maximum value of natural strain which leads to defects in grains [Tsc06]. While the advantages of the bulk metal forming process upsetting are high output rates with small deviations [Lan02] and low waste of material [Lan06], the maximum achievable upset ratio is only about 2.3 in the macro range. Unfortunately, if applied to workpieces in the micro range, the upset ratio reduces to less than 2 [Meß98]. This happens due to buckling effects that occur faster because it is based on the shape inaccuracy of the specimen and of the tool as well as their perpendicularity to each other [Vol13]. To overcome these issues, a new laser-based upsetting process has been developed [Vol08].

The new process consists of two stages: the master forming step, generating a material accumulation; and a subsequent micro cold forming step, in which the specimen reaches its final geometry. In the master forming process, a laser is used to melt a metallic workpiece. Due to the shape balance effect [Vol13], the molten material tends to form a spherical melt pool and remains connected to the workpiece for as long as the surface tension exceeds the gravitational force. This sphere-like shape is conserved even after solidification and the generated material accumulation is called a preform. The process is limited by the maximum thermal upset ratio, which is achieved just before the molten material detaches from the workpiece. In the second process stage, rotary swaging or die forming are used to form the solidified preform into the desired geometry.

The laser-based free form heading process is applicable for two workpiece geometries: within or at the end of a thin rod, and on the rim of a thin sheet. The main investigation areas of the first process stage are the radiation strategy, the energy balance including the laser melting as well as the solidification of the fluid melt pool, and the reproducibility. Afterwards, during the second process stage, the formability, including the yield stress and the natural strain, are significant to reach the final part geometry. Given that handling of micro parts is challenging [Chu11], linked-part combinations are required to process the usually high number of micro parts to different manufacturing machines.

2.2.1 Laser Rod End Melting

The laser rod end melting process consists of two stages. The first step is the thermal upset process, which generates the material accumulation at the end of the rod. Here, the energy equilibrium and the solidification influence the reproducibility and the microstructure of the preform. In the second step, die forming or rotary swaging leads to the final geometry of the part. With regard to industrial applications, the manufacturing of linked part production is also implemented.

2.2.1.1 Thermal Upset Process

To avoid the disadvantages of conventional upsetting processes, a preform at a cylindrical, metallic rod with diameter $d_0 \leq 1.0$ mm can be generated by laser rod end melting. This alternative process, which is shown in Fig. 2.1, is based on applying a laser heat source to the end of the rod, thereby melting the material. The size of the melt pool depends on the amount of energy absorbed by the workpiece, and can be controlled by adjusting the laser beam power and pulse duration. Due to the surface tension, the melt forms spherically to minimize its surface area. By melting the material, the length of the stationary rod reduces and the spherical melt pool, which stays connected to the rod, moves upwards during the process. Thus, the thermal upset ratio given by $u = \frac{l_0}{d_0}$ increases, where l_0 denotes the molten length and d_0 the rod diameter. The thermal upset ratio is limited by the volume of the melt by which the gravity force exceeds the surface tension causing the melt to drip off

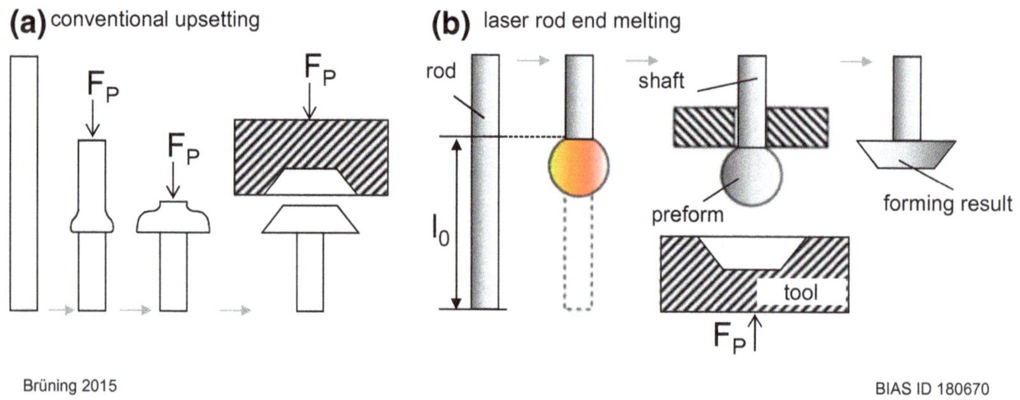

Fig. 2.1 Upsetting **a** multi-stage process conventional upsetting, **b** alternative process laser-based free form heading with only two process stages [Brü16b]

the rod. With decreasing d_0 it is possible to achieve high upset ratios $u \gg 100$; for example, $u \approx 500$ for $d_0 = 0.2$ mm [Ste11]. Thus, the mechanical upset ratio of less than 2 in micro range can be significantly exceeded by the thermal upset ratio of the new process.

To avoid oxidation during the process, a shielding gas atmosphere with Argon is created. After switching off the laser heat source, the melt solidifies rapidly. Therein, the cooling rate has a significant effect on the microstructure and, thus, on the formability of the generated preforms. The final geometry of the part is reached during the subsequent forming process, such as rotary swaging or die forming. In this regards, good reproducibility, high geometrical accuracies, and small reject rates show the advantages for industrial applications in the micro range.

2.2.1.2 *Process Stages and Radiation Strategy*

The master forming process can be subdivided into three process stages: the radiation stage, an intermediate or dead phase, and the cooling stage. To melt the material during the radiation stage, two strategies are possible for the generation of preforms at a rod: coaxial rod end melting and lateral rod end melting. The idea of the coaxial approach is to orientate the rod and the laser heat source coaxially to each other and keep both stationary [Vol08]. In the initial setup of this process design, the rod is positioned in the focus plane of the laser, which is pointing to its head surface, see Fig. 2.2a. In contrast, the lateral process design is based on positioning the laser laterally to the rod and deflecting it along the material using a laser scanner [Brü15a], cf. Fig. 2.2b. Both strategies have advantages, as will be described in the following sections. After switching off the laser, the melting process continues due to the surplus energy that is already added to the rod. This process stage is called the intermediate or dead phase and its duration depends on the amount of energy overheating the melt. When all of the remaining energy is

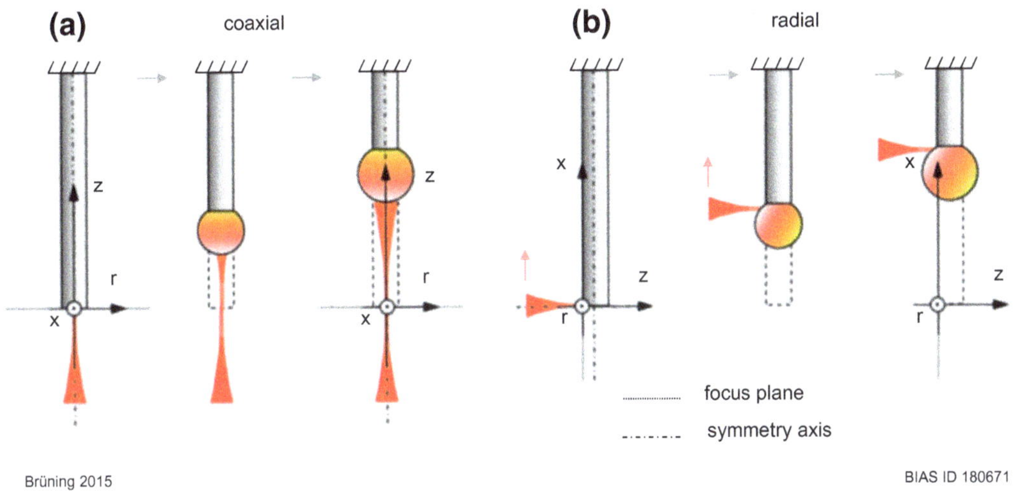

Fig. 2.2 Radiation strategy: **a** coaxial orientated laser beam, **b** lateral orientated laser beam [Brü16b]

consumed, the solidification process starts, which essentially determines the formability of the produced specimen.

2.2.1.3 Modeling and Simulation of the Master Process

Due to its importance for the subsequent forming step and the final result, a detailed analysis of the master forming process and the quantification of the impact of the process parameters on the generated preform is required. Basically, the master forming process and occurring effects can be described by analytical models and result from the similarity theory [Brü12a]. However, a full continuum mechanical model and a accurate corresponding finite element method (FEM) is needed to perform the required throughout process analysis. The modeling and simulation framework can then also be used to find suitable parameter ranges.

From a mathematical point of view, the process can be modeled by coupling the two-phase Stefan problem [Ell81] and the incompressible Navier-Stokes equations with a free capillary surface [Bän01]. Within the thermal problem, laser heating, thermal conduction in the melt and the bulk, cooling described by the Stefan-Boltzmann law, and also the forced convection caused by shielding gas have to be considered to capture all of the relevant effects. The dynamics in the melt are dominated by the movement of the solid-liquid interface and the thermal advection. The resulting partial differential equation (PDE) system is coupled and non-linear [Bän13].

Several finite element approaches have been developed to numerically simulate the process, which are suited to take into account the different radiation strategies in a computationally efficient way. A 2D rotational symmetric approach is used to simulate the process based on a coaxial process design [Jah12a]. Therein, an arbitrary Lagrangian-Eulerian method (ALE) [Bän13] is coupled with an approach based on considering the enthalpy in the workpiece. The throughout analysis of the numerical method is given in [Lut18]. Other than the coaxial process, the lateral process design requires a full 3D simulation [Jah12a]. As a result of the findings in [Lut18], an approach merely based on an enthalpy method was implemented [Jah14b]. To analyze the cooling process and take into account the interaction of melt and shielding gas, the extended finite element method (XFEM) can be used. This method belongs to the class of unfitted approaches and it allows us to decouple the physical domain from the discretization by extending the approximation space. This has the advantage that it allows for a more precise simulation because not only the workpiece but also its surroundings can be considered. In contrast, conventional finite element approaches are based on representing the physical domain by the computational mesh and, therefore, can only consider the impact of the workpiece surrounding via boundary conditions.

After having been validated by comparing the numerical results to the experimental data, such as in regards to the shape of the preform [Jah12], the solidification process [Jah14b], and the energy balance with respect to the self-alignment capability of the process [Jah12a], the finite element simulations are used for parameter identification and for further process analysis.

2.2.1.4 Energy Impact and Heat Dissipation Mechanisms

As previously mentioned, the results of the master forming step are primarily dependent on the amount of energy available for the melting process [Brü12b]. This energy depends not only on the total amount of energy $E = P_L t_L$ applied to the workpiece but also on the composition of the energy, meaning the choice of the laser power P_L, and the radiation time t_L to generate a certain energy E. This happens because the laser power has a significant influence on the absorption rate and the heat dissipation mechanisms. In addition, the divergence of the laser beam has to be taken into account. All these aspects are highly dependent on the radiation strategy and the chosen process parameters.

For laser-based processes, the absorption rate is typically given by the Fresnel equations. For the process at hand, the absorption rate is then approximately 0.38 [Wal09]. However, by significantly increasing the laser beam power P_L, and thus its intensity, a vapor capillary can be created in the melt pool. In this capillary, which is called keyhole, laser rays are reflected multiple times, which leads to abnormal absorption and results in an increasing absorption rate of up to 0.80 (see e.g. [Hüg09]).

A first approximation of the amount of energy E_S needed to generate a preform of volume V_S can be derived using an adiabatic model consisting of a linear relation of absorbed energy and preform volume [Vol08]. However, since this model neglects the fact that energy is permanently dissipated by radiation and convection and, moreover, transported into the non-melting part of the rod by conduction, it only provides a rough estimate. Finite element approaches are used for a precise

Table 2.1 Process parameters for preform generation

Rod diameter	0.40 mm
Laser power	130 W
Pulse duration	100 ms
Absorption coefficient	0.38
Shielding gas	Argon
Deflection velocity	30 mm/s

analysis of the process and the prediction of the impact of different radiation strategies or process parameters on the results.

The heat fluxes for the coaxial and lateral radiation strategy are compared by using the process parameters specified in Table 2.1. The corresponding heat fluxes, which are the absorbed laser energy and cooling caused by the Stefan-Boltzmann law and convection, are and visualized in Fig. 2.3. Therein, one can see that the energy available for the process in both situations differs significantly and, therefore, causes different preform volumes. The main reason for this is that in the coaxial design, the laser heat source always affects the molten part of the rod so that the energy has to be transported through it to melt more material. Thus, the overall temperature of the melt rises permanently, thereby causing more heat dissipation so that the efficiency of the process with respect to the applied energy decreases. Moreover, the divergence of the laser beam has a high impact on the result when using the coaxial radiation strategy. While the rod is initially positioned in the focal plane of a defocussed laser beam that is pointing to its head surface, the melting process shortens the rod. Consequently, the melt leaves the laser's focus plane and the defocussing of the laser beam causes a significant drop of the absorption rate. In contrast, the deflection of the laser beam by a scanner in the lateral process design can guarantee that the laser energy is always applied to the rod near the solid-liquid interface. Furthermore, defocussing is no longer an issue since the lateral scanner position keeps the focus on the rod surface. For the lateral process design, optimal deflection velocities can be determined which depend on the rod diameter [Brü14b]. The results, visualized in Fig. 2.4, show that this velocity is proportional to the laser power and anti-proportional to the square of the rod diameter.

Besides the radiation strategy and the laser beam intensity, a closer look at the process analysis reveals that small changes of the terms of the energy losses can already lead to different results. This happens because of the heat dissipation mechanisms, which depend, partly in a highly non-linear way, on the difference of ambient temperature and the temperature of the preform. In [Jah14], a throughout study quantifying heat dissipation mechanisms for laser based rod end melting is given. Therein, it has been shown that energy is primary dissipated during the radiation stage by the cooling due to the Stefan-Boltzmann law while heat conduction and convection have only a small impact. Because the cooling due to the Stefan-Boltzmann law involves the difference $T^4 - T_a^4$, with T denoting the temperature at the surface and T_a is the ambient temperature, using a high laser power usually results in higher temperatures, which causes more dissipation of energy.

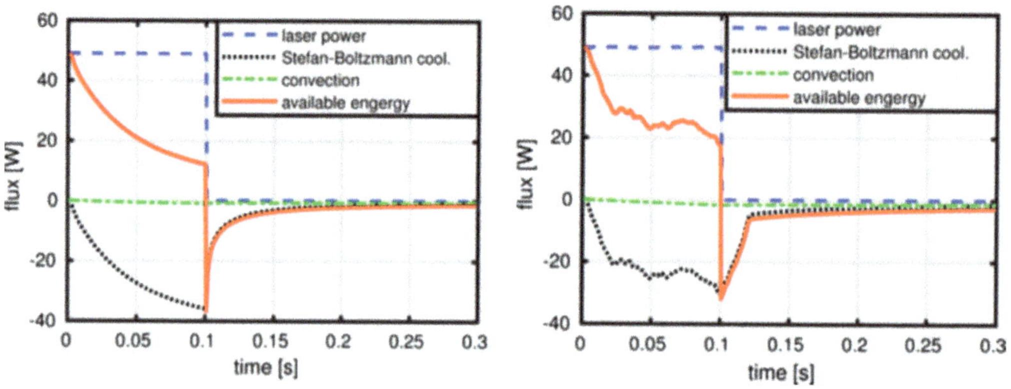

(a) Energy fluxes using coaxial radiation.   (b) Energy fluxes using lateral radiation.

Fig. 2.3 Comparison of energy fluxes for the coaxial and lateral radiation strategy

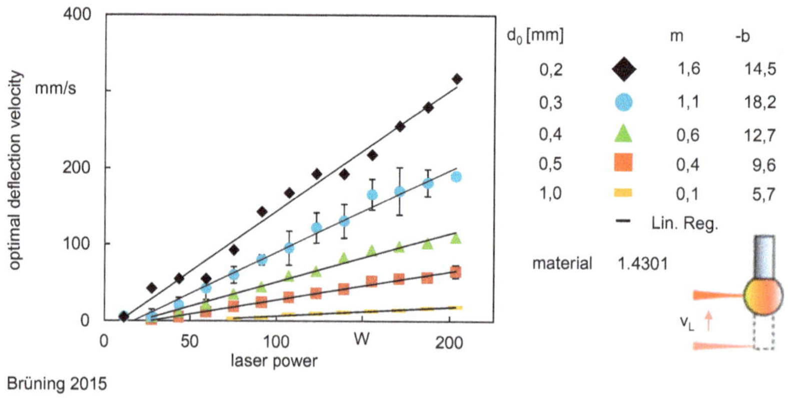

Fig. 2.4 Optimal deflection velocity for different rod diameters [Brü16b]

2.2.1.5 *Solidification and Microstructure*

After switching off the laser, the cooling process starts. This process stage, consisting of the dead phase after switching off the laser and the subsequent solidification process, is crucial for the industrial application of this process. One reason for this is that its duration is the dominating factor of the total process time [Brü13c]. Hence, decreasing the cooling time is mandatory to upscale the process and rapidly generate a high number of preforms. Furthermore, this process stage also defines the microstructure of the preform and, therefore, its formability.

Unfortunately, the duration of the cooling process cannot be considered independently from the radiation stage because it highly depends on the temperature distribution in the melt. In particular, the duration intermediate stage after switching off the laser can be controlled by choosing adapted process parameters. Suitable process values can be determined by combining the finite element simulation with the concept optimization via simulation [Wan07].

In contrast to the melting process, where most energy is dissipated by radiation, the cooling process is governed by convection. Consequently, it can be primarily controlled by changing the shielding gas atmosphere. The duration of the cooling and solidification process has been quantified in [Brü13a], where it has been shown that the cooling time is highly dependent on the diameter of the preform.

The cooling process affects the generated microstructure of the preform, which essentially depends on the temperature gradient at the solid-liquid interface and its velocity [Kur98]. In general, a high interface velocity and a rather small gradient is desirable because it leads to a globular microstructure. A rough estimate of the microstructure can be obtained by an analytical model [Brü16b]: assuming a constant heat dissipation due to radiation and convection, the dependency of the secondary dendrite arm spacing S_D on rod diameter d_0 and the preform diameter d_{PF} is given by Eq. (2.1)

$$S_D = 5.5 \cdot B^{\frac{1}{3}} \cdot \left[\frac{\frac{1}{\pi} \rho V_{PF} H_M}{\dot{q}_{HC} \frac{d_0^2}{4} + \varepsilon \sigma_S \left(d_{PF}^2 - \frac{d_0^2}{4} \right) \left(T_m^4 - T_\infty^4 \right) + \alpha \left(d_{PF}^2 - \frac{d_0^2}{4} \right) (T_m - T_\infty)} \right]^{\frac{1}{3}} \tag{2.1}$$

with the solidification and material specific constant B and the solidification time, described by the density of the material ρ, the volume of the preform V_{PF} and the melting enthalpy H_M in the numerator and the different heat flows in the denominator. Here, the heat conduction $\dot{q}_{HC}$, the emissivity ε, the Stefan-Boltzmann constant σ_S, the melting temperature T_m, and the ambient temperature T_∞ are substituted in the calculation. The model for the expected solidification interval and the secondary dendrite arm spacing shows good agreement with the experiments for smaller solidification intervals, as shown in Fig. 2.5 [Brü16b].

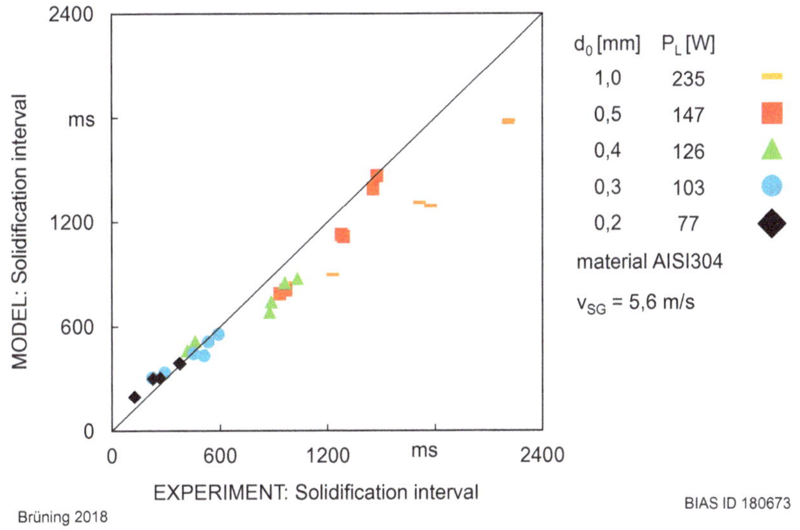

Fig. 2.5 Comparison of solidification interval between model and experiment for different rod diameters [Brü16b]

2.2.1.6 Reproducibility

The laser rod end melting process is very stable with respect to small deviations in the process design and choice of process parameters. For example, the coaxial process has a self-aligning capability which means that the generated preform volume is independent of deviations in rod diameter as long as the pulse energy and the radiation time are scaled accordingly to results from the theory of similarity [Brü12a]. In principle, the coaxial design is prone for a decreasing energy efficiency due to defocussing because of the static position of the focus plane in this design. Therefore, less melting volume can be gained for increasing thermal upset ratios, because the rod shortening increases the distance between focus plane and melt surface, see [Brü15a]. However, this issue can be resolved using a feeding system for the rod. Then, the rod is continuously kept within the focus plane of the laser and, as a result, the thermal upset ratio can be increased to a maximum of $u \approx 500$ [Ste11b]. This means that the energy efficiency is doubled [Ste11a]. Despite this improvement, there are still deviations in the thermal upset ratio compared to the lateral radiation strategy. While the fluctuations of specific melting volume are lower for coaxial radiation, the absolute volume in the lateral process design is still generally higher, making this radiation strategy more energy efficient [Brü15a]. The reasons for this are the different effects of the heat dissipation mechanisms, which were previously described.

With regard to the positioning of the laser beam onto the rod in the lateral process design, Brüning showed in [Brü13e] that positioning deviations up to 100 μm in vertical direction do not essentially affect the resulting preform diameter. Moreover, if the total axial and radial positioning is considered, then the deviations need to be significant less than 100 μm, thus a good initial situation for the subsequent forming process is reached.

As described, the absolute preform volume is proportional to the applied pulse energy [Brü12b]. Differences between the theoretical melting volume and the final volume of the preform can be explained by material losses during the laser process. Especially during the abnormal absorption, vaporization caused by keyhole formation can take place introducing high dynamics into the process. This dynamic can result in spatter formation and, hence, in material losses. Using the lateral radiation strategy, relative mass losses between 0.5 and 2.6% related to the thermal upset ratio and the rod diameter can be measured [Brü16b]. The volume of the preform correlates with the diameter of the solidified preform. Using the coaxial radiation strategy, an increase of preform diameter leads to decreasing eccentricity with a decreasing standard deviation. If the lateral radiation strategy is considered, then the eccentricity, see Fig. 2.6, is lower in a range of 40–60 μm but with a slight fluctuation and a standard deviation of about 10 μm as stated in [Brü16a].

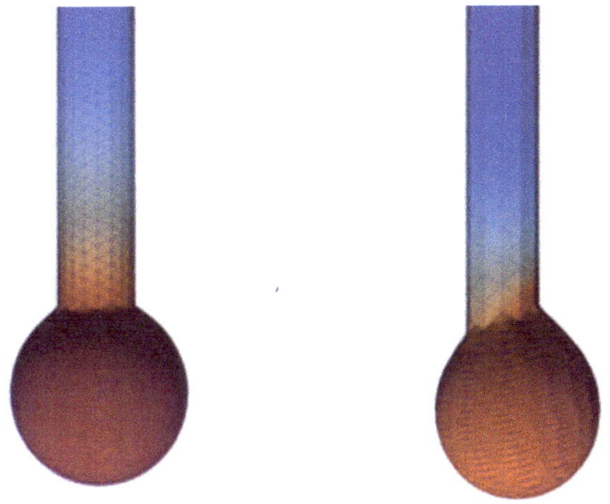

(a) Coaxial radiation. (b) Lateral radiation.

Fig. 2.6 Comparison of simulation results for the laser rod end melting process with coaxial and lateral radiation strategy

2.2.1.7 Formability

The preforms achieve their final geometry during the subsequent forming process. Therefore, the yield stress, the natural strain, and the form filling behavior are of interest. In [Brü13c], it is shown that for preforms with diameters between 1.1 and 2.1 mm, it is possible to reach a maximum average logarithmic natural strain up to 1.7 without damaging the forming tool. Further investigations based on simulations of the forming process show that the influence of the friction coefficient on average natural strain can be neglected [Brü14c]. Here, the distribution of natural strain is inhomogeneous, primarily because the center of the preform interacts with the tool and, thus, increases the centered natural strain. Experiments show that a preform with an initial diameter of 420 m having a dendritic grain structure results in a convex formed specimen with a height of 27 m, even though the surfaces of the open die tool have been planar, as shown in Fig. 2.7. This investigation leads to a maximum averaged natural strain of 2.75 without any occurring defects. Further forming resulted in a plastic deformation of the tools due to strain hardening of the specimen.

The yield stress level increases linearly with the increasing absolute value of average natural strain. Furthermore, a higher secondary dendrite arm spacing leads

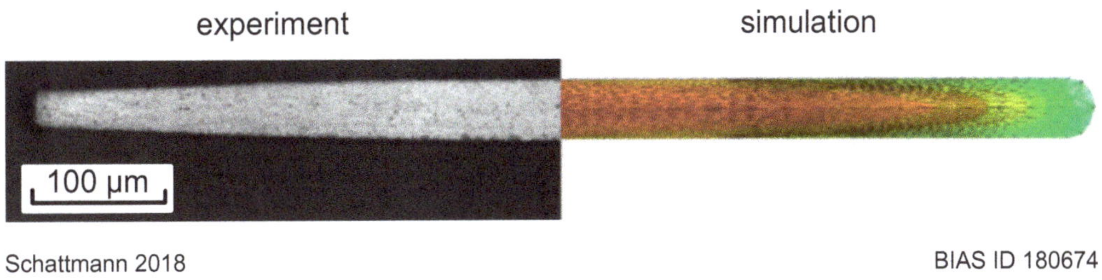

Fig. 2.7 Part after open die forming (simulation of averaged logarithmic natural strain: blue 0, red −2.5; the model consists of rigid surfaces)

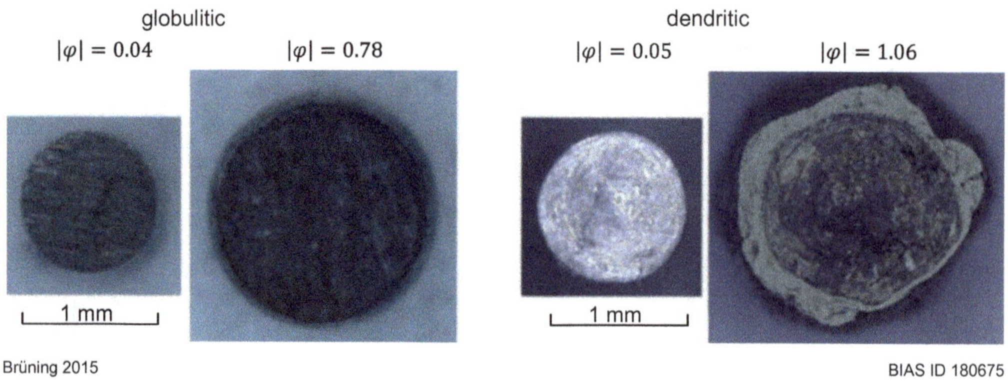

Fig. 2.8 Upsetting of cylinders in micro range with different microstructure depending on average natural strain ϕ, [Brü16b]

to a decreasing yield stress level, see [Brü14a]. Further yield curves arose using an universal testing machine [Vol13]. Here, the yield strength of the preforms that is investigated by upsetting experiments of cylinders is comparable to that of the same material and size with different microstructure [Brü13d]. However, the forming behavior changed with the different microstructure, such that the final geometry after open die forming is different. Within the experiments using globular microstructure inside the cylinder, only barreling occurs. Using the dendritic microstructure of the preform, barreling occurs also, but the size of the front surface is increasingly in homogeneously, as shown in Fig. 2.8 [Brü16b].

During master forming, the eccentricity of the preform with respect to the rod is defined. During the experiments, deviations up to 100 μm depending on the rod diameter can arise. However, the subsequent forming stage also includes a natural calibration step, which reduces the eccentricity to less than 30 μm and which is assumed to be related to the deviation of the coaxial alignment between the upper and the lower die. Furthermore, if an oversized volume of preform is used, then the forming process can be successful without any burr formation. This is possible because the material flows not only inside the cavity but also in direction of the rod. Thus, preform volumes of up to 125% of the die volume can fill the cavity completely without any burr, see [Brü15b].

Finally, it is possible to mold fine surface structures after the forming stage, such as the surface roughness of the die, see [Brü14a], down to feature sizes of 500 nm which is investigated in [Brü15c]. In Fig. 2.9 the different geometries of truncated cone formed parts are shown. Here, the average of natural strain of the manufactured parts is well recognizable.

2.2.1.8 *Linked Part Production*

Given that the handling of the micro parts is always challenging, different linked part systems are investigated to improve the possibilities for mass production and

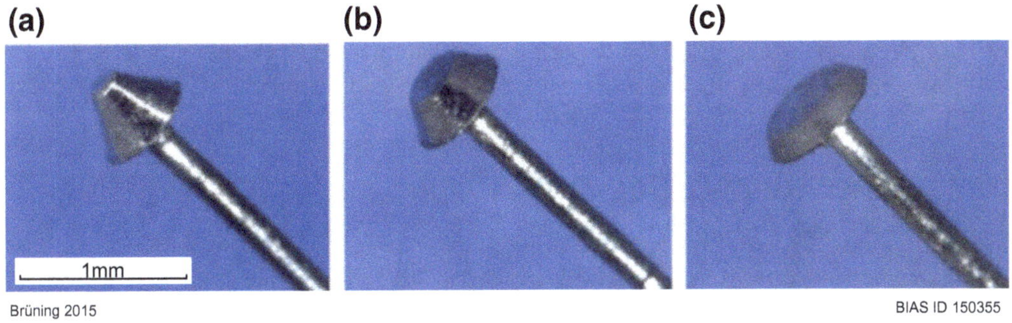

Fig. 2.9 Part after closed die forming using different averages of natural strain [Brü16b]

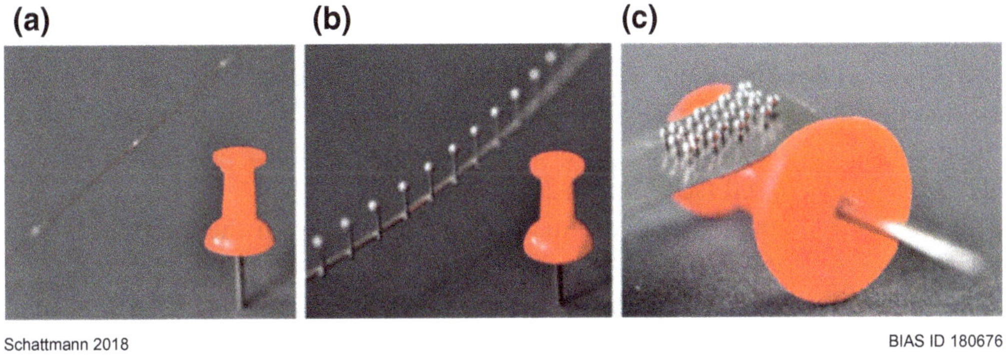

Fig. 2.10 Linked-part production: **a** line-linked, **b** comb-linked, and **c** plate-linked

upscaled output rates during manufacturing process. This includes line-linked parts and comb-linked parts.

For line-linked parts, the thermal upsetting process is performed using a continuous rod, see Fig. 2.10a. The stationary laser process takes place in the middle of a rod. Further rod material is fed into the process zone, so that a material accumulation and thereby the preform arises. A variation of feeding times and laser power result in different preform geometries, which can be formed from round to flat [Sch17b]. Investigations show that the energy efficiency and the cycle time can be improved so that more than 222 parts can be generated per minute [Sch17d].

Because the geometry of the preform is changed slightly due to the surface tension, the comb-linked strategy can be an alternative. In Fig. 2.10b it is shown that the workpiece rod is connected to a conveyor rod. After cutting the workpiece rod, the preform can be generated as usual [Brü13b]. This comb-linked strategy was tested for wire diameters from 0.2 up to 0.5 mm. Depending on the preform diameter, the solidification time varies. Thus, the dead time between laser process and feeding process should be adjusted, see [Jah13].

This linked strategy can be extended by placing the workpiece rods on a carrier plate [Sch17a], as shown in Fig. 2.10c. This opens new possibilities of applications, thus a metallic detachable connection can be introduced. Here, a shear stress of 69 N/cm^2 can be transmitted by a metallic hook-and-loop fastener [Brü14c].

2.2.2 Laser Rim Melting

The laser melting process can also be applied to the rim of metal sheets. Using the lateral process design, a laser beam is deflected with a defined machining depth along the rim of a thin metal sheet [Woi15], as shown in Fig. 2.11. The material starts melting and a molten drop arises which stays connected to the sheet because of surface tension. By deflecting the laser beam along the rim, more material is molten and the melt pool moves along the rim. Due to the physical dimensions of the workpiece, the rear part of the melt solidifies rapidly so that the melt pool always has a specific length which depends on the process parameters and the thickness of the sheet. In general, the process can be used to generate continuous cylindrical preforms and, thus, preforms with irregularities at sheet rims (see Fig. 2.12b) are avoided (see Fig. 2.12a).

In [Brü17], the influence of the melt pool dimensions on the continuous generation of cylindrical preforms was investigated. Here, a length-to-height ratio of 3.0 ± 0.4 of the melt pool was achieved for a blank thickness of 0.2 mm. Furthermore, an analytical model taking into account the surface area of the melt pool not only supports this result but is also able to predict the maximum allowable melt pool length to gain a continuous cylindrical melt pool. Otherwise, instabilities occur, which are comparable to the humping effect during laser welding. In Fig. 2.13, it is apparent that the local frequency of periodical maxima in a continuous preform is influenced linearly by the laser power and by the deflection velocity for sheet thicknesses of 25 up to 100 μm. Here, a high accordance to the results of humping during laser welding by Neumann [Neu12] is viewable. This happens despite the fact that Neumann conducted experiments with a laser power in

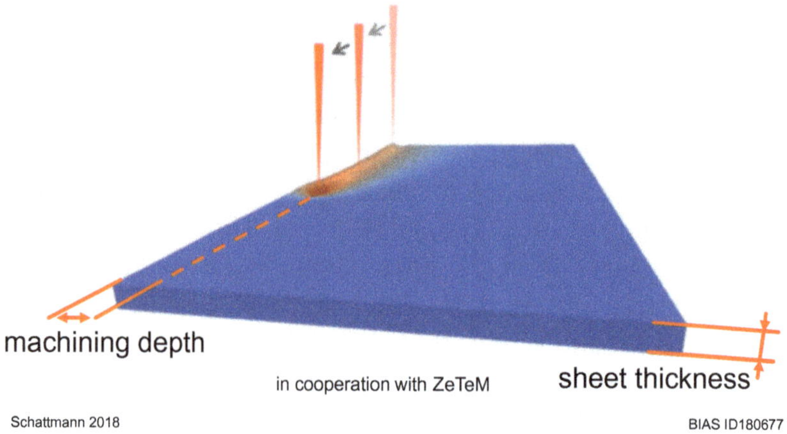

Fig. 2.11 Process laser rim melting including simulated heat distribution

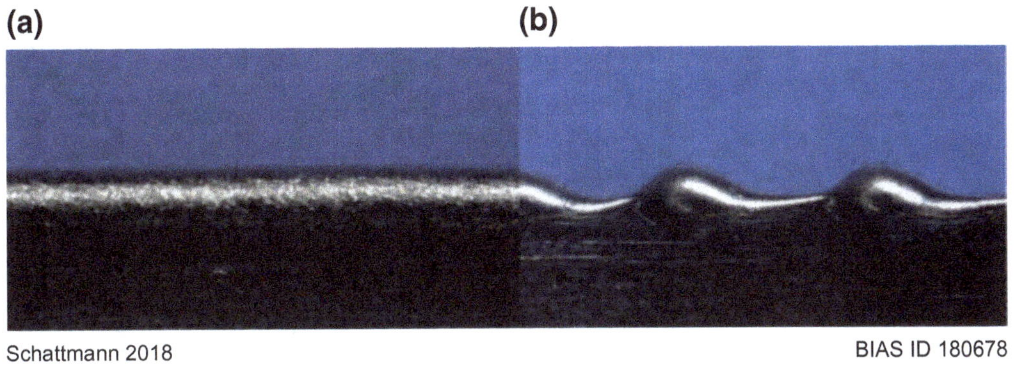

Fig. 2.12 Continuous preform on a metal sheet with a thickness of 70 µm: **a** cylindrical, **b** irregular [Sch17c]

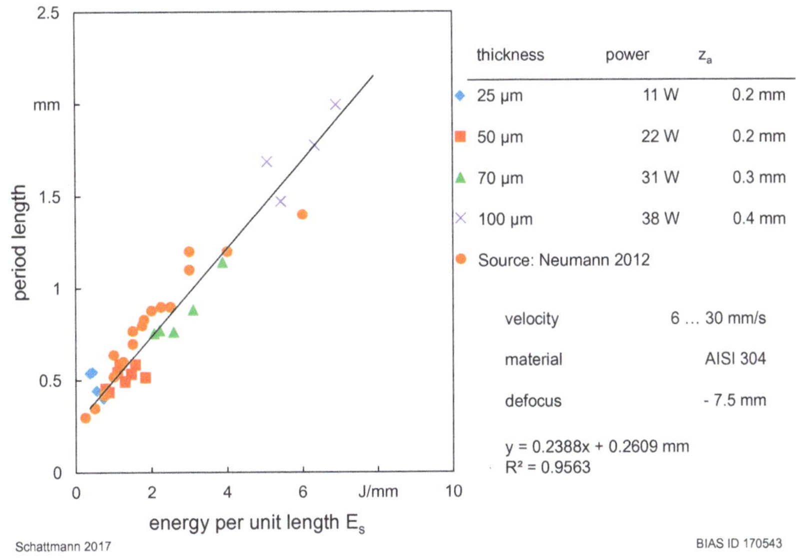

Fig. 2.13 Comparison between humping during laser welding [Neu12] and irregularities during laser rim end melting

the range of some kW and much higher travel speed than used in the current rim melting experiments.

The process can also be applied to closed rims, such as voids in metal sheets. Using the results obtained for the generation of continuous cylindrical preforms, it is for example possible to generate a preform with a height of 0.4 mm at a void of a metal sheet with a thickness of 200 µm, cf. Fig. 2.14a for the simulation result. Thus, it is possible to cut a thread of M2 inside [Sch17c]. Figure 2.14b shows the corresponding scanning electron microscope (SEM) image of one thread with a thickness of 400 µm.

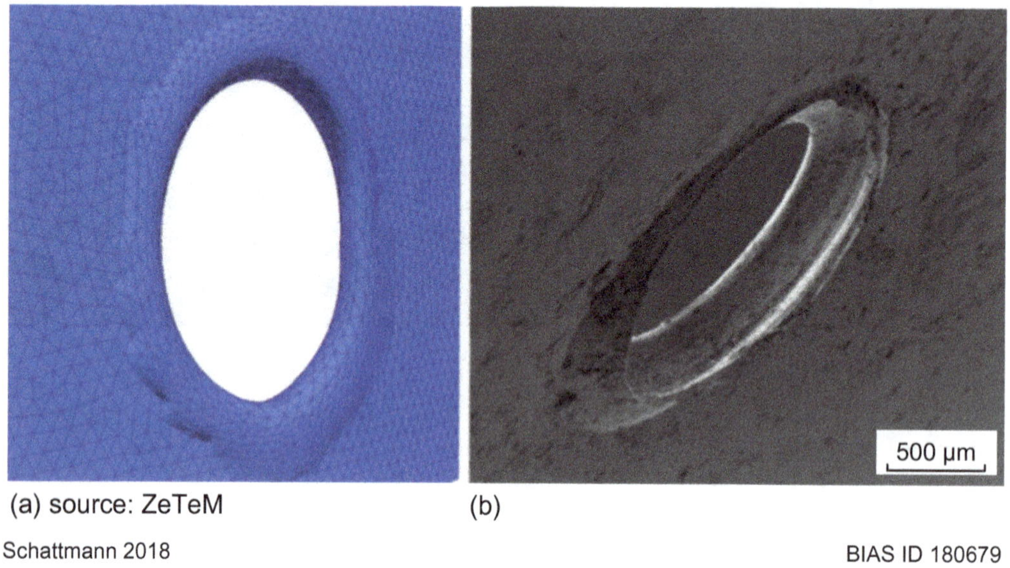

Fig. 2.14 Laser rim melting at void: **a** Simulation, **b** a M2 thread is cut into a continuous cylindrical preform [Sch17c]

The wide range of possibilities for the laser-based free form heading process is appreciated, which is expanded from the rod to the metal sheet, so that more new applications can be exploited and industrial integration can be expected.

2.3 Rotary Swaging of Micro Parts

Eric Moumi*, **Christian Schenck, Marius Herrmann and Bernd Kuhfuss**

Abstract The incremental forming process of micro rotary swaging can be divided into two process variations. During infeed rotary swaging the diameter of a workpiece is reduced over the entire feed length of the workpiece. However, during plunge rotary swaging the diameter is reduced within a limited section by radially feeding the dies. Both variations feature limited productivity due to their incremental nature. An approach to increase the productivity is by increasing the feed rate. However, due to the fixed kinematics, a higher feed rate results in a larger volume of deformed material per stroke, which could lead to failures particularly due to inappropriate material flow. In infeed rotary swaging at a high feed rate, the flow of the workpiece material against the feed direction can result in bending or breaking of the workpiece. In both infeed and plunge rotary swaging, the workpiece material can flow radially into the gap between the dies and provoke the formation of flashes on the workpiece. For plunge rotary swaging the radial flow is also a motion against the feed direction. Measures to control both the radial and the axial material flow to enable high productivity micro rotary swaging are presented. An adjusted clamping device enables an increase of the productivity by a factor of four due to a reduction of the axial forces generated by the undesired axial material flow in infeed rotary swaging. The radial material flow during plunge rotary swaging can be controlled by intermediate elements. Thus, an increase of productivity by a factor of three is possible. Furthermore, the geometry after plunge rotary swaging is strongly influenced by the workpiece clamping and fixation, and the material flow can be controlled by applying low axial forces to the workpiece on one or both sides of the forming zone.

Keywords Forming · Productivity · Quality

2.3.1 Introduction

Micro rotary swaging is an incremental open die forming process to reduce the cross-section of axisymmetric workpieces. Concentrically arranged dies are oscillating with a given stroke frequency f_{st} and a stroke height h_T. The dies strike simultaneously on the workpiece to deform it gradually from the initial diameter d_0 to the final diameter d_1 [Kuh13]. The main components as well as the kinematics (white arrows) are presented in a front view of a rotary swaging machine in Fig. 2.15.

The nominal diameter d_{nom} is defined as the inside diameter of the forming zone of a closed die set. The two main variants of the process are infeed rotary swaging and plunge rotary swaging. During infeed rotary swaging the workpiece is axially

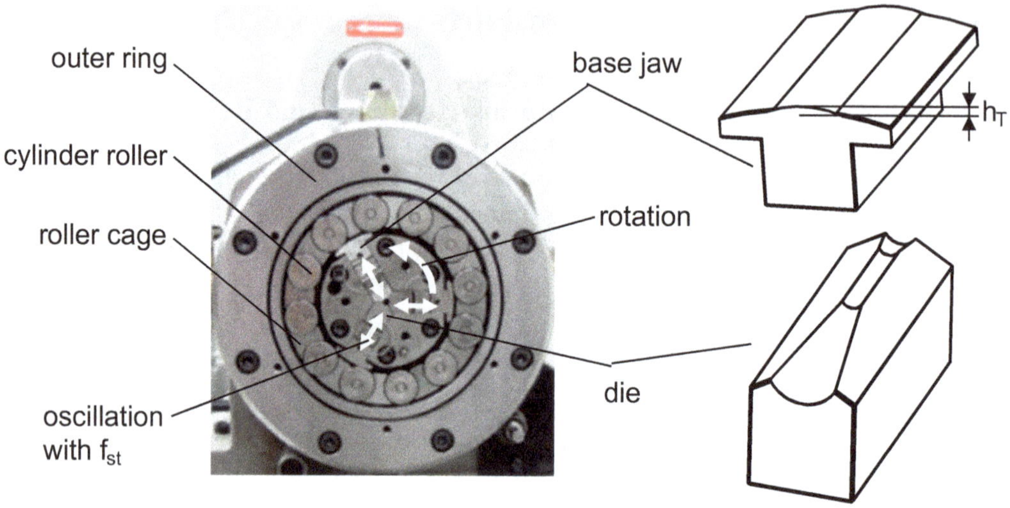

Fig. 2.15 Front view of a rotary swaging machine without end cover

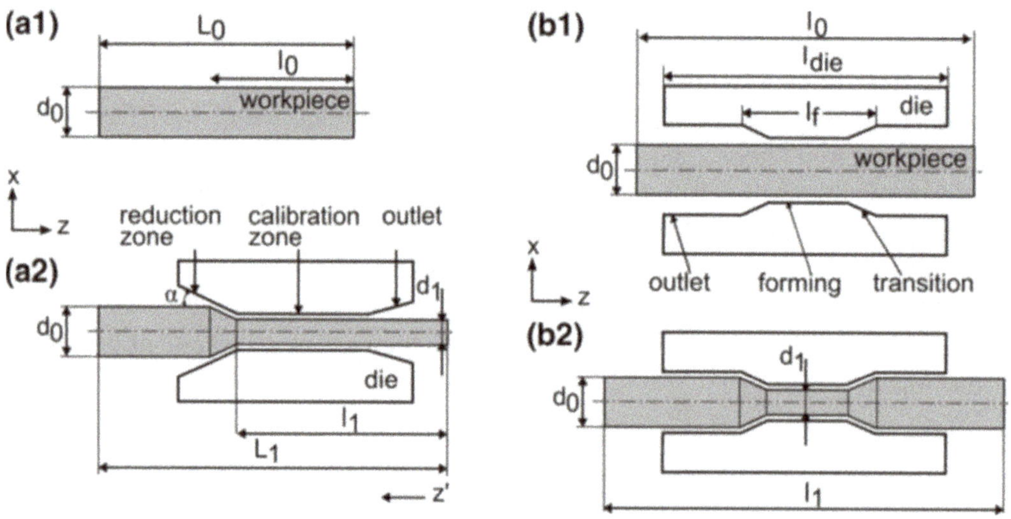

Fig. 2.16 Rotary swaging variants: **a** infeed; **b** plunge [Mou18b]

fed with the feed rate v_f into the swaging unit and is reduced over a length l_0 (see Fig. 2.16a). However, during plunge rotary swaging the workpiece is processed locally as the dies are fed radially with a feed rate v_r while oscillating (see Fig. 2.16b).

Besides the presented developments in micro rotary swaging, investigations on rotary swaging by other authors were made in the macro range and focus on the workpiece properties after forming and on new applications for the process. For example, Alkhazraji et al. found that both tensile and fatigue strength are enhanced in workpieces with ultra-fine grains made by rotary swaging [Alk15]. The process was also analyzed for joining tubes [Zha14] or for plating axisymmetric metallic parts [Sch16].

With regard to high productivity in the micro range, infeed rotary swaging can already facilitate several hundred parts per minute by a continuous process, for example as a preliminary stage for extrusion [Ish18]. But there is still potential for further productivity increases. Plunge rotary swaging enables the manufacturing of variable types of geometries. However, the productivity is considerably low [Kuh09a], due to the low feed rates v_r used to prevent an undesired flow of the workpiece material into the gaps between the dies. Still, for both process variants, high productivity can be achieved by controlling the material flow and the positioning of the workpiece.

The degree of incremental deformation φ_{st} is a common value in rotary swaging to describe and quantify the deformation. It is defined by the initial diameter d_0 and the effective feed per stroke of the die h_{st} (Eq. 2.2). For infeed rotary swaging h_{st} equals the axial workpiece feed per stroke l_{st}, which itself depends on the die angle α and on the ratio of the workpiece feed rate v_f, over the stroke frequency f_{st} (Eq. 2.3). For plunge rotary swaging, h_{st} depends on the ratio of the feed rate of the dies v_r over the stroke frequency f_{st} Eq. 2.4.

$$\varphi_{st} = \ln\left(\frac{d_0 - 2h_{st}}{d_0}\right)^2 \tag{2.2}$$

$$h_{st}\_infeed = l_{st} \tan \alpha = \frac{v_f}{f_{st}} \tan \alpha \tag{2.3}$$

$$h_{st}\_plunge = \frac{v_r}{f_{st}} \tag{2.4}$$

The previous equations show that the degree of incremental deformation φ_{st} does not consider the final diameter d_1. However, this parameter is involved in the volume per stroke V_{st}, which represents the material volume formed by the dies at each stroke. For an idealized forming, V_{st} for infeed rotary swaging can be determined with Eq. 2.5 and for plunge rotary swaging with Eq. 2.6. For plunge rotary swaging the deformed volume of the workpiece is assumed to have a cylindrical shape. In Eq. 2.6 k stands for k-th incremental step.

$$V_{st}\_infeed = \pi l_{st}\left(\frac{d_0^2}{4} - \frac{d_1^2}{4}\right) \tag{2.5}$$

$$V_{st}\_plunge(k) = l_{form}\frac{\pi}{4}\left[(d_0 - kh_{st})^2 - (d_0 - (k+1)h_{st})^2\right] \tag{2.6}$$

In order to improve the micro rotary swaging process, it is necessary to understand and control the material flow. Three methods to improve the rotary swaging in the micro range based on the material flow control are presented. In infeed swaging a higher axial feed rate is enabled by using an adjusted workpiece clamping [Mou18a]. In plunge rotary swaging the application of an elastic tube prevents flash formation [Mou15b, Mou15b] and with the application of external axial forces the part geometry can be controlled [Mou18b].

2.3.2 Process Limitations and Measures for Their Extension

During micro rotary swaging different limitations to the process occur [Kuh08a]. Furthermore, process parameters like the workpiece material, the die geometry, and the tribological conditions influence these limitations.

Typical limitations of the process can be divided into two categories (see Fig. 2.17). The limitations in the first category lead to a termination of the process. These are bending, torsion or breaking of the workpiece. The limitations in the second category are those that affect the workpiece quality. Examples are flashes, high surface roughness [Wil13] and shape deviation [Kuh09b]. A guiding element with a hole slightly larger than d_1 for the workpiece at the inlet in the swaging head is used to prevent premature limitations due, for example, to vibrations occurring in the process [Wil15]. Furthermore, the use of a guiding element, which is necessary and specific for the micro range, reduces the risk of the workpiece getting into the tool gap during the process and thus enables the application of higher strokes.

During rotary swaging material flow takes place predominantly in the axial and radial directions. The axial material flow occurs in both the positive and negative directions. Higher material flow against the axial feed direction induces an axial load that bends the workpiece when the critical bending load $F_{b,crit.}$ is reached. This

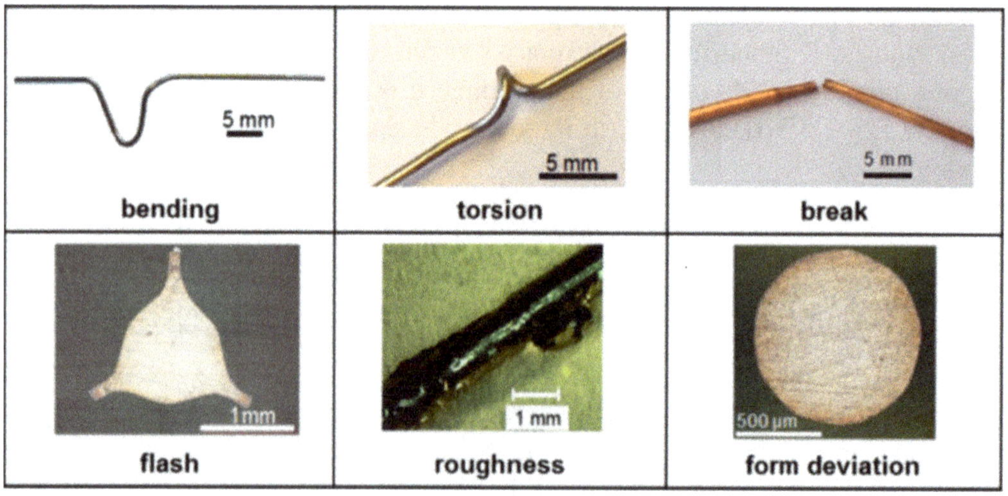

Fig. 2.17 Limitations of micro rotary swaging

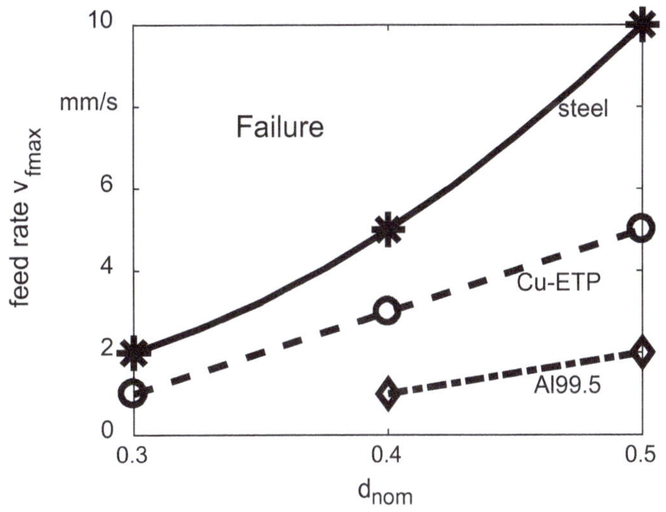

Fig. 2.18 Range of feed rates for forming different materials with different dies without failure. $d_0 = 1.0$ mm Lubricant: Condocut KNR 22

is a specific problem in micro rotary swaging due to the low stiffness of the slim workpieces. The application of a guiding channel for the workpiece minimizes the risk of bending.

Depending on the workpiece material and the diameter reduction, different maximal feed rates can be applied. Infeed rotary swaging experiments with three materials (AISI 304 steel, Cu-ETP, Al99.5) were carried out and the maximum feed rate with which 5 samples could be processed without process failure was determined. Workpieces with an initial diameter $d_0 = 1.0$ mm and a length of 110 mm were reduced by different tool sets with nominal diameters of 0.3, 0.4 and 0.5 [Kuh08b].

Forming was possible within a larger window of feed rate and thus higher V_{st} for materials with higher Young's modulus (see Fig. 2.18). Forming with feed rates above the boundary (lines) for the specific material will lead to failure on the workpiece. From the failures in Fig. 2.17, predominantly bending and torsion occur in infeed rotary swaging.

The tribological conditions and the workpiece material influence the material flow. Especially in micro rotary swaging, due to the slim workpieces, the surface-to-volume ratio is very high, whereby the tribological conditions have a strong influence. During infeed rotary swaging of workpieces from alloy 304 and from Al99.5 with and without lubrication of the workpiece, opposite behaviors of the achievable feed rate are observed [Kuh12b]. While under the dry condition (without lubrication) lower axial feed rates than with lubricant can be achieved for alloy 304, the forming of Al99.5 at higher true strain becomes possible without lubricant. In some cases, Al99.5 rods can be formed only without lubricant, i.e. a reduction from $d_0 = 1.5$ mm to $d_1 = 0.5$ mm and even to 0.4 mm.

However, during forming of aluminum without lubricant, adhesive wear can occur on the tools due to the affinity between the workpiece and tool material. By using tools with amorphous diamond-like carbon coatings, adhesive wear can be avoided and the tool life considerably increased. The tools were coated using magnetron sputtering. Pin-on-disk tests have shown that such coating results in a considerable reduction of friction and wear against aluminum.

For high volume per stroke V_{st}, inadequate material flow in the radial direction can lead to flashes on the workpiece. Because the material cannot flow completely in the axial direction, the generated surplus flows against the direction of the closing motion of the dies, which means into the gaps between the dies. This can be observed both in infeed rotary swaging and in plunge rotary swaging. Furthermore, the inadequate material flow can cause inaccurate geometries (incomplete filling of cavities or asymmetry) or destroy the workpiece irreversibly.

2.3.3 Material Flow Control

2.3.3.1 High Productivity in Infeed Swaging

In order to control the material flow against the axial feed direction and thus prevent process limitations like bending or even breaking, a spring-loaded clamping device for the workpiece was developed [Mou18a]. The clamping device allows an axial evasion of the workpiece. With a rotation lock, free rotation of the workpiece can also be prevented. Thus, the device enables the axial feed rate in infeed micro rotary swaging to be increased by more than four times compared to a fixed clamping. Feed rates higher than $v_f = 100$ mm/s were achieved. Particularly, incremental feed length l_{st} higher than the final diameter d_1 could be fed into the swaging unit, which is only realizable in the micro dimensions. However, the increased deformation per stroke due to higher feed rates leads to a higher forming, and so to high loads in the swaging head. The load leads to an elastic deformation of the outer ring and therefore causes a diameter profile.

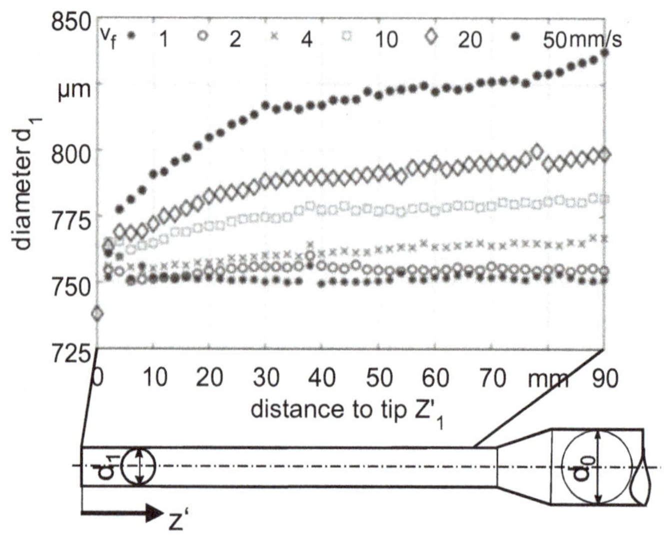

Fig. 2.19 Diameter evolution at different feed rates for AISI 304 steel $d_0 = 1.0$ mm $d_{nom} = 0.75$ mm $h_T = 0.2$ mm. Lubricant: Condocut KNR 22

Figure 2.19 shows the final diameters after a reduction of rods from AISI 304 steel with an initial diameter of $d_0 = 1.0$ mm using a die set with a nominal diameter of $d_{nom} = 0.75$ mm and a calibration zone of $l_{cal} = 20$ mm. For low feed rates ($v_f < 4$ mm/s), d_1 is constant over the workpiece length and close to d_{nom} [Mou18a]. At high feed rates, however, the diameter evolves over the workpiece axis z'. The final diameter d_1 increases with the feed rate and becomes inhomogeneous. An increase of d_1 from the tip towards d_0 can be noticed. The region with the inhomogeneity is many times larger than the calibration zone of the dies. Similar findings are also found for infeed rotary swaging of bars in the macro range, but the inhomogeneous part is much shorter.

Haug explained the conical geometry of the tip of the workpiece with the stiffness behavior of the swaging head and especially for high feed rates with a material flow in the die gaps (flashes) [Hau96]. However, in the presented experiments in the micro range no flashes occurred. Haug reduced the length of the cone by increasing the preloading of the dies [Hau96].

In micro rotary swaging the evolution of the final diameter for high feed rates can be influenced by an in-process adjustment of the radial position of the dies (see Fig. 2.20). A combined axial workpiece feed and a radial die feed with constant feed rates of $v_r = 0.008$ mm/s and $v_f = 20$ mm/s resulted in a more homogeneous diameter evolution.

The stiffness of the swaging head plays a key role in the diameter evolution in the micro range and it is therefore necessary to analyze it during the process. This can be done by monitoring either the elastic deformation of the outer ring of the machine with strain gages [Kuh11] or by monitoring the die closing with a high-speed camera. With the strain gauge measurements, an increase of the elastic deformation of the swaging head with the feed rate of $v_f \geq 2$ mm/s is observed. For lower feed rates, the deformation remains constant as in the idle state before forming. An explanation is the increase of the deformed volume per stroke V_{st} with the feed rate. As a consequence, the outer ring of the swaging head expands more and the dies do not close completely, as can be seen in Fig. 2.21b compared with

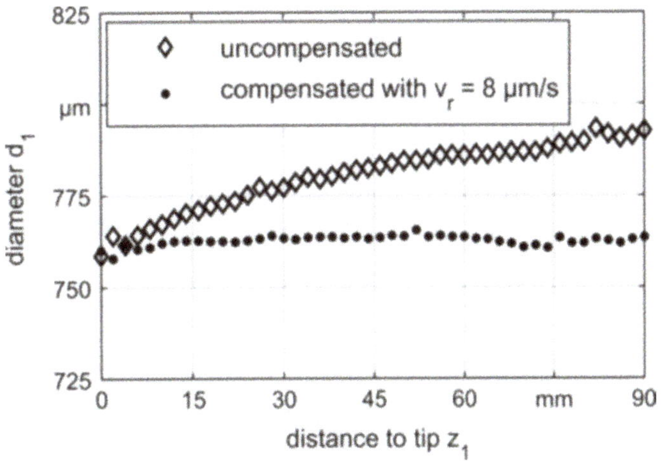

Fig. 2.20 Uncompensated diameter evolution and compensated with $v_f = 20$ mm/s, $v_r = 0.008$ mm/s Material: AISI 304 steel $d_0 = 1$ mm $d_{nom} = 0.75$ mm

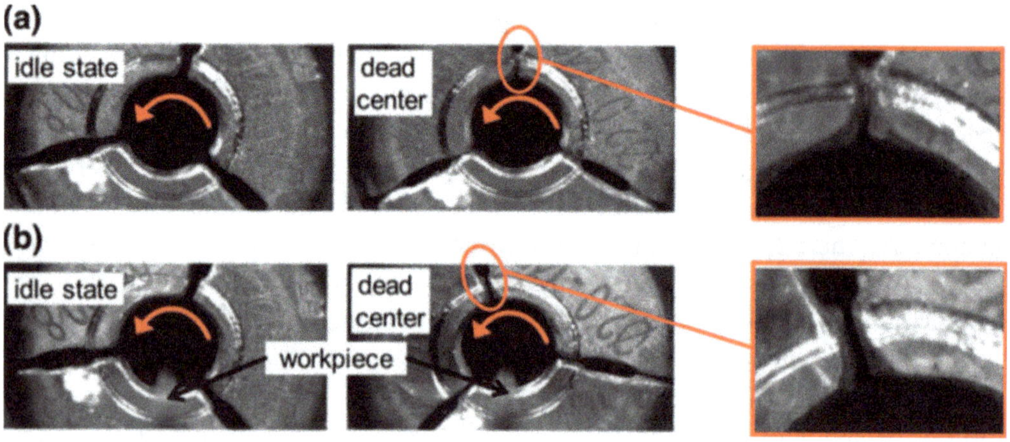

Fig. 2.21 High-speed camera images of the dies: **a** in the idle state; **b** during forming of an AISI 304 steel with $v_f = 20$ mm/s

the idle state in Fig. 2.21a. Due to the incomplete closing of the dies, the final diameter of the workpiece increases.

A rough approximation of the gap between the dies at $v_f = 20$ mm/s leads to a gap width of 55 μm and about 94 μm at 50 mm/s. The two gaps are larger than the corresponding maximal difference between the final diameter d_1 and the die diameter d_{nom} (50 μm at $v_f = 20$ mm/ and 75 μm at $v_f = 50$ mm/s) (see Fig. 2.19). Although a discrepancy exists between the final diameter and the approximated gap width, which can be attributed to difficult accessibility and the measurement conditions, the determined values point nevertheless to a correlation between both.

At high velocities the incremental volume increases. This has an impact not only on the diameter evolution of the workpiece but also on the surface quality and the roundness. However, for velocities up to 50 mm/s these workpiece properties remain almost constant, as can be seen in Fig. 2.22 for the surface roughness. The mean value of the roughness in that velocity range is 2 ± 0.6 μm. The roughness of the small parts is determined from a cross-section of the workpiece. Using the Total

Fig. 2.22 Influence of feed rate on the surface roughness. Infeed rotary swaging $d_0 = 1.0$ mm $d_{nom} = 0.75$ mm $h_T = 0.2$ mm Lubricant: Condocut KNR 22

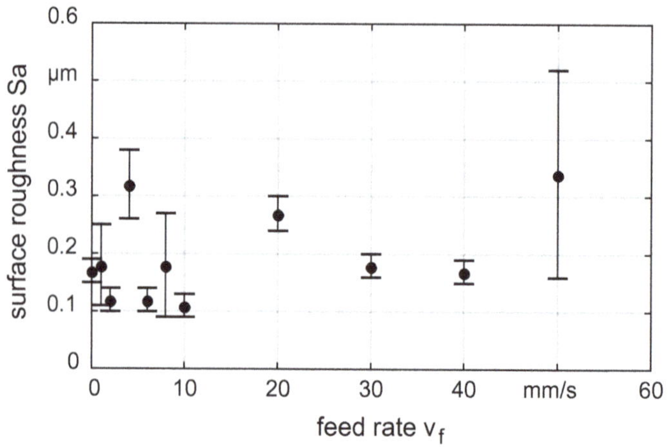

Least Square Fits method, the roughness can be calculated from the extracted contour data.

2.3.3.2 High Productivity in Plunge Rotary Swaging

The main process limitation during plunge rotary swaging is the formation of flashes. Thus, to enable high radial feed rates, the radial material flow needs to be controlled. A new concept was developed to prevent the material from flowing radially into the gaps between the dies by the use of intermediate elements (IE). The IEs could be made of super-elastic metallic or polymeric substances (see Fig. 2.23). However, the elastic behavior of the currently available super-elastic metallic materials is still too low to allow high deformation as in rotary swaging. The use of elastomers in metal forming was summarized in [Thi93], and in [Chu14] a comparable approach for micro patterning was investigated.

Copper rods from Cu-ETP with $d_0 = 1.5$ mm were reduced using dies with $d_{nom} = 0.98$ mm. The stroke was $h_T = 0.3$ mm and different radial feed rates v_r were used. Furthermore, different IEs made of thermoplastic polyurethane with three different hardness grades were tested (low hardness = ILH; middle hardness = IMH and high hardness = IHH) [Mou18a]. For the reduction without IE, a maximum radial feed rate of $v_r = 0.34$ mm/s could be applied before flashes or breaks occurred. The IEs with high hardness enabled an increase of the feed rate up to about 300% (see Fig. 2.24a). The different IEs resulted in different final diameters: for harder IE the diameter increased due to the higher extensions of the outer ring of the swaging head. The extension was provoked by the IE that flowed into the gaps and blocked the closing of the dies. This resulted in an increase of the workpiece's final diameter. The tested IEs were not reusable (see Fig. 2.24b).

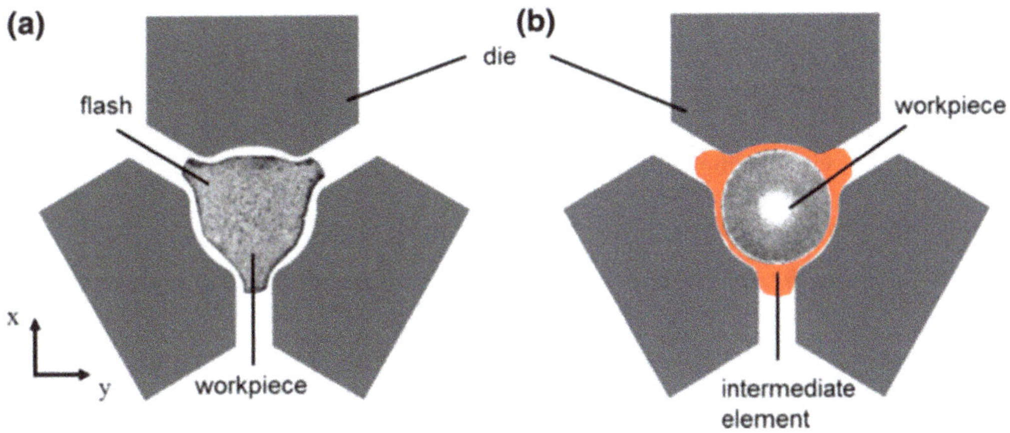

Fig. 2.23 Formation of flashes during rotary swaging: **a** workpiece with flashes; **b** inserted elastic intermediate elements

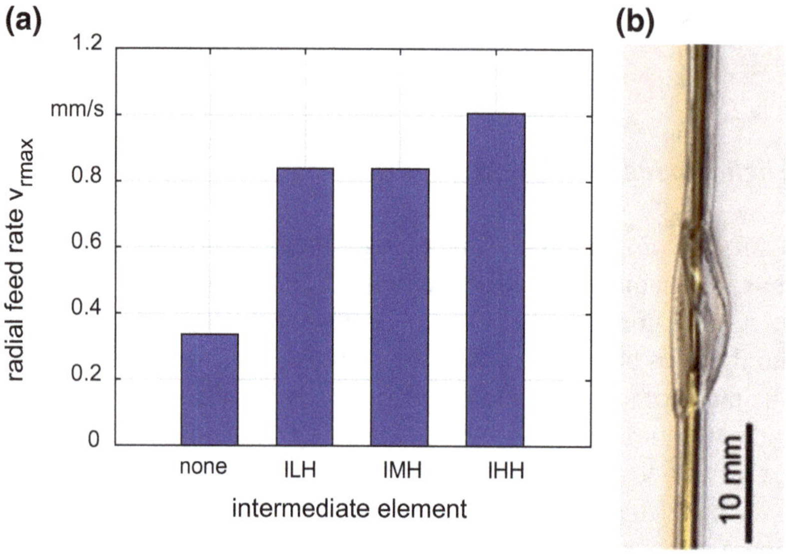

Fig. 2.24 Plunge rotary swaging with elastic intermediate elements: **a** maximal radial feed rate for different intermediate elements; **b** workpiece with intermediate elements after forming

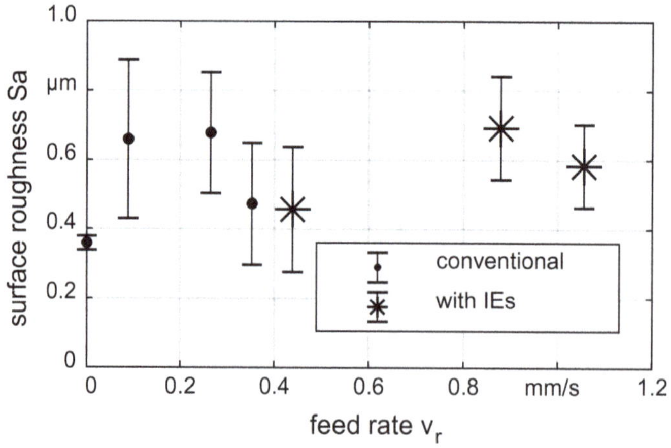

Fig. 2.25 Influence of feed rate and intermediate element on the surface roughness. Plunge rotary swaging $d_0 = 1.0$ mm $d_{nom} = 0.98$ mm $h_T = 0.3$ mm Lubricant: Condocut KNR 22

The surface roughness Sa and the corresponding standard deviation of the produced parts are higher than in the initial rods (Fig. 2.25). However, the increase of the productivity through the IEs does not affect Sa negatively.

2.3.3.3 Application of External Axial Forces in Plunge Rotary Swaging

Controlling the axial workpiece displacement in plunge rotary swaging influences the formation of the final geometry and the axial elongation of the workpiece. A practical way to control the material flow is by applying external axial forces F_a on the workpiece during the forming. These forces are in the order of a few Newton in contrast to the radial forming forces, which are in the range of kN.

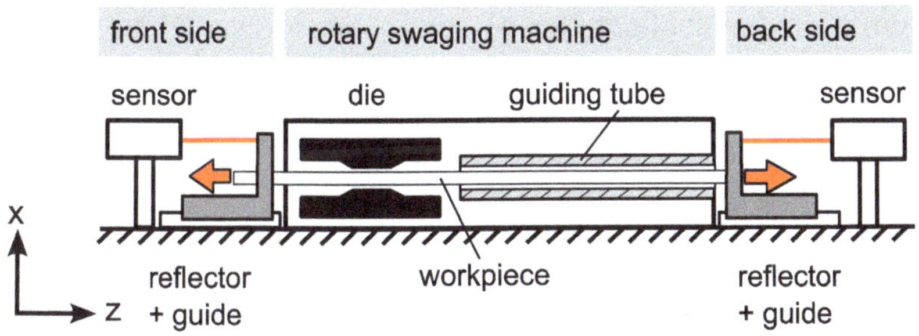

Fig. 2.26 Setup for micro rotary swaging with external axial forces

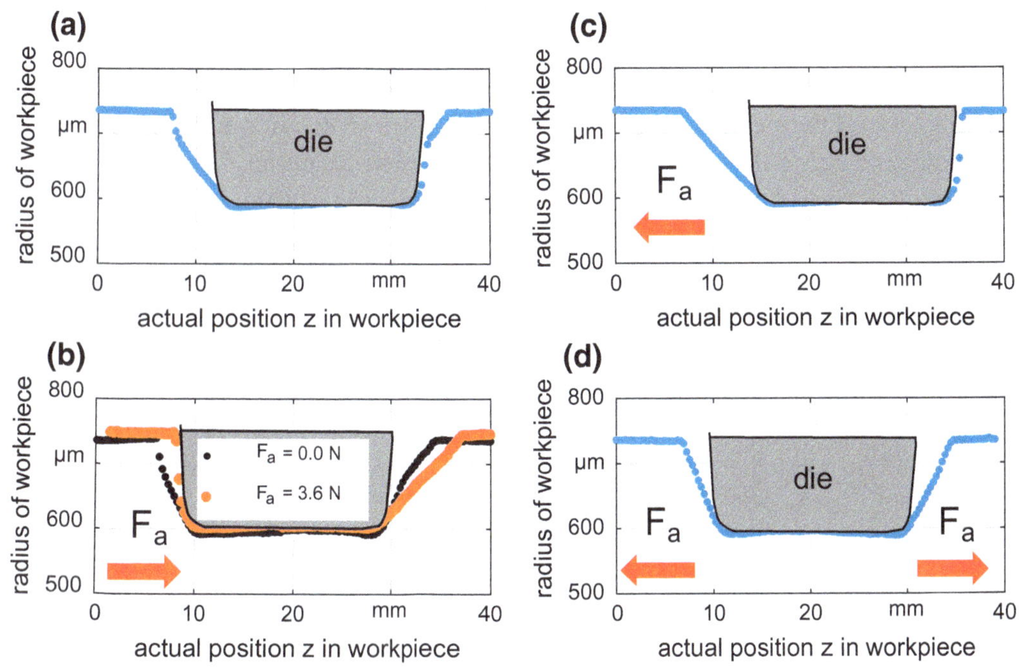

Fig. 2.27 Workpiece geometry after plunge rotary swaging without and with external forces F_a: **a** no F_a; **b** compression force on one side; **c** tension force on one side; **d** tension forces on both sides [Mou18b]

Forming experiments were conducted with alloy 304 with $d_0 = 1.5$ mm and a reduction by $\Delta d = 0.3$ mm by otherwise constant swaging parameters. Four different configurations without (A) and with external forces (B, C and D) were used with the setup illustrated in Fig. 2.26. In configuration B a compressive force was applied in the front, in configuration C a tensile force was applied in the front and in configuration D tensile forces were applied equally in the front and at the back of the rotary swaging machine [Mou18b]. The arrows in Fig. 2.26 represent the tensile forces in configuration D.

The geometry of the parts was investigated after forming. Symmetrical geometries were generated only when the same axial tensile forces were applied on both ends of the workpiece (see Fig. 2.27). Due to the freely moving workpiece in the axial direction for configuration A, the geometry varied strongly. But with external forces only on one end of the workpiece, the geometry shifted in one direction. The material shift can be guided by the external axial forces and, thereby, the produced geometry can be controlled without changing the dies.

2.3.4 Characterization of the Material Flow with FEM

Using the finite element method (FEM), the material flow during the process is investigated. Two-dimensional axisymmetric simulations are commonly used to simulate rotary swaging [Gha08]. The model for infeed rotary swaging consists of two parts: the workpiece (deformable) and the die (rigid) [Mou14a].

One approach to characterize the material flow in rotary swaging is by analyzing the behavior of the neutral zone (NZ). The NZ represents an area in the region of the workpiece being deformed in which the workpiece material features no flow in the axial direction. A position closer to the initial diameter means a higher material flow in the direction of the feed motion. The location of NZ can vary due to the tribology. With a higher friction coefficient, the position of the NZ is shifting in the direction against the feed motion (−z) [Ron06]. Furthermore, the NZ is influenced by v_f as well other process settings, like the final diameter. It was found that the NZ changes its position as well as its shape during a single stroke (see Fig. 2.28). The gray area of the workpiece represents the material flowing against the feed direction, the bright area the material flowing in the feed direction and the boundary between both is the NZ. By an increase of the coefficient of friction value μ, the material flow in the feed direction within one stroke seems to increase. In order to analyze the NZ, a stroke was divided into many small and constant time steps. The number of detected NZ was found not to be equal to the number of time steps but to vary with the friction, the strain and the feed rate. The number of NZ increased with μ and furthermore a shift of the NZ in the negative z-direction was noticed (see Fig. 2.28a and b). The increased number of NZ is a sign of more internal deformation. As the NZ shape can change from concave to convex, the internal deformations are cyclic. Less material flow against the feed direction was observed in experiments without lubricant and thus with higher friction coefficients as in the FEM (see Sect. 2.3.2). For higher axial feed rates a similar trend to that for higher friction was observed, which can be explained by the higher volume per stroke which is deformed (see Fig. 2.28b and c) [Mou14a]. From these results, the improvement of the feed rate and strain range in the forming of the Al99.5 without lubricant can be attributed to the material flow, and more precisely the behavior of the neutral zone.

A further method to characterize the material flow dependent on the process parameters is the history of the plastic strain development. The plastic strain at the outer region of the workpiece is more sensitive to the friction coefficient value (see Fig. 2.29). This can also be seen at the NZ, since the shape is more curved at the surface for higher friction coefficients. But compared to rotary swaged rods in the

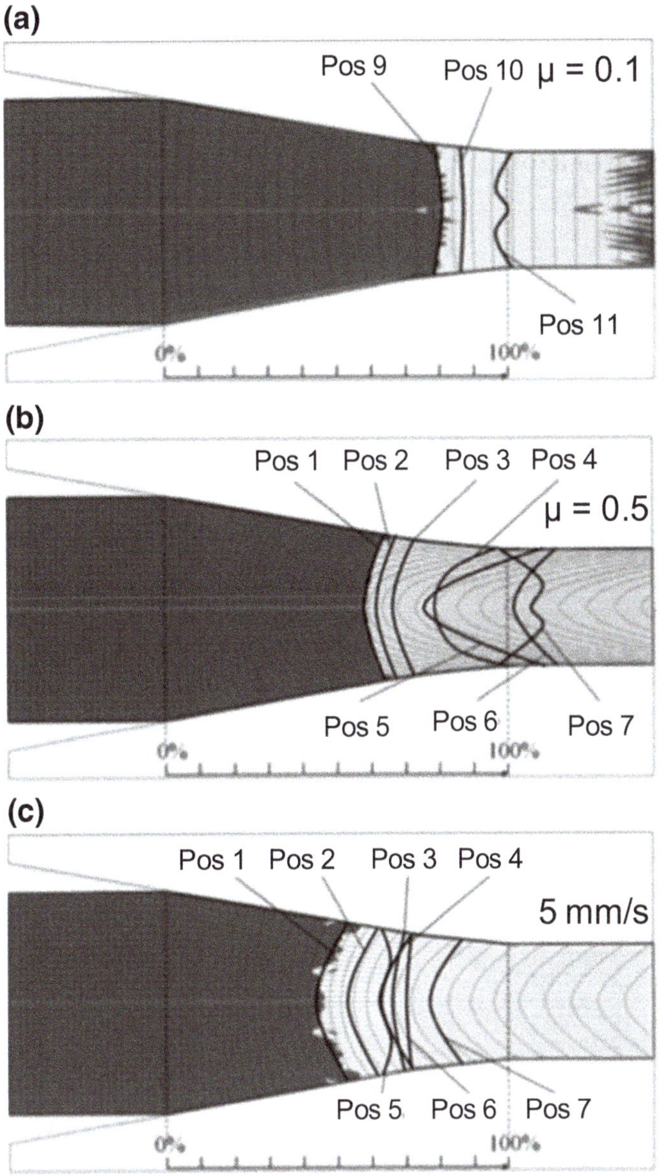

Fig. 2.28 Shape and position of the neutral zone in one single stroke of the forming of Al99.5 rods: **a** $v_f = 1$ mm/s and $\mu = 0.1$; **b** $v_f = 1$ mm/s and $\mu = 0.5$; **c** $v_f = 5$ mm/s and $\mu = 0.5$

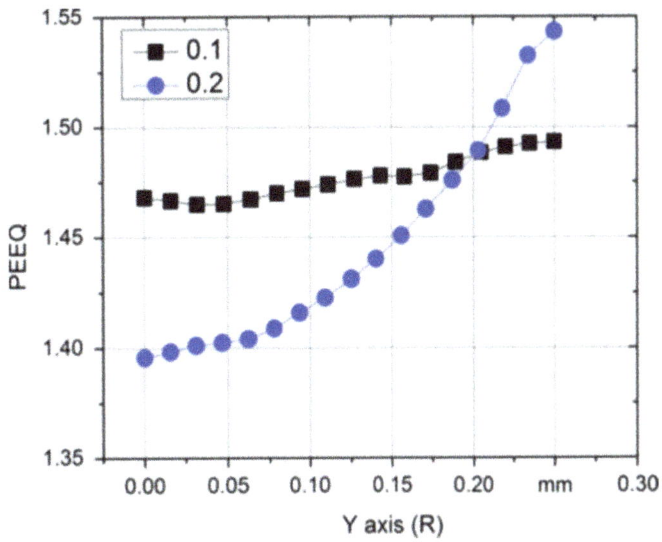

Fig. 2.29 Plastic strain PEEQ over the workpiece radius after deformation for $\mu = 0.1$ and $\mu = 0.2$

macro range, at which only the outer 20% of the diameter is sensitive to friction [Liu17], in the micro range the PEEQ (plastic strain) is influenced over almost the complete radius. This can be explained by the small cross-section of the workpieces in which the surface to volume ratio is very high, thus the tribological conditions have a stronger influence.

2.3.5 Material Modifications

The interaction between the dies and the workpiece during forming leads to material modifications, such as in the mechanical properties and the microstructure. Figure 2.30 presents the Martens hardness in five different regions along a deformed workpiece of alloy 304. The scatter of the Martens hardness after forming (regions I to IV and beginning of V) reaches up to 250 N/mm^2 compared to about 160 N/mm^2 before forming (end of region V). The high scatter occurred because of the inhomogeneity of the microstructure (soft austenite and hard strain-induced martensite) and also the distribution of the hardness across the diameter, as can be

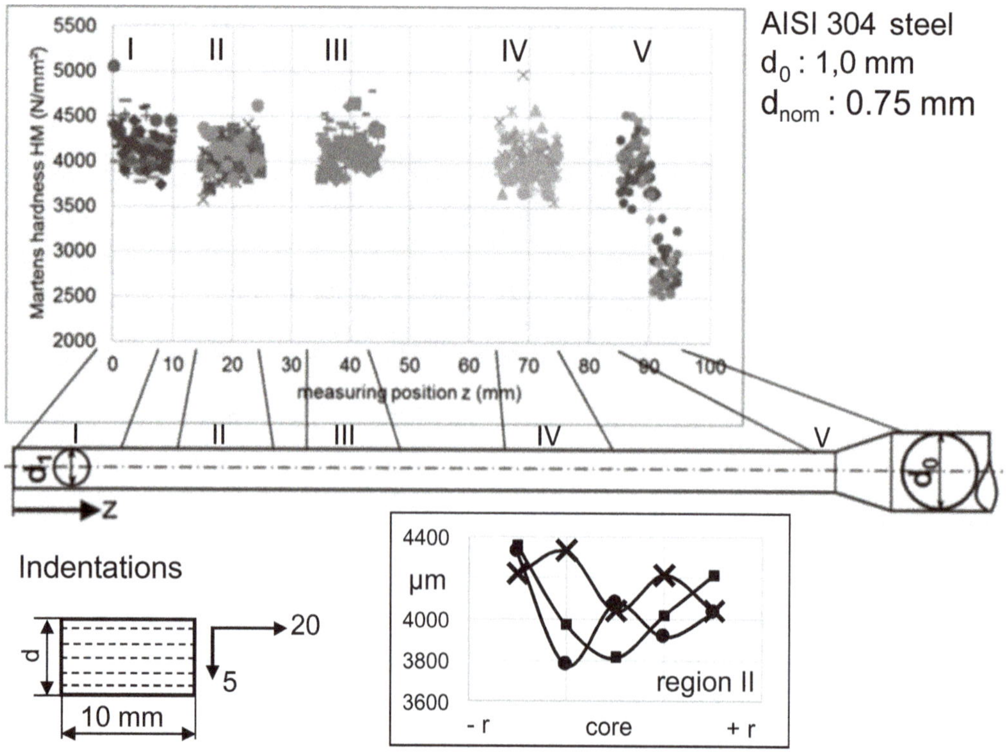

Fig. 2.30 Martens hardness of the workpiece measured after forming in five different regions, detailed sketch of the location and number of indentions, and three hardness profiles over the diameter in region II with hardness values between 3800 and 4400 N/mm^2

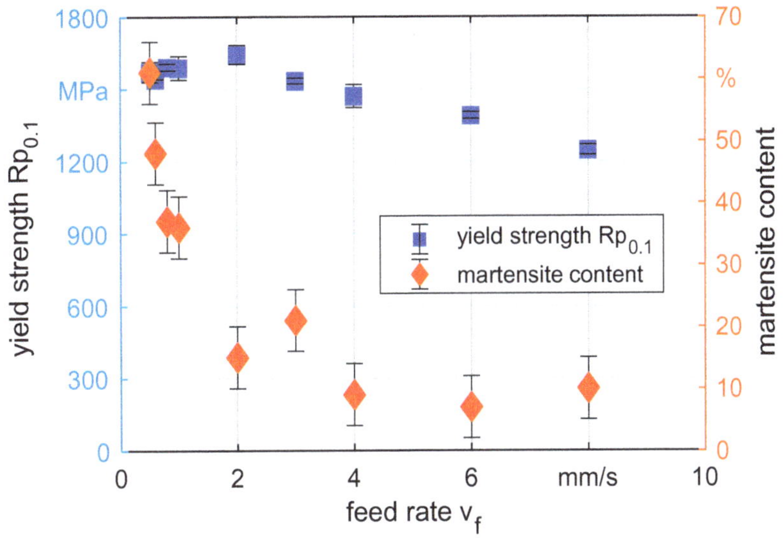

Fig. 2.31 Workpiece properties after forming: yield strength (tensile test) and martensite content (magnetic induction) of AISI 304 steel

seen in detail in Fig. 2.30, where three hardness profiles in region II across the diameter [−r, +r] are presented. The observed distributions reflect inhomogeneous work hardening through the diameter.

Workpieces of AISI 304 steel after rotary swaging present a strong hardening, as the yield strength increases significantly (Fig. 2.31). This increase is generally attributed to work hardening during cold forming. However, further specific phenomena like phase transition as well as dynamic effects influence the development of the material properties.

As is known for AISI 304 steel, martensite usually develops during cold forming. Using a Fischerscope MMS PC that works according to the magnetic induction method, the martensite content can be estimated (Fig. 2.31). The development of martensite with regard to the feed rate in the presented case can be approximated by Eq. 2.7. The equation is derived from the profile in Fig. 2.31.

$$\alpha_M = C * v_f^{-n} \tag{2.7}$$

α_M is high for low v_f and decreases when v_f increases. As the workpieces have the same initial diameter and nominal true strain, this behavior can be attributed to the number of strokes experienced by the workpiece and to an adiabatic heating during forming. At each stroke, new volume fractions of austenite are transformed into martensite, hence more martensite is present at low feed rates. When the feed rate increases, the deformed volume at each stroke also increases, which means the deformation work is higher. As a result, the workpiece temperature increases and the martensitic transformation is reduced. The yield strength, determined by tensile testing, is also high for workpieces manufactured at low feed rates and decreases almost linearly for the feed rate range between 2 and 8 mm/s. While the amount of martensite is reduced from 60 to 10%, the yield strength is reduced only by about 30% within the investigated feed rate range. Therefore, the martensite-induced transformation in micro rotary swaging is sensitive to feed rate changes.

2.3.6 Applications and Remarks

Micro rotary swaging is an eligible convenient process when parts with high strength and high surface quality are needed [Mou14b]. Due to its incremental characteristic, it allows larger cross-section reductions compared to other forming processes. Moreover, not only rotationally symmetrical cross-sections are possible but also polygons. The smaller dimensions of the workpiece compared to the dimensions of available machines open new possibilities for adjusting rotary swaging for micro rotary swaging. Besides the manufacturing in linked parts (Sect. 2.3.2) several workpieces can be formed at the same time (multi-forming) with adequate die geometries. Figure 2.32 presents examples of micro parts generated by micro rotary swaging.

Important factors for generating good parts in micro rotary swaging are a deep understanding of the flow behavior of dissimilar (a) or similar (b) materials; the control of the relative motion between the workpiece and the dies (b), (c); the control of the positioning in process and the tool design (e). The part (a) in Fig. 2.32 shows a composite component made of two different materials (copper in the core, and AISI 304 steel as shell) by infeed rotary swaging [Kuh10]. The softer material in the core guarantees a tight connection of the two components [Kuh12a].

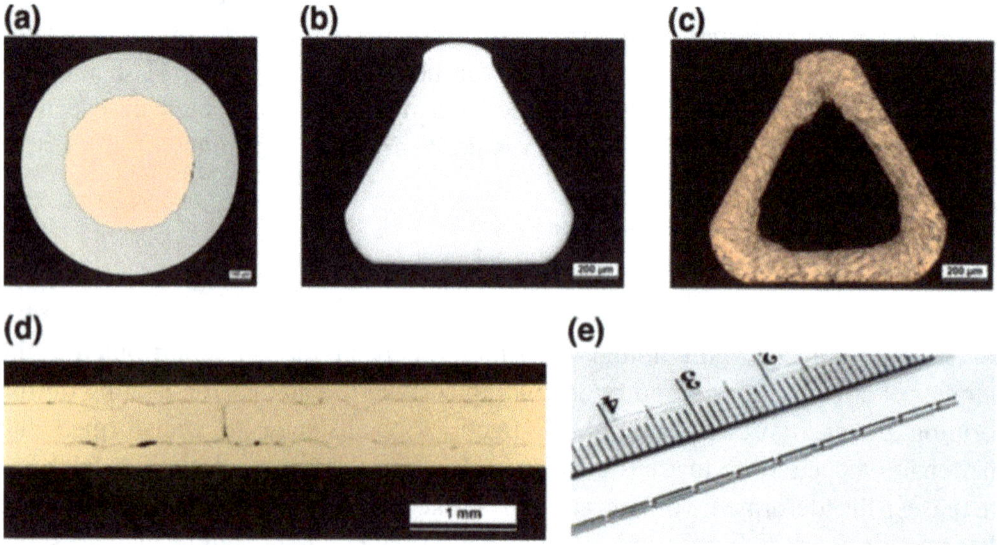

Fig. 2.32 Parts made by micro rotary swaging: **a** composite component consisting of a core from Cu-ETP and a shell from AISI 304 steel; **b**, **c** triangular cross-sections from AISI 304 and Cu-ETP; **d** axially joined parts by forming, **e** multi forming of notches

Part (d) was made by joining two rods [Mou14c] in plunge rotary swaging, using a tube as the connecting component. In both cases (a) and (d), the connection between the joining partners can be a form and a force fit. The correct relative rotational motion between the workpiece and the dies not only allows further material modifications (Sect. 2.4) but also leads to more possible geometry variations. For (b) and (c) this relative rotation was kept to zero and the calibration zones of the dies were flat. Using a mandrel with tubes, polygons like (c) can be manufactured. In (e) nine notches (depth 0.25 mm) manufactured in three steps are presented. Because the notches are small the feed rates remain the same for one or more elements when manufactured at the same time, but the productivity is increased with regard to the number of geometries designed in the die.

By the control of the material flow in micro rotary swaging with different measures, the process can be improved in the sense of productivity, the accuracy of the final product or even material modifications. Using a spring-loaded clamping device, the axial material flow in infeed rotary swaging can be controlled and thus enables about four times higher axial feed rates. With intermediate elements around the workpiece in plunge rotary swaging, an undesired radial material flow into the gaps between the dies can be prevented and, as a result, the radial feed rate can be significantly increased. Furthermore, external forces applied on one or both ends of the workpiece allow the material shift to be guided, thus the final geometry of the product can be influenced.

In infeed rotary swaging, the final diameter can be constant or have a profile over the workpiece length, depending on the feed rate used. For low feed rates, the generated diameter is closer to the nominal diameter of the dies d_{nom} and is constant along the workpiece length, besides the tip. At high feed rates, an inhomogeneity of the diameter over the total part length is produced. The diameter increases with the feed rate and from the tip to the end of the deformed part. The diameter increase can be explained by the machine resilience, which leads to an incomplete closing of the dies during the forming with high feed rates as in the idle state. However, the rising value of the final diameter over the length of the workpiece can be overcome by using another machine with less resilience or by controlling the position of the dies by adding a radial feed during the infeed process.

In plunge micro rotary swaging, intermediate elements around the workpiece enable up to three times higher radial feed rates v_r compared to forming without these elements. This is because flashes on the workpiece can be prevented. However, the final diameter increases slightly, which depends on the stiffness of the elastic intermediate element, with bigger diameters for harder elements. This fact has to be considered during the die design.

A further approach to control the material flow during plunge rotary swaging involves the external forces at the workpiece ends. The material flow reacts very sensitively to these. Low forces at one side can direct the material flow so that it occurs in a preferred direction. A symmetrical material flow is possible if equal external axial forces are applied on both sides of the forming zone. Thus, the final shape of the workpiece can be controlled.

For the process and especially for the material flow in micro rotary swaging, specific micro effects have to be considered. The first particularity are the slim workpieces, which tend to bend, break or even get into the die gaps, so a guiding element needs to be applied during forming. Furthermore, the surface-to-volume ratio is much higher and for that reason the friction condition has a greater effect on the process. The friction influences, for example, the plastic strain in a thicker area from the surface of the workpiece. In addition, the feasible axial feed rates reach such high values that the feed length during one stroke exceeds the workpiece's initial diameter. For this reason, the radial forces are high and the outer ring of the swaging head expands, which leads to an inhomogeneous diameter evolution. The previous results present the potential of rotary swaging for micro manufacturing.

Latin

Variable	Unit	Explanation
C	(−)	Material constant
d_0	mm	Initial diameter
d_1	mm	Final diameter
d_{nom}	mm	Diameter of the forming zone of a closed die set
f_{st}	s^{-1}	Stroke frequency
F_a	N	External axial forces
$F_{b,crit}$	N	Critical bending load
h_T	mm	Stroke height
h_{st}	mm	Radial die feed per stroke
l_0	mm	Initial length of the deformed workpiece part
L_0	mm	Initial length of the workpiece
l_1	mm	Final length of the deformed section
L_1	mm	Final length of the workpiece
l_{cal}	mm	Length of the calibration zone of the die
l_{form}	mm	Length of the forming zone of the die for plunge swaging
l_{st}	mm	Axial workpiece feed per stroke
n	(−)	Hardening behavior
v_f	mm/s	Workpiece feed rate
v_r	mm/s	Die feed rate
V_{st}	mm^{-3}	Volume displaced by the dies per stroke

Greek

α	°	Tool angle
α_M	%	Volume fraction of martensite
Δd	mm	Diameter difference
φ_{st}	(−)	Incremental deformation degree
μ	(−)	Friction coefficient

2.4 Conditioning of Part Properties

Svetlana Ishkina*, **Christian Schenck and Bernd Kuhfuss**

Abstract Micro rotary swaging is an incremental cold forming process that changes the geometry and the microstructure of the swaged parts. Due to the opportunity to produce parts with an accurate diameter and to influence the surface and the mechanical characteristics by infeed rotary swaging, it is possible to prepare micro parts for further forming processes such as extrusion. This targeted conditioning was further enhanced by modifications of rotary swaging tools and kinematics. The formability of the modified workpieces was reflected by the required forming force of the subsequent extrusion process, where not only a force increase but also a force reduction was observed with modifications in rotary swaging. In this section a process chain "rotary swaging – extrusion" for austenitic stainless steel AISI304 is shown.

Keywords Incremental forming · Microstructure · Process modifications

2.4.1 Introduction

Cold forming of complex workpieces is usually accompanied by work hardening and is usually realized by multi-stage forming processes. In addition, difficult to form materials have to be annealed between the forming steps due to the strain hardening. This heat treatment contributes to a substantial rise of costs for equipment and energy supply. One possibility to avoid this intermediate heat treatment and to improve the formability of the workpieces is a targeted conditioning of the workpieces for further forming. This conditioning includes adjusting the forming characteristics of the workpiece, the geometry (diameter) and the surface (roughness, lubrication pockets). Very good formability with such characteristics as good strength in combination with a sufficient ductility provide ultrafine-grained (UFG) materials. Refining of materials can be generated by applying severe plastic shear deformation, e.g. by pressing the workpiece into a die with an angular channel [Val00]. Further improvement of the forming process can be achieved by reducing the friction between the workpiece and the forming die. This can be realized, for example, by modifying the profile and the surface topology of the blanks [Ish17a].

Micro infeed rotary swaging not only changes the geometry and the surface of the swaged parts, but also influences the microstructure and, thus, the mechanical properties of swaged workpieces. Due to the special adjustments of the machine, it was possible to vary the produced diameter [Kuh13] and the roughness of the swaged parts [Ish15c]. Moumi et al. investigated the change of the microstructure after rotary swaging depending on the feed velocity as well as the increase or decrease of the martensite content with the variation of the forming temperature [Mou15a]. They deduced that rotary swaging could provide the opportunity to adjust the forming characteristics of the workpieces for further forming steps [Ish15a]. However, the process was to be enhanced in terms of tool geometry and process kinematics.

2.4.2 Process Chain "Rotary Swaging—Extrusion"

2.4.2.1 Modifications of the Die Geometry

The usage of curve-shaped dies (CSD) is typical of rotary swaging of circular tubes and rods (Fig. 2.33a). These dies feature a curved surface in the reduction zone, where the diameter of the part is reduced, as well as in the calibration zone, where the diameter of the part is defined. However, the simulation analysis with FEM software showed that the shear strain distribution depended on the geometry of the dies [Ish15a] (see Fig. 2.34). This planar model represented the reduction of the work-piece cross-section at the forging zone (the part of the reduction zone where forming occurred) after the first revolution of the dies and at the end of the reduction.

An increase of the shear strain (PE12) was observed in the simulations with flat shaped dies (FSD) compared to the curve-shaped dies. These dies featured a curved surface in the reduction zone and a flat surface in the calibration zone (Fig. 2.33b). Further modifications of the die surface in the calibration zone were introduced with double-flat-shaped dies (DFSD), which featured a flat surface in both the reduction and the calibration zone (Fig. 2.33c).

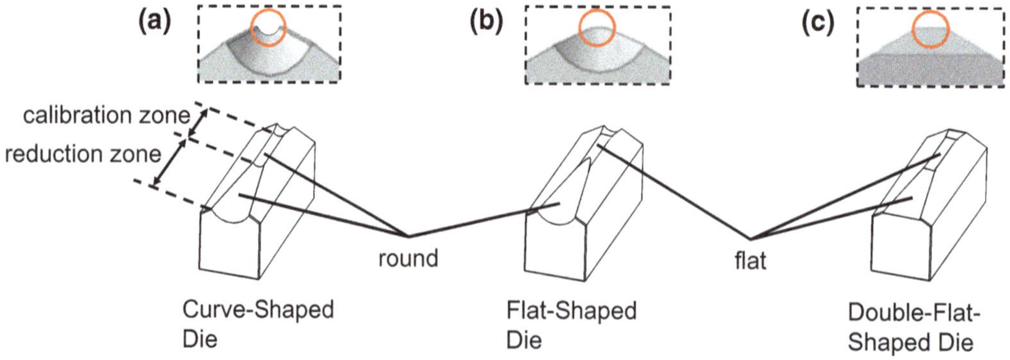

Fig. 2.33 Evolution of the swaging dies from curve-shaped dies (CSD) to double-flat-shaped dies (DFSD) to increase the shear strain in the formed workpieces

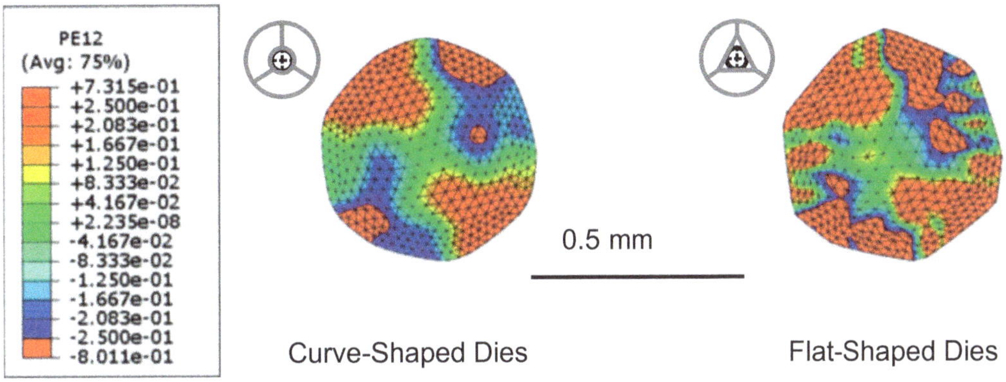

Fig. 2.34 Distribution of the shear strain PE12 at the end of the reduction zone

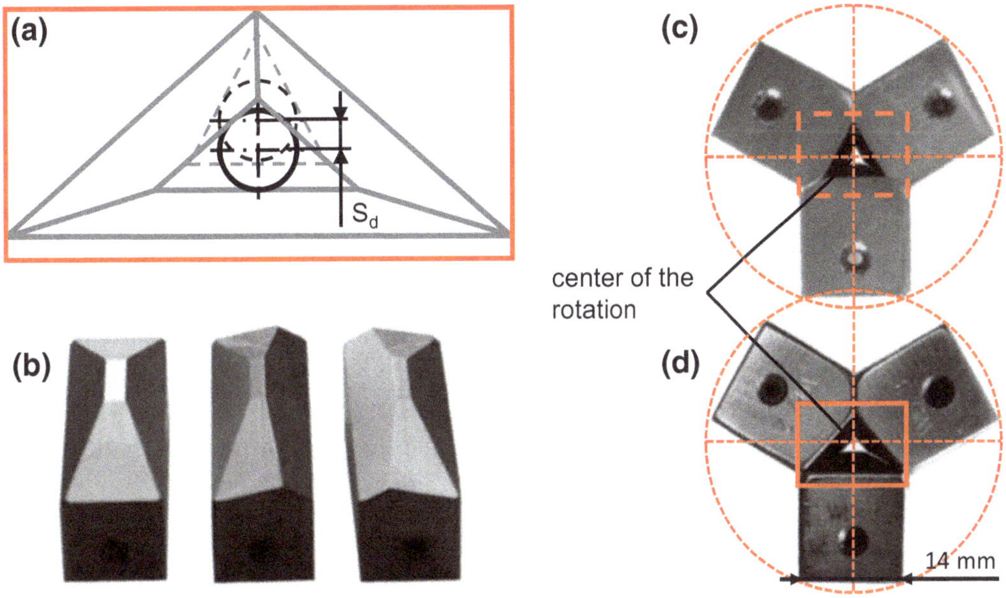

Fig. 2.35 Eccentric rotary swaging: **a** principal shift of center line; **b** eccentric dies; closed double-flat-shaped dies for **c** centric rotary swaging; **d** eccentric rotary swaging

2.4.2.2 Modifications of Process Kinematics

Eccentric Rotary Swaging

Further increase of the shear strain in the workpiece was realized by eccentric rotary swaging. For this aim, new swaging dies were designed (see Fig. 2.35). The aim of these dies was to shift the center line with every stroke of the tools during their rotation (Fig. 2.35a) [Toe18].

Control of the Stroke-Following Angle

Angular driving of the part allowed control of the stroke-following angle $\Delta\phi$, which was defined as the difference between the rotation angle of the workpiece and the rotation angle of the dies during two consecutive strokes. When the angular velocity

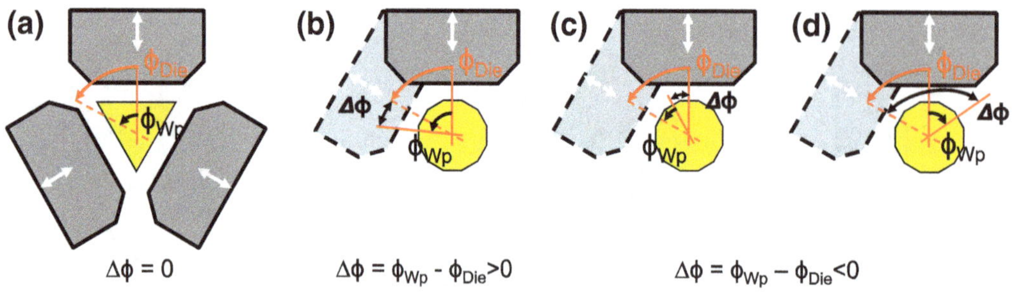

Fig. 2.36 Variation of the stroke-following angle $\Delta\phi$: **a** without; **b** positive; **c** and **d** negative

of the tools was kept constant at $\omega_{Die} = 89.23\ ^{rad}/_{s}$ and the part was not rotating ($\omega_{Wp} = 0\ ^{rad}/_{s}$), the stroke-following angle $\Delta\phi$ between two successive strokes (stroke frequency $f_{st} = 102\ s^{-1}$) resulted in $\Delta\phi = 51°$ (see Eq. 2.8):

$$\Delta\phi = \frac{360° * (\omega_{Die} - \omega_{Wp})}{2\pi * f_{St}} \tag{2.8}$$

In the case of the same rotation speed of the workpiece and the flat-shaped dies, the stroke-following angle $\Delta\phi$ equals zero and the dies impact the workpiece at all times at the same circumferential line (Fig. 2.36a). The result of this forming process is a triangle. In other cases, the workpiece can rotate either in the same direction or opposite to the dies rotation. When they rotate in the same direction, a positive stroke-following angle $\Delta\phi$ (Fig. 2.36b) is achieved when the workpiece rotates faster than the dies. If the workpiece rotates more slowly, the stroke-following angle $\Delta\phi$ is negative (Fig. 2.36c). If the workpiece rotates in the other direction (Fig. 2.36d), the stroke-following angle $\Delta\phi$ remains negative as well.

2.4.2.3 *Extrusion*

Ten samples were examined in each series of experiments. For a comparison of the forming characteristics, the samples "NS" (not swaged) in the initial stage were turned from non-swaged parts. The diameter of the extrusion bore was $d_{1m} = 1.30$ mm (Fig. 2.41a). The diameter of the lower bore was $d_{2m} = 1.00$ mm. All parts were radially preloaded in a metal sleeve. Forming was preceded with a manual pre-lubrication of the extrusion die. The press force was obtained by a piezo-based force sensor to characterize the formability of the swaged parts (Fig. 2.37b).

In order to regard the polygonal geometry and small variations in the diameter of the samples d_1, the forming force F was related to the cross-sectional area A of the respective swaged workpieces. All curves had a similar gradient angle in the elastic range (at the beginning of the process) and differed during the plastic deformation.

2.4.2.4 *Experimental Design*

The forming parameters for all experiment series are summarized in Table 2.2.

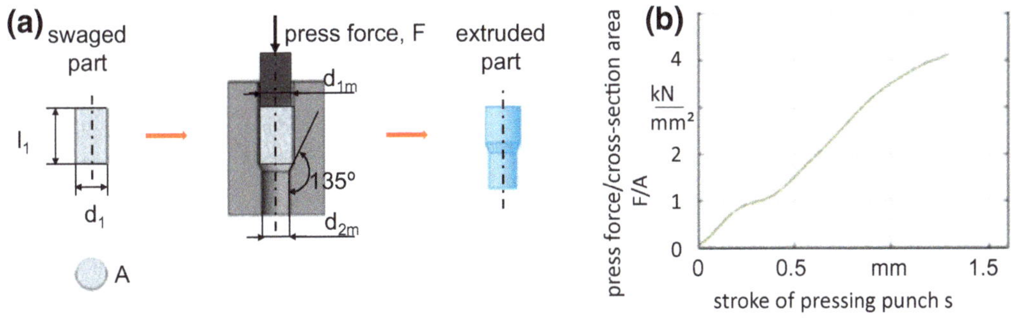

Fig. 2.37 Forward extrusion process: **a** schema; **b** characterization curve

Table 2.2 Forming parameters: rotary swaging

Test number	Die	Feed velocity v_f, mm/s	Stroke- following angle $\Delta\phi$	Shift of the center line S_d, mm
I	CSD	1.0	51°	0
I-a		2.0		
II	FSD	1.0		
II-a		2.0		
III	DFSD	1.0		
III-a		0.5	60°	
III-b			51°	
III-c			40°	
III-d			30°	
IV	EFSD	1.0	51°	0.3
IV-a		2.0		

For the extrusion, all the swaged parts with the diameter $d_1 = 1.28 \pm 0.01$ mm were cut into small parts with a length of $l_1 = 3.0$ mm. The samples were pressed with a constant pressing velocity of v = 0.1 mm/s.

2.4.3 Results and Discussion

The analysis of the characterization curves reflected the possibility of conditioning of part properties by means of modifications in rotary swaging. The influence of the modifications in the swaging die design on the extrusion force is shown in Fig. 2.38. The stroke s of the pressing punch varied with the actual initial length of the samples. Curve NS represents the unswaged material during extrusion. The analysis revealed that rotary swaging with curve shaped dies increased the required extrusion force (curve I above curve NS). In contrast, using flat-shaped and double-flat-shaped dies decreased the required forces significantly (curves II and III

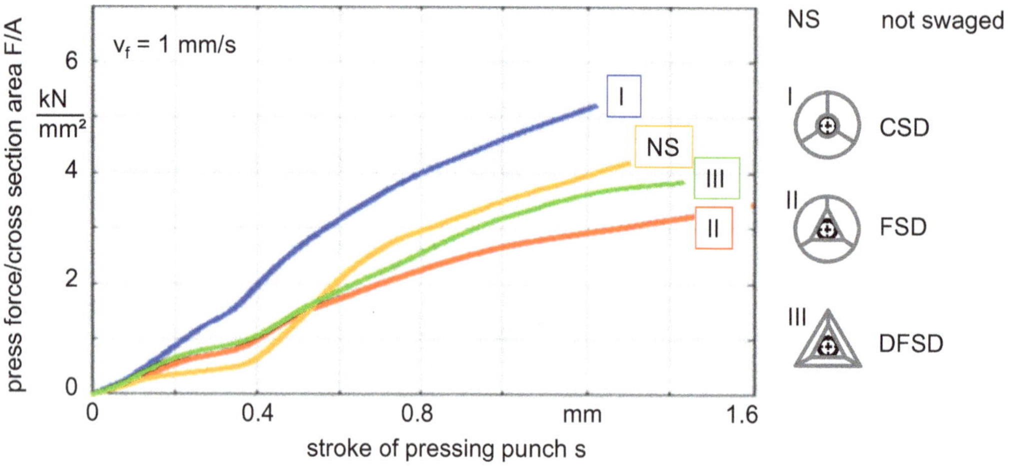

Fig. 2.38 Characterization curves based on tool design. I: curve-shaped dies (CSD), II: flat-shaped dies (FSD), III: double-flat-shaped dies (DFSD), NS: not swaged

are below curve NS). This difference can be explained by various aspects, for example a change of the microstructure, a reduction of the grain size or martensite and carbide formation [Ish17b].

The microstructures are shown in Fig. 2.39. Independent of the die design, a clear grain refinement in the radial direction resulted from the swaging process. However,, the grain boundaries after swaging with curve-shaped dies (Fig. 2.39b) were more easily detectable than after swaging with flat- or double-flat-shaped dies (see Fig. 2.39c and d).

Another aspect is the roughness and the topology of the surface. The surface roughness of the initial stage samples revealed a value of $S_a = 0.92 \pm 0.16$ µm. The geometry (cross-section) of swaged stainless steel (alloy AISI304) specimens using the curve-shaped dies as well as the flat- or double-flat-shaped dies featured a circular geometry. However, the value of surface roughness S_a of the latter two grew to $S_a = 1.50 \pm 0.35$ µm and $S_a = 1.11 \pm 0.61$ µm compared to the conventional swaged workpieces with $S_a = 0.41 \pm 0.05$ µm (settings II, III and I). Hence, the workpieces swaged with curve-shaped dies tended to provide smoother surfaces, but the flat- or double-flat-shaped dies promoted pocket formation on the sample surface. Consequently, different die designs, curved or flat, led to significantly different forming properties [Ish18].

Although flat-shaped dies allow a significant reduction of the yield stress [Ish17b, Ish15a] and work hardening, an increase of feed velocity from $v_f = 1$ mm/s (settings I and II) to $v_f = 2$ mm/s (settings I-a and II-a) decreased this improvement. The required extrusion force increased when forming with flat-shaped dies (curve II-a higher than curve II) (Fig. 2.40), while the feed velocity did not affect the extrusion force when using curve-shaped dies (curve I equals curve I-a).

The roughness of the workpieces swaged with flat-shaped dies was influenced by the feed velocity v_f as well (decreased to $S_a = 0.94 \pm 0.23$ µm), while swaging

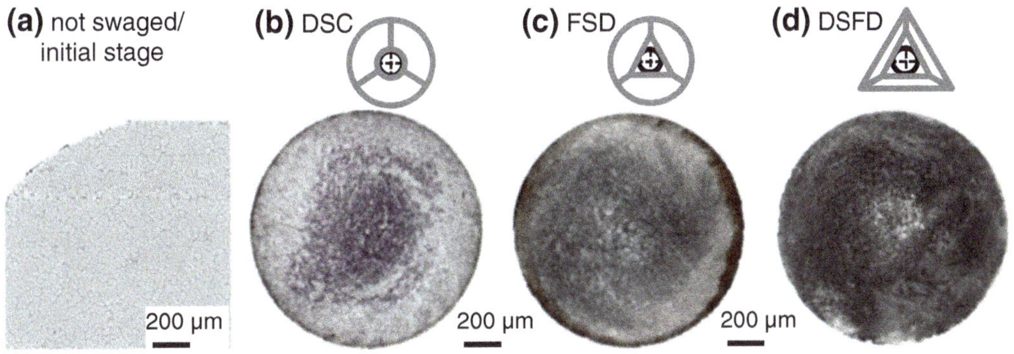

Fig. 2.39 Microstructure (cross-section) of the workpieces **a** before forming; and after rotary swaging at the setting parameters **b** I, **c** II and **d** III

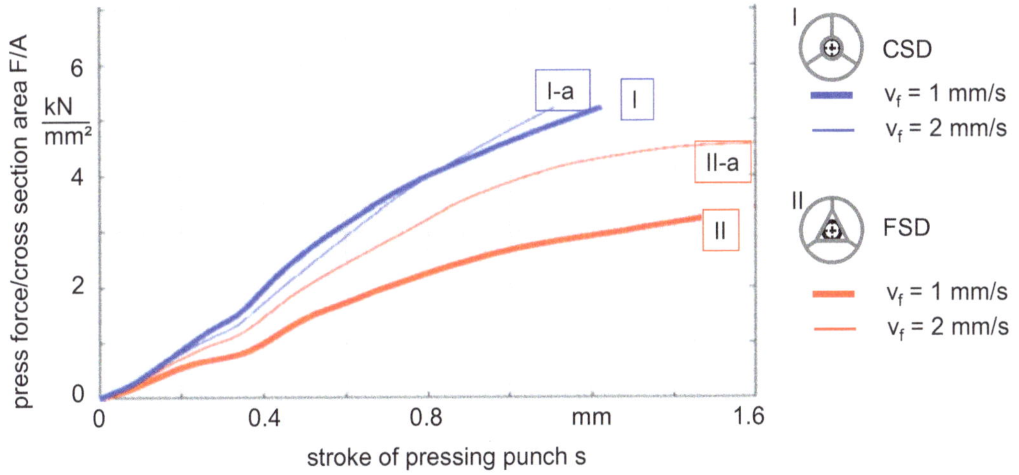

Fig. 2.40 Characterization curves based on feed velocity v_f (I and I-a: curve-shaped dies (CSD), II and II-a: flat-shaped dies (FSD))

with curve-shaped dies at a higher feed velocity $v_f = 2$ mm/s changed this insignificantly ($S_a = 0.58 \pm 0.18$ µm) (Fig. 2.41).

These results correspond to the martensite content, which was more influenced by the feed velocity if flat-shaped dies were applied [Ish17b]. Additionally, the number of strokes was higher with lower feed velocity. The number of strokes influenced the cross-sectional shape more when using flat-shaped dies, due to the pockets on the surface (see Fig. 2.45). The microstructure was refined more by the higher degree of shear strain [Ish15a].

Another major influence on the extrusion characteristics was the modification of the process kinematics. The use of the eccentric flat-shaped dies changed the process kinematics by means of radial displacement of the center-line during the process and thus resulted in a higher shear strain. Due to a significant reduction of the grain size and the martensite content compared to the rotary swaging using

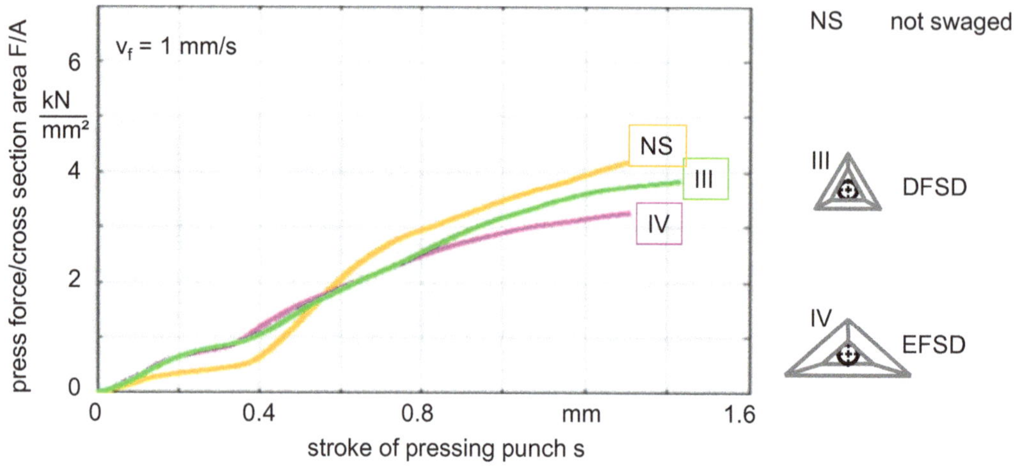

Fig. 2.41 Characterization curves based on process kinematics: eccentric flat-shaped dies (EFSD, setting IV) compared with double-flat-shaped dies (DFSD, setting III)

double-flat-shaped dies [Ish17b], a reduction of the required press force during extrusion could be observed (Fig. 2.41, curve IV below curve III).

Eccentric rotary swaging influenced also the development of the microstructure. As a result, the transformed martensite showed a strong reduction compared to the amount of martensite after conventional rotary swaging. Thus, pre-forming of the workpieces in this special way affected work softening, reduction of yield stress, and increase of the plastic strain [Ish17b]. The grain boundaries were detectable in the core area of the processed workpieces while they were not visible any more in the outer regions (Fig. 2.42b). Furthermore, the specimens exhibited typical eddy patterns and a spiral grain orientation, which were more pronounced with a feed velocity of $v_f = 2$ mm/s (setting IV-a) (see Fig. 2.42c). Moreover, the shape of these workpieces was a polygon with eight corners. Using this method, the surface roughness was also influenced by the feed velocity. The workpieces swaged at a feed velocity of $v_f = 1$ mm/s (setting IV) featured a roughness of

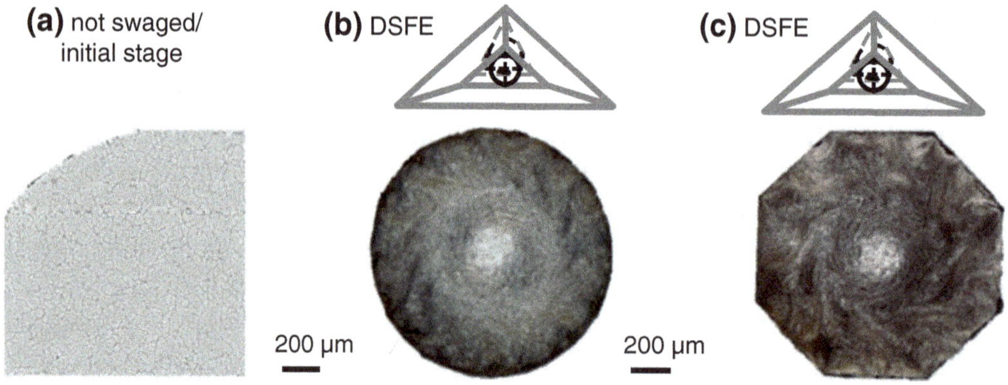

Fig. 2.42 Microstructure (cross-section) of the workpieces **a** before forming; and after eccentric rotary swaging at feed velocity of **b** $v_f = 1$ mm/s (setting IV); **c** $v_f = 2$ mm/s (setting IV)

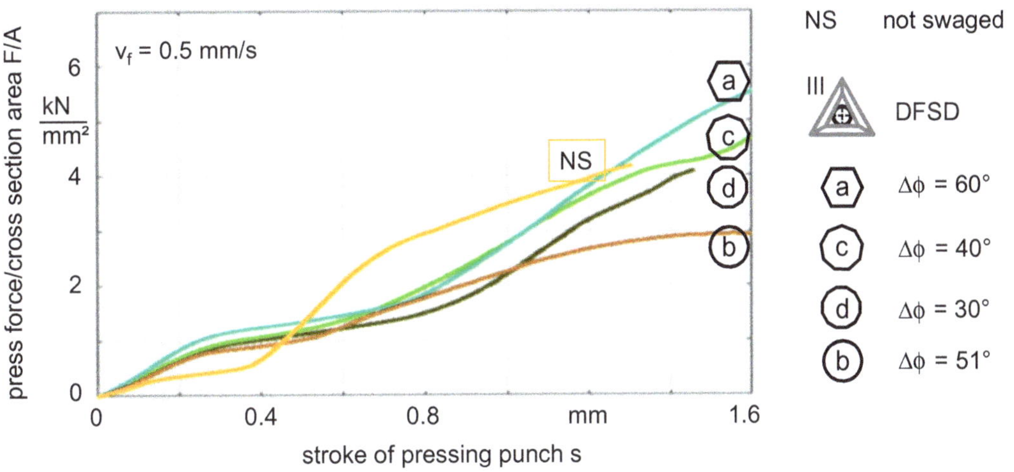

Fig. 2.43 Characterization curves based on process kinematics: stroke-following angle $\Delta\phi$ using double-flat-shaped dies (settings III-a to III-d)

$S_a = 1.09 \pm 0.26$ μm, which rose at the increased feed velocity of $v_f = 2$ mm/s (setting IV-a) to a value of $S_a = 1.36 \pm 0.48$ μm.

A variation of the stroke-following angle $\Delta\phi$ allowed the production of parts with polygonal geometries. The shape of the conventional swaged workpieces using curve-shaped dies was actually circular. The usage of flat-shaped dies or double-flat-shaped dies in combination with the targeted adjustment of the stroke-following angle $\Delta\phi$ delivered polygonally shaped parts with several facets. The shaping correlates to Eq. 2.9, where n is the number of corners:

$$n = 360^\circ/\Delta\phi \qquad (2.9)$$

With the increasing number of facets the characterization curves became flatter (Fig. 2.43).

In dependence on the stroke-following angle $\Delta\phi$, not only the microstructure but also the surface roughness was influenced. A stroke-following angle $\Delta\phi = 60^\circ$ (setting III-a) delivered a hexagon (see Fig. 2.44a), with a surface roughness of $S_a = 0.45 \pm 0.16$ μm. The round shaped workpieces formed by $\Delta\phi = 51^\circ$ (setting III-b) featured the highest surface roughness value for these experiments with $S_a = 0.72 \pm 0.26$ μm. Workpieces swaged by $\Delta\phi = 40^\circ$ (setting III-c) developed the shape of a nonagon (see Fig. 2.44c), and a surface roughness of $S_a = 0.66 \pm 0.36$ μm. All dodecagons were swaged by $\Delta\phi = 30^\circ$ (setting III-d) (see Fig. 2.44d), and had a roughness of $S_a = 0.53 \pm 0.17$ μm.

The workpieces swaged with curve-shaped dies revealed a circular cross-section and a smooth surface (Fig. 2.45a). Swaging with the stroke-following angle of $\Delta\phi = 51^\circ$ (non-rotating workpiece, setting III-b) using double-flat-shaped dies led also to a round geometry (Fig. 2.44b), but many small bevels (see Fig. 2.45b),

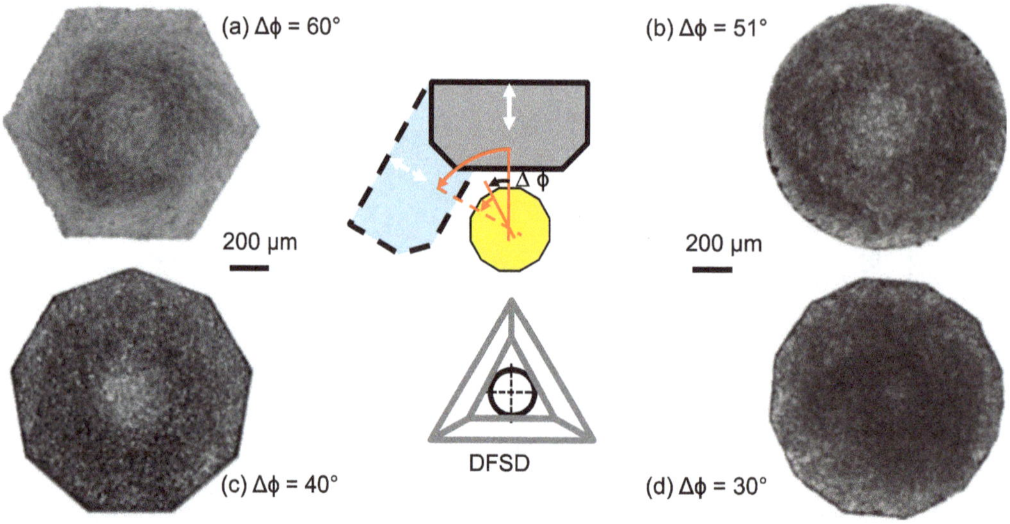

Fig. 2.44 The influence of the stroke-following angle on the geometry and the microstructure (cross-section) of the swaged workpieces using double-flat-shaped dies at feed velocity of v_f = 0.5 mm/s (settings III-a to III-d)

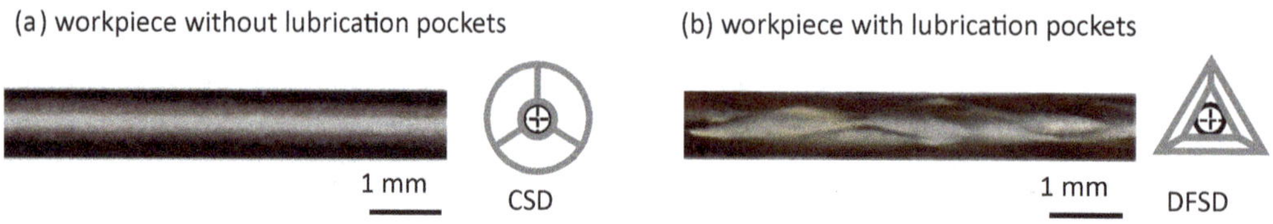

Fig. 2.45 Influence of the swaging dies on the creation of lubrication pockets by non- rotating workpiece $\Delta\phi = 51°$

which are dependent on the feed velocity v_f. The swaging dies did not strike along the same line of the workpiece any more and the resulting shape formed was not a polygon. The surface roughness increased to the value of $S_a = 0.72 \pm 0.26$ µm.

2.5 Influence of Tool Geometry on Process Stability in Micro Metal Forming

Lewin Rathmann*, Lukas Heinrich and Frank Vollertsen

Abstract Deep drawing is a well-suited technology for the production of hollow metallic micro parts due to its excellent qualities for mass production. However, the downscaling of the forming process leads to new challenges in tooling and process design, such as high relative deviation of the tool geometry and increased friction. In order to overcome these challenges and use micro deep drawing processes in an efficient production system, a deeper understanding of the micro deep drawing process is necessary. In this investigation, an overview of the substantial influences on process stability in micro deep drawing and a method for description, called "tolerance engineering", are presented.

Keywords Micro forming · Deep drawing · Tool geometry

2.5.1 Introduction

Deep drawing is a well-suited technology to produce such parts due to its excellent qualities for mass production. However, the downscaling of the forming process leads to new challenges in tooling and process design. Cheng et al. investigated how the interaction of free surface roughening and the size effect affect the micro-scale forming limit curve [Che17]. They modified the original Marciniak-Kuczynski model by introducing free surface roughening to describe the decrease of the micro-scale forming limit curve as a result of the geometry and grain size effects.

Saotome et al. showed that the process window decreases with increasing ratio between the punch diameter and sheet thickness when thin foils are used [Sao90]. In preparation for the deep drawing process, Gau et al. expanded the process window for forming micro cups with cup height to outer diameter ratios by annealing stainless steel 304 sheets at temperatures not less than 900 °C for more than 3 min [Gau13]. In 2014, a novel technique in micro sheet metal forming was presented by Irthiea et al. [Irt14]. They used a flexible die to produce micro metal cups from stainless steel 304 foils. The die consisted of a cylindrical tank filled with a rubber pad. Forming parameters like anisotropy of SS 304 material and friction conditions at various contact interfaces were investigated experimentally as well as in simulation with this setup.

Messner et al. demonstrated that increasing punch forces can be expected in the micro range due to increasing friction [Mes94]. Furthermore, Michel et al. showed that the yield point and stress decrease with decreasing foil thickness [Mic03]. Besides material and friction, the geometry of the forming tools has a significant impact on the process forces, stress states and the process window of the deep drawing process.

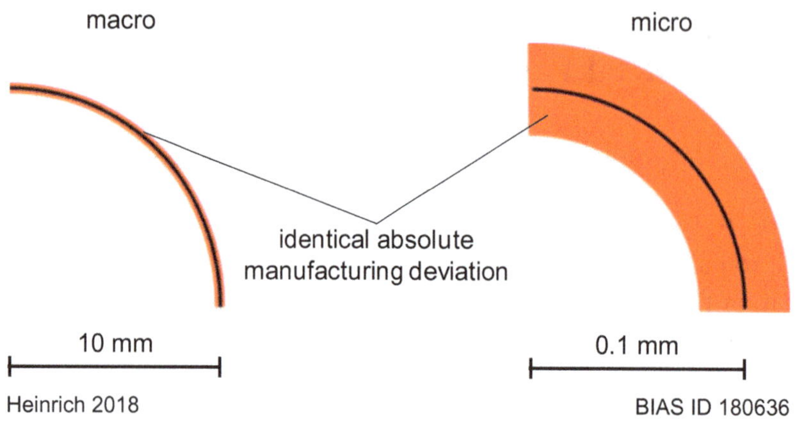

Fig. 2.46 Relative deviations in manufacturing of a forming tool due to identical absolute deviations in macro and micro range

Aminzahed et al. present in [Ami17] a piezoelectric actuator as a new approach in a deep drawing operation to draw rectangular foils. They also studied blank holder effects on thickness distributions, punch force and spring-back and used experiments and simulations to obtain results.

Wagner et al. demonstrated that the punch radius has a significant influence on the process force and that higher punch radii are beneficial for the process [Wag05]. Thus, the quality of the produced parts depends on the choice of suitable geometric parameters of the forming tools. In conventional deep drawing processes, manufacturing deviation in the sub-millimetre range in tooling do not affect the deep drawing process since these deviations are neglectable in size compared to the dimensions of the tools. Moreover, small deviations are compensated by the formability of the workpiece material. Due to size effects, the tribology changes in the micro range, which leads to smaller process windows in micro deep drawing [Vol09]. If scaled down, the deep drawing process becomes more sensitive to process deviations. This is due to the relative deviations of the tool geometry, caused in tool manufacture. These increase with decreasing size in the micro range because the accuracy of manufacturing reaches its limits [Hu09]. Figure 2.46 illustrates the effect of larger relative manufacturing deviations in the micro range. Due to these effects, micro forming processes cannot be simply transferred to the micro range and further investigation is needed. Therefore, the aim of this study is to determine the influence of tool geometry on micro forming processes to allow a specific process design and improved process stability as well as a quantitative assessment of the effect of wear- and production-related deviations of tool geometry.

2.5.2 Experimental Setup

For the deep drawing experiments described in this investigation, blanks were cut out using a picosecond pulsed laser with a wavelength of 1030 nm in order to prevent burr formation at the edge of the blank. The diameters were checked with a Keyence VHX 1000 digital microscope. For the tools, ledeburitic powder-metallurgical steel 1.2379 (X153CrMoV12) was used. The relevant geometric parameters of the tools used in the experiment were measured using a laser scanning microscope Keyence VK 9700. The drawing process was carried out on a single-axis micro forming press with a maximum punch force of 500 N and a constant punch velocity of 10 mm/s using HBO 947/11 as lubricant. The punch force was measured with a Kistler 9217A piezo load cell with an accuracy of 0.01 N. The punch stroke was measured with a Heidenhain LS477 linear scale with an accuracy of 1 μm. The press was driven by a servo motor controlled by a NI 9514 servo drive interface. The blank holder acts passively. It uses its own weight and is supported by two springs. The blank holder pressure can be adjusted by changing the spring tension. The blank positioning was realized by an automated positioning system. The blank was positioned with a pneumatic gripper that is driven by a linear direct drive cross table. The position of the blank was then measured using an Allied Vision G 917 B monochrome CCD camera with a resolution of 9 MP equipped with a telecentric lens with built-in coaxial illumination and a magnification of 0.75. With this setup the blanks can be positioned within a radius of 10 μm from the center of the drawing die.

2.5.3 Numerical Models

The Finite Element Method (FEM) is a proven tool for routine calculation tasks and is used in plant and mechanical engineering as well as in vehicle construction. This method enables largely realistic statements in the stages of concept finding and the development of components and structures using computer-assisted simulations of physical properties. This results in a significant reduction of product development time [Kle15]. For FEM analysis, the software Abaqus 6.14 was used. A 3-dimensional model for micro deep drawing was used. All tools are defined as analytical rigid shell objects and the blank was defined as a deformable body using the 8-node linear brick 3D-stress element C3D8R for the mesh. Within the sheet thickness four elements were used. In order to save computational time, only one half of the blank was considered. For the elastic plastic material model of the foil, tensile tests were performed to determine the required flow stress curves.

2.5.4 Circular Deep Drawing

The punch force presents an important parameter to assess the process stability of deep drawing processes. In [Beh13] the influence of variation of the tool geometry on the punch force is given. By using experimental data and FEM simulation, each geometry variation is classified regarding its impact on the punch force and therefore on the process stability, as shown in Fig. 2.47.

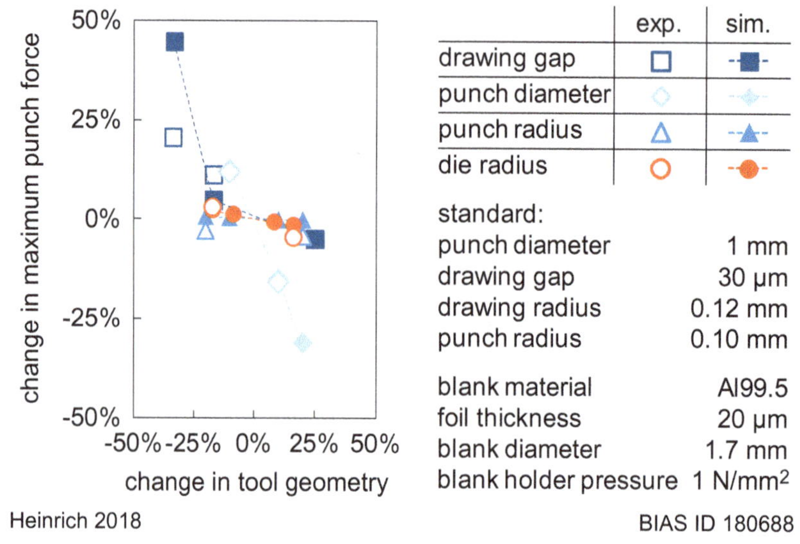

Fig. 2.47 Influence of tool geometry variation on the maximum punch force in circular micro deep drawing

Figure 2.47 shows that the change of punch diameter and drawing gap resulted in the greatest impact on the punch force and therefore should be carefully controlled during tool manufacturing. Changes in the punch radius, drawing die radius and edge shape, on the other hand, proved to be of minor importance.

The limiting drawing ratio LDR describes the maximum value of the drawing ratio that can be achieved under the given process conditions. A larger achievable LDR acts as an indicator of a more stable deep drawing process and therefore makes it attractive for industrial use. In [Beh16] the influence of the blank material and tool geometry variation on the LDR is described. It is shown that, regardless of the tool geometry, the austenitic steel 1.4301 enables the largest process window compared to the foil materials Al99.5 and E-Cu58, and therefore proves to be most stable in micro deep drawing processes, as shown in Fig. 2.48.

For all the materials used, the die clearance and die radius have the biggest influence on the LDR. While the die radius should be increased, a decrease of the drawing clearance below 1.25 times the foil thickness is beneficial to the drawing process.

The investigation of the influence of a tool geometry deviation on the part geometry is mentioned in [Hei17]. In this case, the influence of an uneven drawing radius deviation was investigated for different drawing ratios using experimental data and FEM simulation. The deviation of the die radius is described by the ratio of curvature (ROC), which is defined as the ratio between the maximum and minimum curvature of the die edge. This sort of variation has a significant influence on the part shape, resulting in uneven cup height after the drawing process due to stronger stretching of the cup wall at positions with a smaller die radius. This effect increases with higher drawing ratios. Furthermore, it is shown that by using FEM simulation an initial blank displacement sufficient to result in an even cup height can be found.

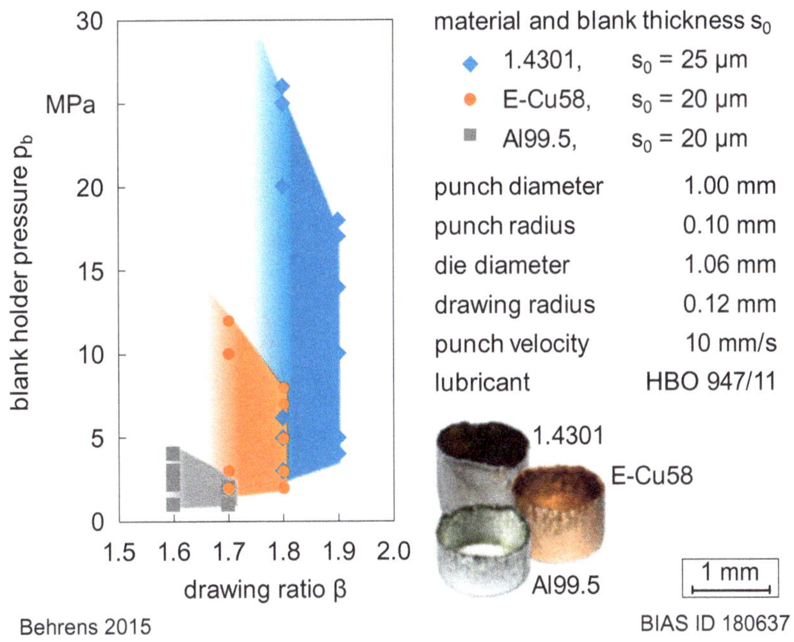

Fig. 2.48 Process window for different materials in micro deep drawing

However, the effect of an uneven drawing radius cannot be compensated entirely. While an appropriate blank displacement results in an even cup height when an uneven drawing radius is used, a significant difference in strain remains in the part, as shown in Fig. 2.49. This is explained by the different mechanisms that result in an uneven cup geometry. While blank displacement leads to material surplus on one side of the cup, an unsymmetrical radius deviation on the other hand generates uneven stretching of the cup wall. It can be concluded that the effect of an uneven radius geometry proves to be of major importance to produce accurately shaped micro cups.

The investigation of the interactions between individual sub-processes in multi-staged micro deep drawing and their impact on the overall process is time-consuming and cost-intensive when done only experimentally. Thus, a simulation model of a two-stage micro deep drawing process is developed in [Tet15] and showed that this model can display the first and second stroke in one forming simulation in accordance with the experiments conducted. Figure 2.50 shows the overview of the two-stage micro deep drawing process simulation model. It consists of two pairs of dies and punches with different diameters and a blank holder impinging the blank with pressure. After the first stroke, the second punch clamps the cup on the radius of the second die while the first punch and die move backwards. Then, the second punch with a smaller diameter starts moving, which draws the first cup into the second die, which results in a second cup with a smaller diameter and higher cup walls.

This model is the basis for further investigations with two-stage simulation models with varied tool geometries. This enables an identification of the influence

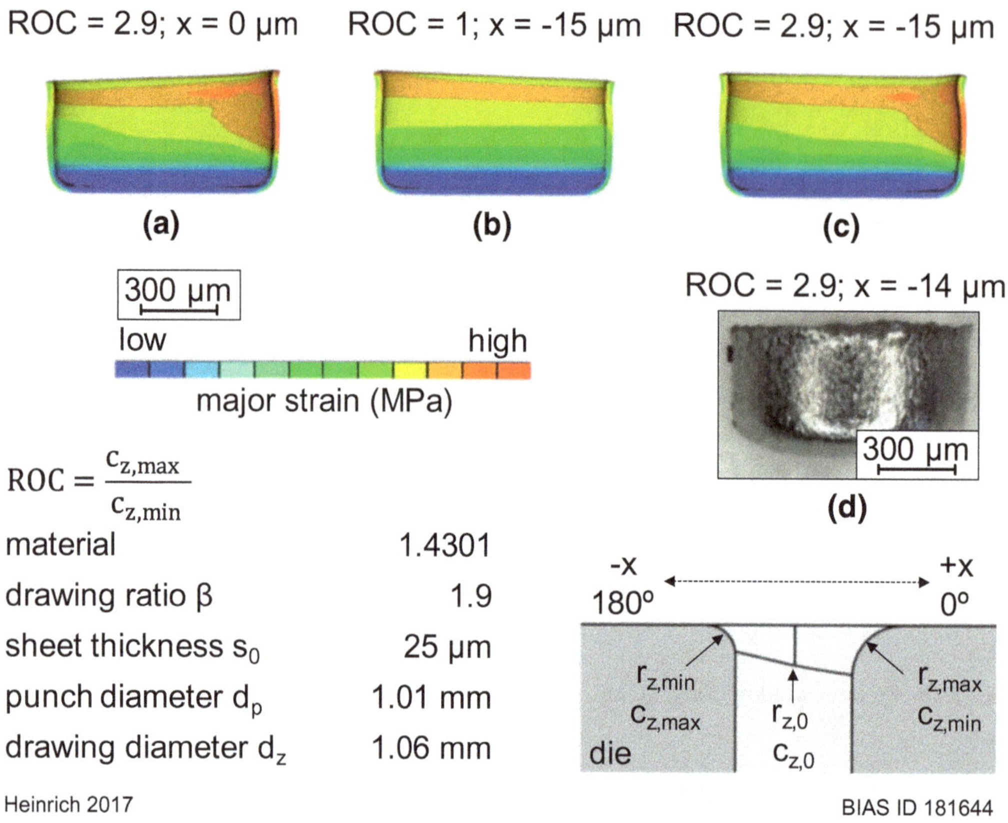

Fig. 2.49 Influence of uneven die radius described by the ratio of curvature ROC and blank displacement x on micro deep drawing. Uneven die radius and centered blank (**a**), even die radius and blank displacement (**b**), uneven radius compensated by blank displacement in FEM (**c**) and experiment (**d**)

of opportune or inopportune in series connected tool geometry. The investigated characteristics are the curvature of the rib, the mean wall thickness alteration and the maximum punch force. The geometry is varied in the range of ±3.5% around the initial tool geometry to investigate these influences. This variation represents unpreventable deviations that occur when fabricating real tools. It also makes it possible to use a maximum and minimum tool geometry in the simulation model. Hence, these options are expressed in two adjustments. On the one hand, (+) expresses the positive deviation of +3.5% to larger tool dimensions from the initial tool geometry, while (−) represents the negative deviation of −3.5% to lower dimensions. For example, a die with smaller dimensions (−) than the ideal is used in the first stroke and combined with a die with greater dimensions (+) than the ideal during the second stroke. As identified in [Beh13], the drawing gap variation has a significant influence on the punch force. Consequently, the punch diameter is held constant while the drawing gap is varied by using the two mentioned adjustments for the die diameters during the simulative investigations. Moreover, the die radius is varied in the same range as the diameter.

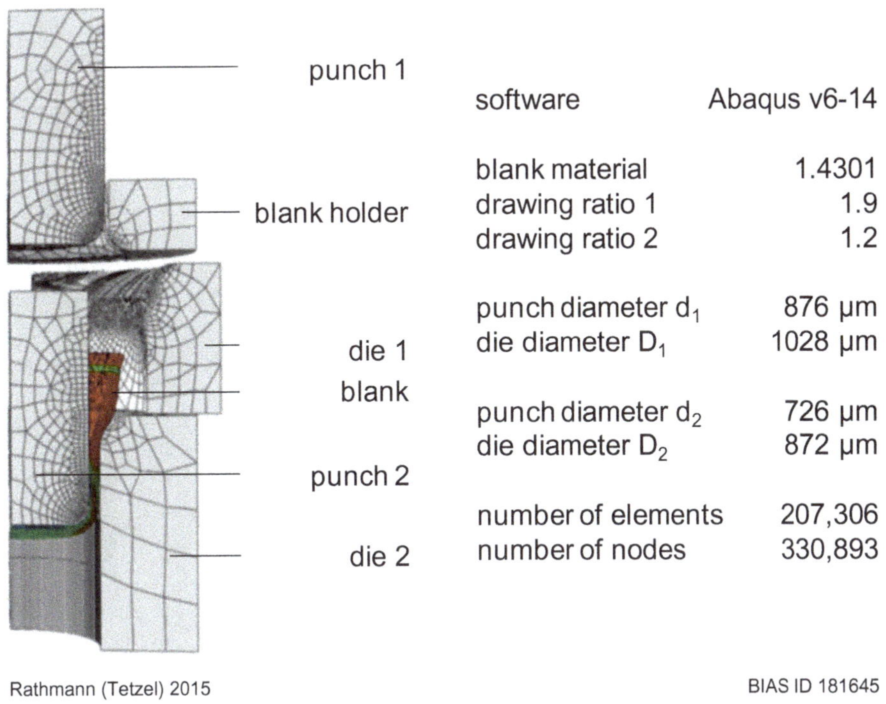

Fig. 2.50 Simulation model of a two-stage micro deep drawing process

The first characteristic is the curvature γ along the cup wall. The curvature is a result that can be found in both the simulation and experiment and results from an oversized drawing gap [Klo06]. It describes lip forming and bulge at the rib of a cup, which are defects along the cup wall and thus criteria for whether a cup is usable or not. This curvature is the maximum distance between a tangent and a point A. Both are located at the outer cup wall. The tangent goes through two points that mark the two local maximum peaks given by the lip forming and the rib of the cup wall. The point A is between these two peaks and marks the local minimum of the cup's wall topography. Figure 2.51 gives a pictorial representation of the curvature.

The influence of the adjustment of the first and second die diameters on the curvature γ is shown in Fig. 2.52. The abscissa labels the die diameter deviation from the ideal geometry for a minimum (−) and maximum value (+), while the ordinate displays the curvature in μm. Furthermore, the main influence of the adjustment of D_2 on γ is represented by a black dotted line. The main influence of D_2 means that the influence of the adjustment of D_1 on γ is not considered. Two courses happen and show that a large die diameter in the first stroke and a large one in the second stroke reduce the curvature. For an explanation, it is important to differentiate between the outer and inner cup wall. A close look at these two sides reveals that the second stroke relocates the curvature from the outer to the inner wall. Responsible for this is the second die radius, which shifts the curvature to the inner wall by straightening the outer. The greater the curvature is after the first stroke, the better this mechanism works. If the curvature is low after the first stroke,

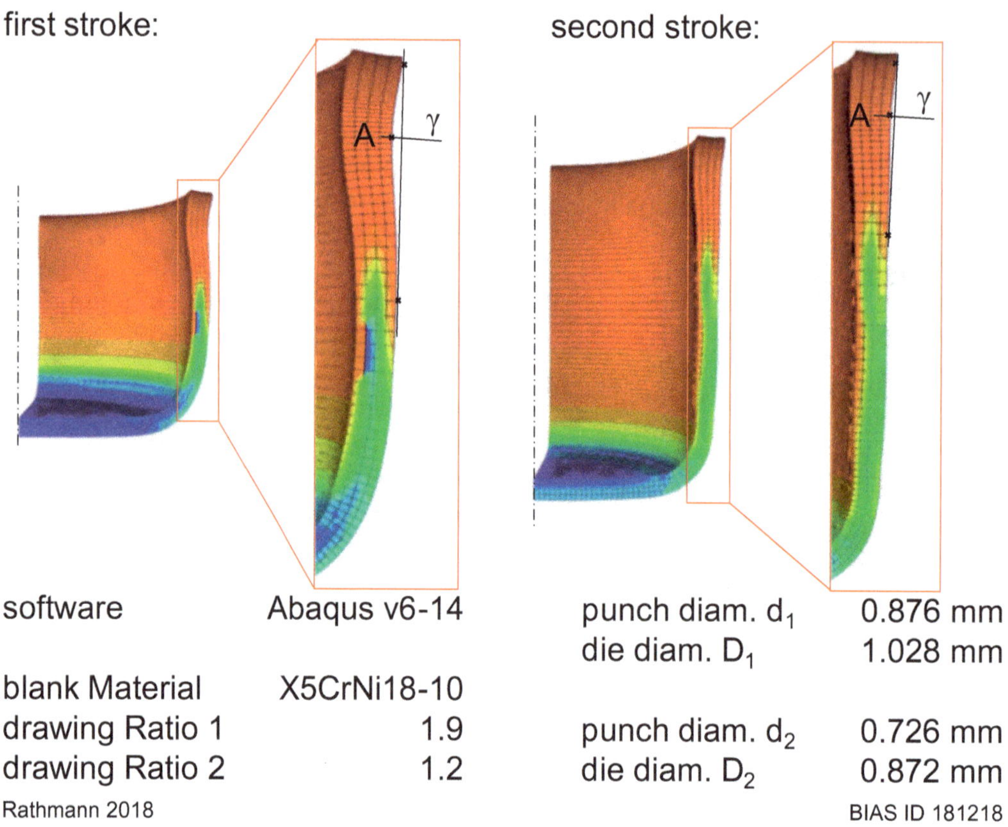

Fig. 2.51 Pictorial representation of the curvature γ at the cup wall due to a large drawing gap

this mechanism does not occur and the measured curvature after the second stroke gets worse.

Another deep drawing quality feature is a homogeneous wall thickness over the entire cup wall. As a measure for this, the mean wall thickness alteration Δ is used. The development is shown in dependency on the adjustment of die diameter D_1 in Fig. 2.53. As in Fig. 2.52, the abscissa labels the die diameter deviation from the ideal geometry for a minimum (−) and maximum value (+), while the ordinate displays the mean wall thickness alteration in μm. Furthermore, the main influence of the adjustment of D_2 on Δ is represented by a black dotted line. It can be seen that the mean wall thickness alteration is held almost constant by choosing a small diameter D_1 in the first deep drawing step and a small D_2 in the second one. Choosing a small die diameter leads to small drawing gaps in both deep drawing steps. The small drawing gap effects a levelling on deviations of the large wall thickness. Especially in the area of the punch radius at the cup wall, a thinner wall can be found. Otherwise, the wall thickness increases at the end of the rib due to the missing influence of the blank holder when the blank emerges under it. Between those zones, the cup wall thickness is in the range of the initial blank thickness. If this maximum deviation is reduced by applying a large die diameter in the second stroke, the smaller drawing gap reduces the mean wall thickness alteration and

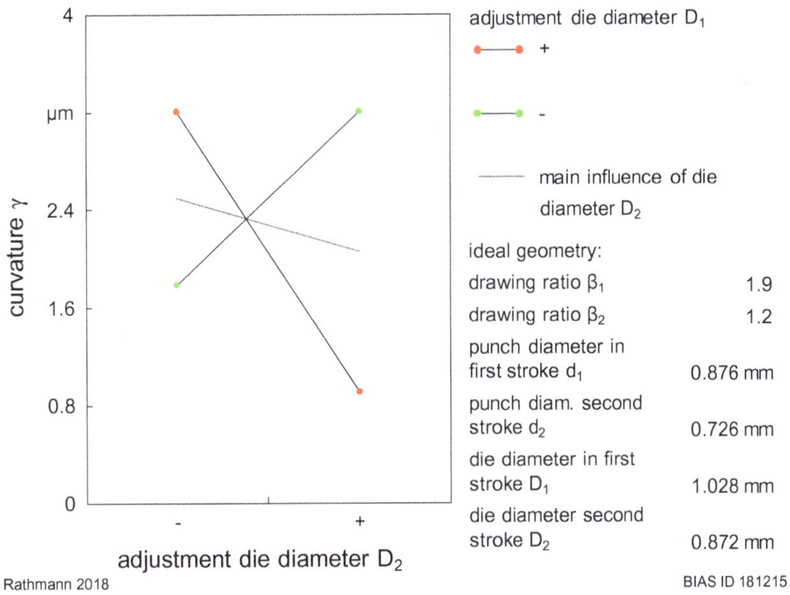

Fig. 2.52 Influence of the adjustment of die diameter D_2 in second stroke on curvature γ with varied adjustments of die diameter D_1 in first stroke

improves the homogeneity of the wall thickness distribution. Nevertheless, a small thickness deviation remains due to the drawing gap being larger than the blank thickness, leaving enough space for minimum alterations to occur.

The punch force is one of the most important considerations in metal forming. This force correlates with the tool wear, surface quality and the parameters of the conferred forces [Beh13]. Therefore, the influence of the adjustment of die diameter D_2 on the maximum punch force $F_{p,max}$ is shown in Fig. 2.54. As in Fig. 2.52 the abscissa labels the die diameter deviation from the ideal geometry for a minimum (−) and maximum value (+), while the ordinate displays the maximum punch force in N. Furthermore, the main influence of the adjustment of D_2 on $F_{p,\ max}$ is represented by a black dotted line. The diagram shows that a large diameter D_1 in the first stroke as well as a large diameter D_2 in the second stroke lead to stronger decreasing maximum punch forces than choosing a small D_1 and a large D_2. The decrease in the force is due to the reduced contact between the tools and workpiece. This contact does not only appear among the radii of the die and punch and the blank, but it also counts for the inner walls of the die and the blank. Contact connected with relative movement between a contact pair is always accompanied by friction. In the micro range, the punch force is affected by friction between the tool and workpiece [Beh13]. Therefore, a large diameter D_2 decreases the punch force, because it leaves enough space for the blank to avoid contact with the inner die walls. Additionally, a larger diameter D_2 reduces the contact pressure between the blank and die radius. This pressure arises from the contact conditions between the die radius and blank, ending in greater friction. And finally, a lower drawing ratio decreases the punch force. The maximum punch forces for the second stroke in dependency on D_1 differ strongly, because of the remaining stresses in the cup

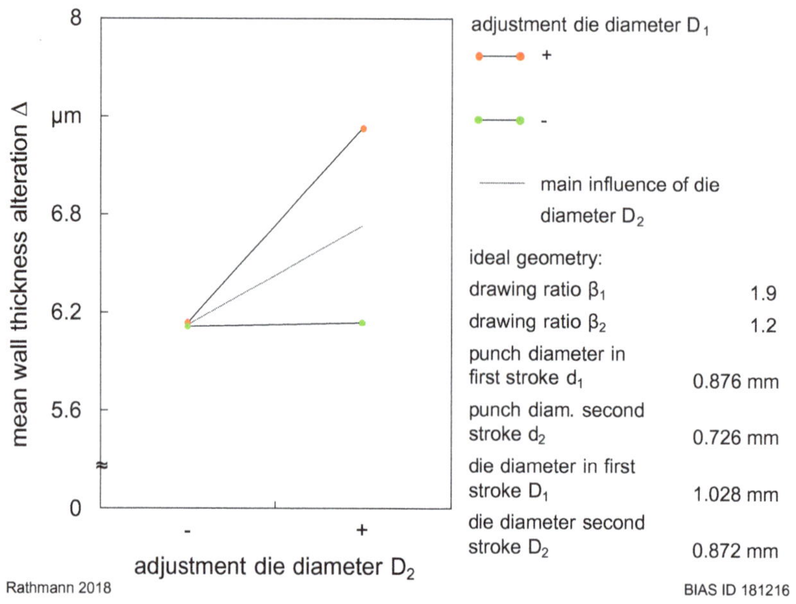

Fig. 2.53 Influence of the adjustment of die diameter D_2 in second stroke on mean wall thickness alteration Δ with varied adjustments of die diameter D_1 in first stroke

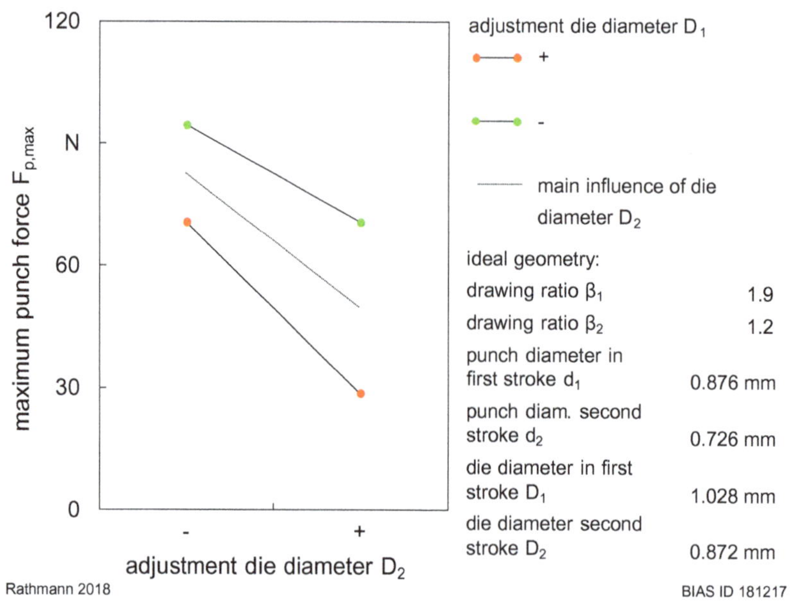

Fig. 2.54 Influence of the adjustment of die diameter D_2 in second stroke on maximum punch force $F_{p,max}$ with varied adjustments of die diameter D_1 in first stroke

wall. A smaller D_1 causes greater stresses than a larger D_1. The punch must additionally overcome more stress during the second stroke. This increases the maximum punch force.

2.5.5 Deep Drawing of Rectangular Parts

Compared to deep drawing of circular parts, there is a significant increase in complexity when rectangular parts are formed. As additional steps such as handling, or trimming become increasingly difficult when scaling processes to the micro range, it was the aim to produce accurately shaped parts only through deep drawing. A commonly used method to achieve near net shaped parts in deep drawing is blank shape optimization. However, due to size effects this method cannot be transferred unadjusted for the optimization of micro deep drawing processes. For example, there is a significant increase in friction with decreasing process dimensions. This effect has a fundamental influence on the shape of the final deep drawn part. In [Hu11] it is shown that, if the material model and friction coefficient are adjusted to the actual condition in the micro range, it is possible to determine the optimal blank shape in a size-dependent FEM simulation and accordingly produce near net shaped parts in rectangular micro deep drawing, as shown in Fig. 2.55.

Furthermore, the work carried out showed that material characteristics are even more important than friction and that the results of the FEM simulation are in good

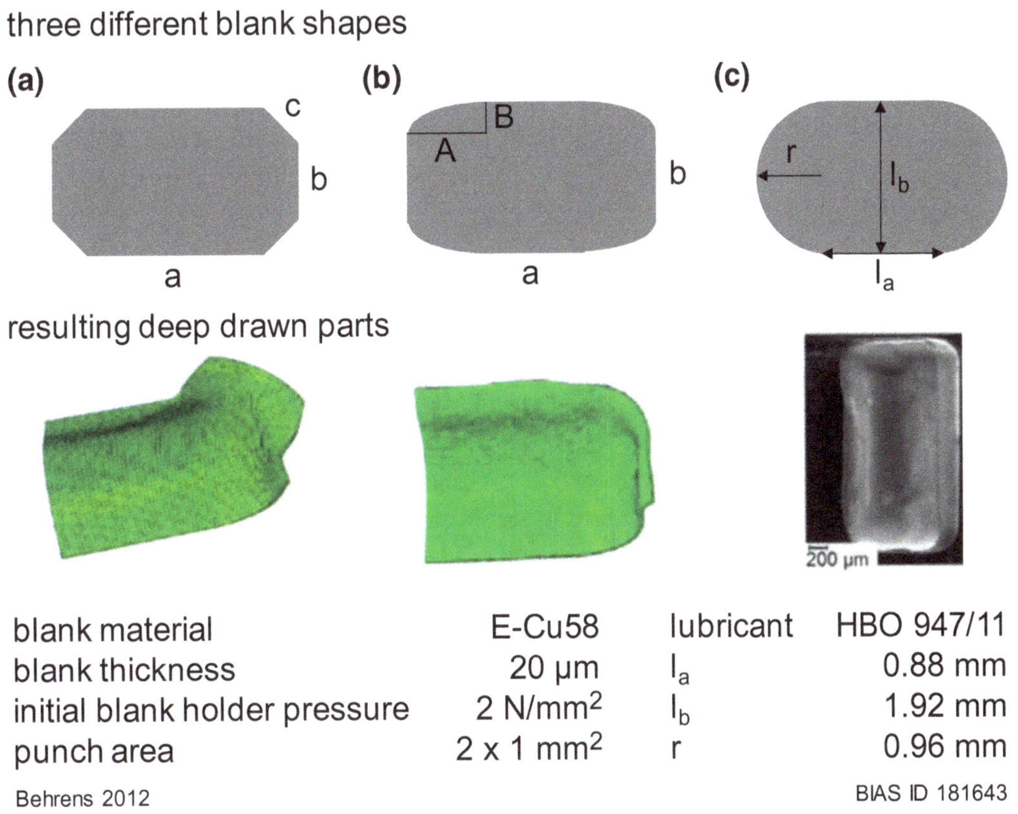

Fig. 2.55 Optimization of blank shape in FEM (**a**) and (**b**) and experiment (**c**) for deep drawing of rectangular blanks

agreement with experimental findings. Using Al99.5, rectangular parts with no remaining flange can be formed. In [Hu11] these findings can be transferred to blanks made of E-Cu58 and 1.4301. The influence of tool geometry deviation on the punch force in rectangular micro deep drawing has been investigated in Behrens et al. [Beh15]. It is demonstrated that increasing the corner radius or decreasing the die radius increases the resulting punch force and therefore has a negative effect on the drawing process. Additionally, the results show that the influence of a die radius variation on the punch force is far more prominent in rectangular deep drawing compared to the forming of circular parts, which can be explained by the different composition of the punch force.

2.5.6 Forming Limit

In order to enable the possibility of predicting component failure in deep drawing like bottom fracture using FEM simulation in addition to experimental investigations, it is necessary to integrate the material failure as an input variable into the simulation. Conventional methods like hydraulic or pneumatic bulge tests are available to determine forming limit curves even for thin foil materials. With these methods, only positive minor strains are measurable. Therefore, a scaled-down Nakajima test was developed and described in Veenaas et al. [Vee15]. Using an evaluation method adapted to the micro range, complete forming limit curves (FLC) can be determined. Especially for deep drawing processes, negative minor strains and the left side of a forming limit diagram are more important [Vee15]. It was possible to determine forming limit strain curves for the micro range for the foil materials Al 99.5, E-Cu58, 1.4301 and for PVD-sputtered Al–Zr, both for the positive and for the negative side of minor strain, as shown in Fig. 2.56.

2.5.7 Change of Scatter

Tolerance engineering is an approach for assessing influences in multi-stage processes but is also applicable in a single-stage process. The aim is the possibility of extrapolating the permissible tool geometry deviations in every process step from geometrical requirements on the final product geometry prescribed by operators. This becomes necessary because each process step increases or decreases the actual deviation of the workpiece. In order to describe this change of deviation, so-called "tolerance functions" are formulated. These functions answer the question, what is the necessary tolerance or admissible wear to get sound parts. For a process chain, the outcome can be calculated by a series of tolerance functions.

An example of data useful for tolerance engineering is given in Brüning [Brü15c] and is described by using upsetting of preforms with cone shaped cavities. In this example, four different dies are compared regarding their forming results.

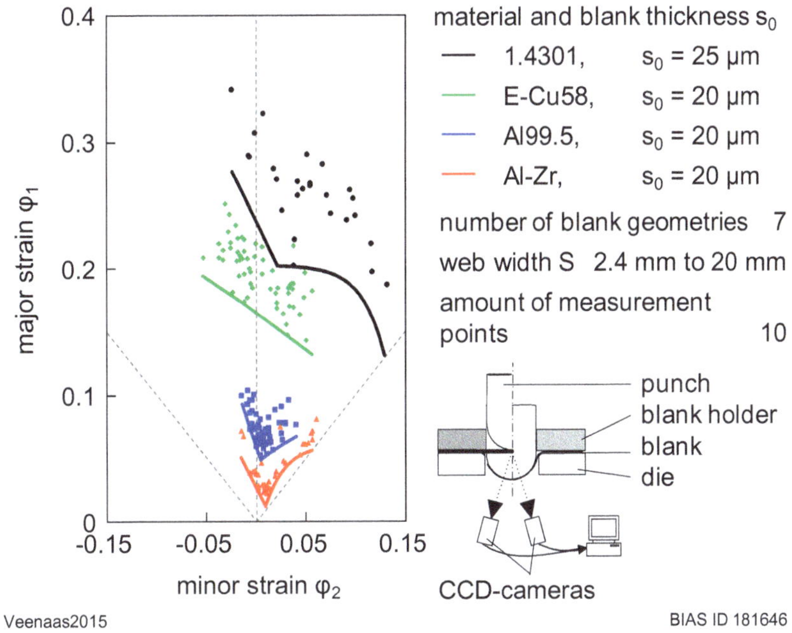

Fig. 2.56 Forming limit diagram and the lower forming limit curves for different foil materials

These results depend on the radii of the dies and preforms after upsetting as well as on the relative preform volumes. It is shown that the preforms generated by the laser rod end melting process can be calibrated well within a single-stage cold forming operation. Furthermore, for average natural strains $\varphi^* \leq 0.7$, the tolerable eccentricity of the preform is determined to 25 µm and increases with decreasing relative volume of the preform. This volume can scatter within 3% without a negative influence on the filling of the cavity. The influence of the calibration step on the eccentricity of the preform after the laser rod end melting process for one die is shown in Fig. 2.57.

The figure shows that the scattering field of the deviations regarding the position in the z- and r-direction is subjected to a shift. This shift is in the direction of the origin. Additionally, the scattering field of deviations after the calibration step is much smaller than the one after the laser rod end melting process. As Fig. 2.57 shows, it is possible to improve the scattering field and the position of its centroid by decreasing the field's width and length and relocating the center point in the direction of the origin.

According to tolerance engineering, it is now important to know the allowed deviation of the laser rod end melting process so that the subsequent process step is able to calibrate those deviations and create a sound part. Therefore, matrices are used to transform the deviations from the cold forging process into the allowed deviations of the previous step by multiplying them with damping factors and subtracting the shift of the mean deviations of both processes. An exemplary equation is shown in the following Fig. 2.58.

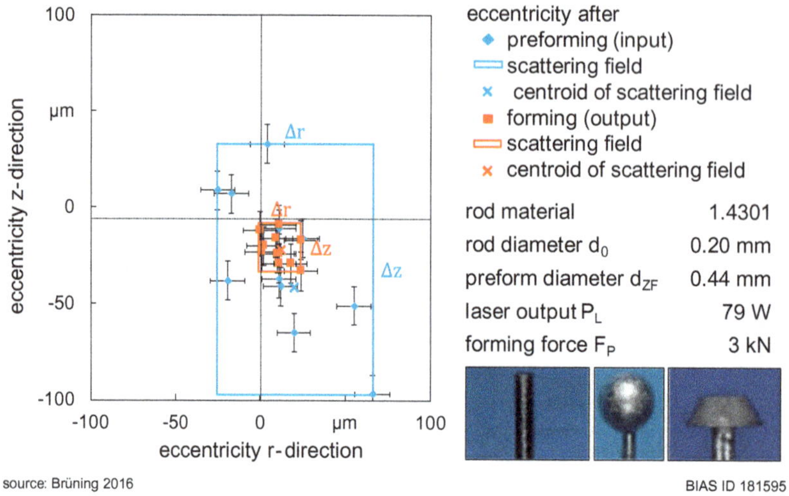

Fig. 2.57 Influence of a calibration step on the eccentricity of the preform after the laser rod end melting process

inverse calculation for i+1th step:

$$\underbrace{\left[\begin{pmatrix} r_i \\ z_i \end{pmatrix} + \begin{pmatrix} \pm\Delta r_i/2 \\ \pm\Delta z_i/2 \end{pmatrix}\right]^T}_{\text{maximum allowed deviation of the preforming process}} = \underbrace{\left[\begin{pmatrix} r_{i+1} \\ z_{i+1} \end{pmatrix} + \begin{pmatrix} \pm\Delta r_{i+1}/2 \\ \pm\Delta z_{i+1}/2 \end{pmatrix}\right]^T}_{\text{deviation after forming process}} \underbrace{\begin{pmatrix} b_{11}^{i+1} & 0 \\ 0 & b_{22}^{i+1} \end{pmatrix}}_{\text{damping factors}} - \underbrace{\left[\begin{pmatrix} r_i \\ z_i \end{pmatrix} - \begin{pmatrix} r_{i+1} \\ z_{i+1} \end{pmatrix}\right]^T}_{\text{shifting of the centroid of the scattering field}}$$

b_{11}^{i+1}, b_{22}^{i+1}: damping factors

with
$b > 1$: decreasing
$b < 1$: increasing
the scatter due to process i+1

process step	parameter	b_{11}	b_{22}
cold forging	eccentricity	2.63	2.75
thermal upsetting (single factor)	diameter	3.07	3.07

Rathmann 2018 BIAS ID 181596

Fig. 2.58 Exemplary equation for tolerance engineering

On the left side of the equation, the mean deviation (r_i z_i) and the scatter of the laser rod end melting process ($\pm\Delta r_i/2$ $\pm\Delta z_i/2$) are mentioned. The mean deviation results from the movement velocity of the laser in the lateral direction to the rod, while the scatter develops, for example, from the flow of the inert gas during preforming. On the right side of the equation, the mean deviation of the cold forging (r_{i+1} z_{i+1}) and the scatter of this process ($\pm\Delta r_{i+1}/2$ $\pm\Delta z_{i+1}/2$) is mentioned. In the case of the mean deviation, it results from positioning deviations of the upper and lower die in relation to each other, and the scatter develops from differences in the form filling of the die by the preform. The factors b_{11}^{i+1} and b_{22}^{i+1} are damping factors with values greater than one if the (i+1)th process step decreases the scatter, or with values greater than zero and less than one if the (i + 1)th process step increases the scatter. These values result from the reverse calculation, which means the calculation of the scatter of the (i − 1)th process step from the deviations of the prior step. The factors have to be evaluated prior to the calibration process by using simulations or experiments. In Fig. 2.58 two examples of the parameter b_{11} and b_{22} are given. In the first row the parameters for the cold forging process after the laser rod end melting process are mentioned. In this case, the parameters show that the scattering field is decreased. The second row includes the parameter for the scattering field of the preform diameter after a thermal upsetting. As can be seen, the values for the parameters are also greater than one and therefore describe a decreasing of the scatter. Finally, the shifting of the centroid of the scattering field from the cold forging to the laser rod end melting has to be considered. This shifting is expressed by the difference between the two centroids.

If a process comprises more than two process steps, a series of tolerance functions is used to calculate the necessary tolerance to get sound parts. An example of such a series of functions is given by the following Fig. 2.59.

As can be seen, the left side of the equation in Fig. 2.59 is the same as in Fig. 2.58 and so is the part of the equation in the tall square brackets, but with different exponents and indices due to the three-stage process chain. This last part still describes the transformation of the (i + 2)th scattering field of the final product into the (i + 1)th scatter of the previous step. Now, the process chain consists of three process steps and the transformation of the scatter from the (i + 1)th step into the ith step has to be considered. The part between the tall square brackets can be understood as the (i + 1)th centroid and its scattering field, and is now multiplied with the damping factors for this (i + 1)th step and relocated by the last

inverse calculation for a process chain with three steps:

$$\left[\binom{r_i}{z_i} + \binom{\pm\Delta r_i/2}{\pm\Delta z_i/2}\right]^T = \left[\left[\binom{r_{i+2}}{z_{i+1}} + \binom{\pm\Delta r_{i+2}/2}{\pm\Delta z_{i+2}/2}\right]^T \begin{pmatrix} b_{11}^{i+2} & 0 \\ 0 & b_{22}^{i+2} \end{pmatrix} - \left[\binom{r_{i+1}}{z_{i+1}} - \binom{r_{i+2}}{z_{i+2}}\right]^T\right] \begin{pmatrix} b_{11}^{i+1} & 0 \\ 0 & b_{22}^{i+1} \end{pmatrix} - \left[\binom{r_i}{z_i} - \binom{r_{i+1}}{z_{i+1}}\right]^T$$

b_{11}, b_{22}: damping factors

with

$b > 1$: decreasing

$b < 1$: increasing

the scatter due to process i+2 and i+1

Rathmann 2018 BIAS ID 181596

Fig. 2.59 Inverse calculation for a three-staged process chain

mathematical term. As before, the damping factors have to be evaluated in simulation or experimentally in the first place.

To receive a tolerance function for an arbitrary n-stage process chain, the following steps have to be considered. Firstly, the starting point is always the mathematical description of the center point and the scatter of the nth process step. Secondly, the first of the overall n − 1 shifts takes place. Therefore, the center point and the scatter are multiplied with the damping factors of the (n − 1)th process step. Then, the difference between the center points of the nth and (n − 1)th steps is subtracted. The result is the center point and the scatter of the (n − 1)th step. Thirdly, the (n − 2)th shift has to be done, and so on. This procedure is repeated until the (n − (n − 1))th shift is done. Finally, a series of tolerance functions is received that describes the development of the allowed deviations or the tolerable tool wear through the overall process chain.

References

[Alk15] Alkhazraji, H., El-Danaf, E., Wollmann, M., Wagner, L.: Enhanced fatigue strength of commercially pure Ti processed by rotary swaging. Adv. Mat. Sci. Eng. (2015). ID 301837

[Ami17] Aminzahed, I., Mashhadi, M.M., Sereshk, M.R.V.: Investigation of holder pressure and size effects in micro deep drawing of rectangular work pieces driven by piezoelectric actuator. Mater. Sci. Eng. **C71**, 685–689 (2017)

[Bän01] Bänsch, E.: Finite element discretization of the Navier-Stokes equations with a free capillary surface. Numer. Math. **88**(2), 203–235 (2001)

[Bän13] Bänsch, E., Paul, J., Schmidt, A.: An ALE finite element method for a coupled Stefan problem and Navier–Stokes equations with free capillary surface. Int. J. Numer. Methods. Fluids **71**(10), 1282–1296 (2013)

[Beh12] Behrens, G., Kovac, J., Köhler, B., Vollertsen, F., Stock, H.: Drawability of thin magnetron sputtered aluminium zirconium foils in micro deep drawing. In: ICNFT 2012, Harbin China, pp. 30–35 (2012)

[Beh13] Behrens, G.; Vollertsen, F.: Influence of tool geometry variation on the punch force in micro deep drawing. In: Proceedings of the ESAFORM 2013 Key Engineering Materials, The Current State-of-the-Art on Material Forming, vols. 554–557, pp. 1306–1311. Trans Tech Publications, Zurich-Durnten (2013)

[Beh15] Behrens, G., Ruhe, M., Tetzel, H., Vollertsen, F.: Effect of tool geometry variations on the punch force in micro deep drawing of rectangular components. J. Prod. Eng. **9**(2), 195–201 (2015)

[Beh16] Behrens, G., Trier, F., Tetzel, H., Vollertsen, F.: Influence of tool geometry variations on the limiting drawing ratio in micro deep drawing. Int. J. Mater. Form. **9**(2), 253–258 (2016)

[Brü12a] Brüning, H., Vollertsen, F.: Self-alignment capability of laser based free form heading process. In: Cosic, I. (ed.) Proceedings of 11th International Scientific Conference, pp. 427–430, Serbia (2012a)

[Brü12b] Brüning, H., von Bargen, R., Vollertsen, F., Zoch, H.-W.: Wärmebehandlung laser-generierter Vorformelemente: Anlayse zur Robustheit der Fertigung von Vorformelementen sowie Einfluss einer Wärmebehandlung. wt Werkstatttechnik online (102), 702–708 (2012b)

[Brü13a] Brüning, H., Jahn, M., Schmidt, A.: Process window for forming of micro preforms at different temperatures. In: Merklein, M., Franke, J., Hagenah, H. (eds.) Proceedings of the WGP Congress 2013, pp. 173–180, Germany (2013a)

[Brü13b] Brüning, H., Vollertsen, F.: First approach to generate comb linked parts by laser melting. J. Technol. Plasticity **38**(1), 1–12 (2013b)

[Brü13c] Brüning, H., Vollertsen, F.: Formability of micro material pre-forms generated by laser melting. In: Hinduja, S., Li, L. (eds.) Proceedings of the 37th International MATADOR Conference, pp. 373–376. Springer, London (2013c)

[Brü13d] Brüning, H., Vollertsen, F.: Investigation on flow stress level of spherical preforms generated by laser melting. Mater. Manuf. Process. **28**(5), 519–523 (2013d)

[Brü13e] Brüning, H.: Robustness of the laser melting process. In: Proceedings of 32th International Congress on Applications of Lasers and Electro-Optics (ICALEO13), pp. 852–857 (2013e)

[Brü14a] Brüning, H., Teepe, M., Vollertsen,F.: Surface roughness and size effect in dendrite arm spacing at preforms of AISI 304 (1.4301) generated by laser rod end melting. Procedia Eng. **81,** 1589–1594 (2014a)

[Brü14b] Brüning, H., Veenaas, S., Vollertsen, F.: Determination of axial velocity of material accumulation in laser rod end melting process. In: Fang, F., Brinksmeier, E., Riemer, O. (eds.) Proceedings of the 4th International Conference on Nanomanufacturing (nanoMan 2014), pp. F1–3 (2014b)

[Brü14c] Brüning, H., Vollertsen, F.: Mit dem Laser zum Klettverschluss. Laser **3,** 26–28 (2014c)

[Brü14d] Brüning, H., Vollertsen, F.: Numerical approach to determine natural strain of spherical preforms in open die upsetting. Adv. Mater. Res. **1018,** 325–332 (2014d)

[Brü15a] Brüning, H., Vollertsen, F.: Energy efficiency in laser rod end melting. In: Graf, T., Emmelmann, C., Overmeyer, L., Vollertsen, F. (eds.) Lasers in Manufacturing 2015 Proceedings, p. 138 (2015a)

[Brü15b] Brüning, H., Vollertsen, F.: Form filling behaviour of preforms generated by laser rod end melting. CIRP Ann. Manuf. Technol. **64**, 293–296 (2015b)

[Brü15c] Brüning, H., Vollertsen, F.: Surface accuracy achieved by upsetting of preforms generated by laser rod end melting. In: Qin, T., Dean, T.A., Lin, J., Yuan, S.J., Vollertsen, F. (eds.) 4th International Conference on New Forming Technology, vol. 21, pp. 1–7. EDP Sciences (2015c)

[Brü15d] Brüning, H.: Form filling behaviour of preforms generated by laser rod end melting. CIRP Ann. Manuf.Technol. **64**, 293–296 (2015d)

[Brü16a] Brüning, H., Jahn, M., Vollertsen, F., Schmidt, A.: Influence of laser beam absorption mechanism on eccentricity of preforms in laser rod end melting. In: 11th International Conference on Micro Manufacturing, p. 77 (2016a)

[Brü16b] Brüning, H.: Prozesscharakteristiken des thermischen Stoffanhäufens in der Mikrofertigung. Strahltechnik. BIAS Verlag (2016b)

[Brü17] Brüning, H., Schattmann, C., Krüger, M.: Melt pool geometry in laser blank rim melting to generate continuous cylindrical preforms. Prod. Eng. Res. Develop. **4–5**(11), 425–433 (2017)

[Che17] Cheng, C., Wan, W., Meng, B.: Size effect on the forming limit of sheet metal in micro-scaled plastic deformation considering free surface roughening. In: ICTP 2017, 17–22 September 2017, Cambridge, United Kingdom

[Chu11] Chu,H.: An Automated Micromanipulation System for 3D Parallel Microassembly. Dissertation edition (2011)

[Chu14] Jin, C.K., Jeong, M.G., Kang, C.G.: Effect of rubber forming process parameters on micro-patterning of thin metallic plates. Procedia Eng. **81**, 1439–1444 (2014)

[Eic10] Eichenhüller, B., Engel, U., Geiger, M.: Microforming and investigation of parameter interactions. Prod. Eng. **4**(2–3):135–140 (2010)

[Ell81] Elliot, C.M.: On the finite element approximation of an elliptic variational Inequality arising from an implicit time discretization of the Stefan problem. IMA J. Numer. Anal. **1,** 115–125 (1981)

[Gau13] Gau, J.-T., Teegala, S., Huang, K.-M., Hsiao, T.-J., Lin, B.-T.: Using micro deep drawing with ironing stages to form stainless steel 304 micro cups. J. Manuf. Process. **15**, 298–305 (2013)

[Gha08] Ghaei, A., Movahhedy, M.R., Karimi Taheri, A.: Finite element modelling simulation of radial forging of tubes without mandrel. Mater. Des. **29**, 867–872 (2008)

[Hau96] Haug, R.: Infeed rotary swaging - Ph.D. Thesis. University of Stuttgart (1996)

[Hei17] Heinrich, L.; Kobayashi, H.; Shimizu, T.; Yang, M.; Vollertsen, F.: Influence of asymmetrical drawing radius deviation in micro deep drawing. In: Proceedings of the 36th IDDRG Conference—Materials Modelling and Testing for Sheet Metal Forming. Journal of Physics, vol. 896, p. 012060 (2017) (online)

[Hof17] Hof, L., Heinrich, L., Wüthrich, R.: Towards high precision manufacturing of glass tools by spark assisted chemical engraving (SACE) for micro forming techniques. In: World Congress on Micro and Nano Manufacturing (WCMNM 2017), Kaohsiung, Taiwan (2017). USB-Stick

[Hu09] Hu, Z., Walter, R., Vollertsen, F.: Forming tools for micro deep drawing—influence of geometrical tolerance of forming tools on the punch force in micro deep drawing. wt Werkstatttechnik online, H **11**(12), 814–819 (2009)

[Hu11] Hu, Z., Vollertsen, F.: Investigation on the optimisation of the blank shape for micro deep drawing of rectangular parts. In: Steel Research International, Special Edition: 10th International Conference on Technology of Plasticity (ICTP 2011), VCH Weinheim, pp. 974–978 (2011)

[Hüg09] Hügel, H., Graf, T.: Laser in der Fertigung: Strahlquellen, Systeme, Fertigungsverfahren. Studium. Vieweg + Teubner, Wiesbaden, 2., neu bearb. aufl. edn (2009)

[Irt14] Irthiea, I., Green, G., Hashim, S., Kriama, A.: Experimental and numerical investigation on micro deep drawing process of stainless steel 304 foils using flexible tools. Int. J. Mach. Tools Manuf. **76**, 21–33 (2014)

[Ish15a] Ishkina, S., Kuhfuss, B., Schenck, C.: Grain size modification by micro rotary swaging. Key Eng. Mater. 651–653, 627–632 (2015a)

[Ish15b] Ishkina, S., Schenck, C., Kuhfuss, B.: Beeinflussung der Umformbarkeit von Halbzeugen durch Modifikationen beim Mikrorundkneten. Tagungsband. In: Hopmann, Ch., Brecher, Ch., Dietzel, A., Drummer, D., Hanemann, T., Manske, E., Schomburg, W.K., Schulze, V., Ullrich, H., Vollertsen, F., Wulfsberg, J.-P. (eds.) Kolloquium Mikroproduktion in Aachen, pp. 64–71, 16–17 November 2015, IKV Aachen (2015b). ISBN: 978-3-00-050-755-7

[Ish15c] Ishkina, S., Kuhfuss, B., Schenck, C.: Influence of the relative rotational speed on component features in micro rotary swaging. In: Qin, Y., Dean, T.A., Lin, J., Yuan, S.J., Vollertsen, F. (eds.) Proceedings of the 4th International Conference on New Forming Technology (ICNFT 2015) 6–9 August 2015, Glasgow, Scotland (UK). Published in: Materials Science, Engineering and Chemistry vol. 21, p. 09012 (2015c). 1–7 ISBN: 978-2-7598-1823-5

[Ish17a] Ishkina, S., Schenck, C., Herrmann, M., Kuhfuß, B.: Konditionierung von Halbzeugen durch Mikrorundkneten. Fachbeiträge. In: Vollertsen, F., Hopmann, C., Schulze, V., Wulfsberg, J. (eds.) Kolloquium Mikroproduktion, Bremen, 27–28 November 2017, pp. 99–106. BIAS Verlag (2017a). (online) ISBN: 978-3-933762-56-6

[Ish17b] Ishkina, S., Schenck, C., Kuhfuss, B., Moumi, E., Tobeck, K.: Eccentric rotary swaging. Int. J. Precis. Eng. Manuf. **18**(7), 1035–1041 (2017b)

[Ish18] Ishkina, S., Schenck, C., Kuhfuss, B.: Conditioning of material properties by micro rotary swaging. In: AIP Conference Proceedings 1960, 160013 (2018). https://doi.org/10.1063/1.5035039

[Jah12a] Jahn, M., Schmidt, A.: Finite element simulation of a material accumulation process including phase transitions and a capillary surface. Technical Report 12-03, ZeTeM, Bremen

[Jah12] Jahn, M., Luttmann, A., Schmidt, A.: A FEM simulation for solid-liquid-solid phase transitions during the production of micro-components. In: Cosic, I. (ed.) Proceedings of 11th International Scientific Conference MMA - Advanced Production Technologies, pp. 231–234, Serbia (2012)

[Jah13] Jahn, M., Brüning, H., Schmidt, A., Vollertsen, F.: Herstellen von Mikroverbundstrukturen durch Stoffanhäufen und Laserschweißen. In: Tutsch, R. (ed.) Tagungsband 6. Kolloquium Mikroproduktion, p. A11. Shaker Verlag (2013)

[Jah14b] Jahn, M., Schmidt, A., Bänsch, E.: 3D finite element simulation of a material accumulation process including phase transistions and a capillary surface. Technical Report 14-03, ZeTeM, Bremen

[Jah14] Jahn, M., Brüning, H., Schmidt, A., Vollertsen, F.: Energy dissipation in laser-based free form heading: a numerical approach. Prod. Eng. **8**(1–2), 51–61 (2014)

[Kle15] Klein, B.: FEM. Grundlagen und Anwendungen der Finite-Elemente-Methode im Maschinen- und Fahrzeugbau. Springer Vieweg, Wiesbaden (2015)

[Klo06] Klocke, F., König, W.: Fertigungsverfahren—Umformen. Springer-Verlag, Berlin, Heidelberg (2006)

[Kov18] Kovac, J.; Heinrich, L.; Köhler, B.; Mehner, A.; Clausen, B.; Zoch, H.: Comparison of the tensile properties and drawability of thin bimetallic aluminium-scandium-zirconium/stainless steel foils and monometallic Al–Sc–Zr fabricated by magnetron sputtering. In: Vollertsen, F., Dean, T.A., Qin, Y., Yuan, S.J. (eds.) 5th International Conference on New Forming Technology (ICNFT 2018), MATEC Web Conference, vol. 190, p. 15012 (2018)

[Kuh08a] Kuhfuss, B., Moumi, E., Piwek, V.: Micro rotary swaging: process limitations and attempts to their extension. Microsys. Technol. **14** 1995–2000 (2008a)

[Kuh08b] Kuhfuss, B., Moumi, E., Piwek, V.: Influence of the feed rate on work quality in micro rotary swaging. In: 3rd ICOMM Pittsburgh, USA pp. 86–91 (2008b)

[Kuh09a] Kuhfuss, B., Moumi, E., Piwek, V.: Manufacturing of micro components by means of plunge rotary swaging. In: Proceedings of the 9th Euspen, San Sebastian, pp. 58–61 (2009c)

[Kuh09b] Kuhfuss, B., Piwek, V., Moumi, E.: Vergleich charakteristischer Einflussgrössen beim Mikro- und Makrorundkneten. In: Vollertsen, F., Büttgenbach, S., Kraft, O., Michaeli, W. (eds.) Kolloquium Mikroproduktion, pp. 219–228. BIAS Verlag, Bremen (2009d)

[Kuh10] Kuhfuss, B., Moumi, E., Piwek, V.: Mikrorundkneten. Ein neues Umformverfahren mit viel Potenzial. Industrie Manag. **6,** 14–16 (2010)

[Kuh11] Kuhfuss, B., Moumi, E., Piwek, V.: Load measurement during rotary swaging of micro components using strain gauges. In: 6th International Conference on Micro Manufacturing (ICOMM 2011) Tokyo, Japan, pp. 53–58 (2011)

[Kuh12a] Kuhfuss, B., Moumi, E., Piwek, V., Hork, M., Werner, M.: Reib- und formschlüssige Verbindungen durch Rundkneten, VDI-Konstruktion, Ingenieurwerkstoffe Metall, Jan./Feb. 1/2-2012

[Kuh12b] Kuhfuss, B., Moumi, E., Piwek, V.: Effects of dry machining on process limits in micro rotary swaging. In: 7th ICOMM 2012, pp. 243–247, Evanston, IL, USA (2012b)

[Kuh13] Kuhfuss, B., Moumi, E.: Bulk metal forming. In: Vollertsen, F. (ed.) Micro Metal Forming, pp. 103–133. Springer, Hedielberg (2013). ISBN: 978-3-642-30915-1

[Kur98] Kurz, W., Fisher, D.J.: Fundamentals of Solidification, 4 rev. ed. Trans Tech Publications, Aedermannsdorf (1998)

[Lan02] Lange, K.: Umformtechnik Handbuch für Industrie und Wissenschaft: Band 4: Sonderverfahren, Prozeßsimulation, Werkzeugtechnik, Produktion. Springer, Heidelberg (2002)

[Lan06] Lange, K.: Handbook of Metal Forming. Society of Manufacturing Engineers, Dearborn, Michigan (2006)

[Liu17] Liu, Y., Herrmann, M., Schenck, C., Kuhfuss, B.: Plastic deformation history in infeed rotary swaging process. In: AIP Conference Proceedings 1896, 080013 (2017). https://doi.org/10.1063/1.5008093

[Lut18] Luttmann, A.: Modellierung und Simulation von Prozessen mit fest-flüssig Phasenübergang und freiem Kapillarrand. Dissertation, Universität Bremen (2018)

[Mes94] Messner, A.; Engel, U.; Kals, R.; Vollertsen, F.: Size effect in the FE-simulation of micro-forming processes. J. Mater. Process. Technol. **45**(1–4), 371–376 (1994)

[Meß98] Meßner, A.: Kaltmassivumformung metallischer Kleinstteile: Werkstoffverhalten, Wirkflächenreibung, Prozessauslegung. Meisenbach, Bamberg (1998)

[Mic03] Michel, J.F., Picart, P.: Size effects on the constitutive behaviour for brass in sheet metal forming. J. Mater. Process. Technol. **141**, 439–446 (2003)

[Mou14a] Moumi, E., Ishkina, S., Kuhfuss, B., Hochrainer, T., Struss, A., Hunkel, M.: 2D-simulation of material flow during infeed rotary swaging using finite element method. Procedia Eng. **81**, 2342–2347 (2014a)

[Mou14b] Moumi, E., Kuhfuss, B.: Properties of alloy 304 micro parts processed by rotary swaging. In: Fang, F., Brinksmeier, E., Riemer, O. (eds.) Proceedings of the NanoMan (2014b). (digital)

[Mou14c] Moumi, E., Wilhelmi, P., Kuhfuss, B., Schenck, C., Tracht, K.: Wire joining by rotary swaging. Procedia Eng. **81**, 2012–2017 (2014c)

[Mou15a] Kuhfuss, B., Moumi, E., Clausen, B., Epp, J., Koeler, B.: Investigation of deformation induced martensitic transformation during incremental forming of 304 stainless steel wires. Key Eng. Mater. 651–653, 645–650 (2015a)

[Mou15b] Moumi, E., Kuhfuss, B.: Erhöhung der Umformgeschwindigkeit beim Mikrorundkneten durch Einsatz von Zwischenelementen. In: Hopmann, Ch, Brecher, Ch., Dietzel, A., Drummer, D., Hanemann, T., Manske, E., Schomburg, W.K., Schulze, V, Ullrich, H, Vollertsen, F, Wulfsberg, J.-P. (eds.) Kolloquium Mikroproduktion, pp. 72–78. IKV Aachen (2015b)

[Mou18a] Moumi, E., Schenck, C., Herrmann, M., Kuhfuss, B.: High productivity micro rotary swaging. In: MATEC Web Confrences, vol. 190, p. 15014 (2018a). https://doi.org/10.1051/matecconf/201819015014

[Mou18b] Moumi, E., Wilhelmi, P., Schenck, C., Herrmann, M., Kuhfuss, B.: Material flow control in plunge micro rotary swaging. In: MATEC Web Conference, vol. 190, p. 15002 (2018b). https://doi.org/10.1051/matecconf/201819015002

[Neu12] Neumann, S.: Einflussanalyse beim single mode Faserlaserschweißen zur Vermeidung des Humping-Phänomens, volume 48 of Strahltechnik. BIAS Verlag, Bremen (2012)

[Ron06] Rong, L., Nie, Z., Zuo, T.: FEA modeling of effect of axial feeding velocity on strain field of rotary swaging process of pure magnesium. Trans. Nonferrous Met. Soc. China **16**, 1015–1020 (2006)

[Sao90] Saotome, Y.; Kaga, H.: Micro deep drawability of very thin sheet steels. Adv. Technol. Plasticity **3**, 1341–1346 (1990)

[Sch16] Şchiopu, V., Luca, D.: A new net-shape plating technology for axisymmetric metallic parts using rotary swaging. Int. J. Adv. Manuf. Tech. **85**, 2471–2482 (2016)

[Sch17a] Schattmann, C., Brüning, H., Ellmers, J.-H., Vollertsen, F.: Mikroverbund-Strukturen für die Kaltmassivumformung. UMFORMtechnik Massiv + Leichtbau **1,** 22 (2017a)

[Sch17b] Schattmann, C., Hennick, C., Brüning, H., Pfefferkorn, F.E., Vollertsen, F.: Generation and shape control of line-linked preforms by laser melting. J. Technol. Plasticity **42**(1), 21–32 (2017b)

[Sch17c] Schattmann, C., Vollertsen, F.: Metrisches Gewinde in dünnen Blechen. Mikroproduktion **06,** 60–61 (2017c)

[Sch17d] Schattmann, C., Wilhelmi, P., Schenck, C., Kuhfuß, B., Vollertsen, F.: Erhöhung der Taktrate bei der Herstellung von Zwischenformen im Linienverbund. In: Vollertsen, F., Hopmann, C., Schulze, V., Wulfsberg, J.P. (eds.) Fachbeiträge 8. Kolloquium Mikroproduktion, pp. 227–236. BIAS Verlag, Bremen (2017d)

[Ste11a] Stephen, A., Brüning, H., Vollertsen, F.: Fokuslagensteuerung beim laserbasierten Stoffanhäufen. In: Kraft, O., Haug, A., Vollertsen, F., Büttgenbach, S. (eds.) Kolloquium Mikroproduktion und Abschlusskolloqium SFB 499, pp. 155–160. KIT Scientific Reports (2011a)

[Ste11b] Stephen, A., Vollertsen, F.: Influence of the rod diameter on the upset ratio in laser-based free form heading. In: Hirt,G., Tekkaya, A.E. (eds.) Steel Research Int, Special Edition: 10th Int. Conf. on Technology of Plasticity, pp. 220–223, Weinheim. Wiley-VCH Verlag GmbH (2011b)

[Tet15] Tetzel, H.; Rathmann, L.; Vollertsen, F.: Entwicklung eines Simulationsmodells für mehrstufige Tiefziehprozesse im Mikrobereich. 7. Kolloquium Mikroproduktion (16–17 November 2015, Aachen), pp. 79–86. IKV Aachen (2015)

[Tet16] Tetzel, H., Rathmann, L., Heinrich, L.: Simulation accuracy of a multistage micro deep drawing process. JPT J. Technol. Plasticity **41**, 1–11 (2016)

[Thi93] Thiruvarudchelvan, S.: Elastomers in metal forming - a review. J. Mat. Proc. Tech. **39**, 55–82 (1993)

[Toe18] Toenjes, A., Ishkina, S., Schenck, C., von Hehl, A., Zoch, H.-W., Kuhfuss, B.: Eccentric rotary swaging variants. In: Proceedings of the 5th International Conference on New Forming Technology (ICNFT 2018), 18–21 September 2018, Bremen, Germany (DE), MATEC Web of Conferences, vol. 190, p. 15003 (2018). https://doi.org/10.1051/matecconf/201819015003

[Tsc06] Tschätsch, H.: Metal Forming Practise: Processes—Machines—Tools. Springer-Verlag, Heidelberg (2006)

[Val00] Valiev, R.Z.; Islamgaliev R.K.; Aleksandrov, I.V.: Bulk nanostructured materials from severe plastic deformation. Prog. Mater. Sci. **45**, 103–189 (2000)

[Vee15] Veenaas, S., Behrens, G., Kröger, K., Vollertsen, F.: Determination of forming limit diagrams for thin foil materials based on scaled Nakazima tests. In: Progress in Production Engineering, WGP Kongress, pp. 190–198 (2015)

[Vol08] Vollertsen, F., Walther, R.: Energy balance in laser-based free form heading. CIRP Ann. Manuf. Technol. **57**(1), 291–294 (2008)

[Vol09] Vollertsen, F., Biermann, D., Hansen, H.N., Jawahir, I.S., Kuzman, K.: Size effects in manufacturing of metallic components. CIRP Ann. Manuf. Technol. **58**, 566–587 (2009)

[Vol13] Vollertsen,F.: Micro Metal Forming. Lecture Notes in Production Engineering. Springer, Heidelberg (2013)

[Wag05] Wagner, S., Felde, A.: Niederfrequent schwingender Niederhalter beim Tiefziehen kleiner Bauteile. Abschlussbericht DFG-Schwerpunktprogramm 1074, Verlagshaus Mainz GmbH Aachen, pp. 364–371 (2005)

[Wal09] Walther, R.: An enhanced model for energy balance in laser-based free form heading. J. Micromechanics Microengineering **19**(5), 054007 (2009)

[Wan07] Wang, G.G., Shan,S.: Review of metamodeling techniques in support of engineering design optimization. J. Mech. Des. **129**(4), 370 (2007)

[Wil13] Wilhelmi, P., Moumi, E., Foremny, E., Kuhfuss, B., Schenck, C.: Feeding and positioning of linked parts in micro production chains. In: Archenti, A., Maffei, A. (eds.) Proceedings of NEWTECH 2013, vol. 2, pp. 75–84 (2013)

[Wil15] Wilhelmi, P., Moumi, E., Schenck, C., Kuhfuss, B.: Werkstofffluss beim Mikrorundkneten im Linienverbund In: Hopmann, Ch., Brecher, Ch., Dietzel, A., Drummer, D., Hanemann, T., Manske, E., Schomburg, W.K., Schulze, V., Ullrich, H., Vollertsen, F., Wulfsberg, J.P. (eds.) Kolloquium Mikroproduktion, pp. 72–78 (2015)

[Woi15] Woizeschke, P., Brüning, H., Vollertsen, F.: Lasergenerierte Blechkanten. Laser Mag. **5**, 22–24 (2015)

[Zha14] Zhang, Q., Jin, K., Mu, D.: Tube/tube joining technology by using rotary swaging forming method. J. Mat. Proc. Tech. **214**, 2085–2094 (2014)

Chapter 3
Process Design

Claus Thomy, Philipp Wilhelmi, Ann-Kathrin Onken, Christian Schenck, Bernd Kuhfuss, Kirsten Tracht, Daniel Rippel, Michael Lütjen and Michael Freitag

C. Thomy (✉)
BIAS—Bremer Institut für Angewandte Strahltechnik GmbH, Bremen, Germany
e-mail: info@bias.de

P. Wilhelmi (✉) · A.-K. Onken · C. Schenck · B. Kuhfuss · K. Tracht
Bremen Institute for Mechanical Engineering (bime), University of Bremen, Bremen, Germany
e-mail: wilhelmi@bime.de

D. Rippel (✉) · M. Lütjen
BIBA—Bremer Institut für Produktion und Logistik GmbH, University of Bremen, Bremen, Germany
e-mail: rip@biba.uni-bremen.de

M. Freitag
Faculty of Production Engineering, University of Bremen, Bremen, Germany

F. Vollertsen et al. (eds.), *Cold Micro Metal Forming*, Lecture Notes in Production Engineering, https://doi.org/10.1007/978-3-030-11280-6_3

3.1 Introduction to Process Design Claus Thomy

The design of micro production processes and systems faces a variety of significant challenges today. This is due both to economic and to technological challenges. Generally speaking, in most industries micro production tends to be characterized by mass production, meaning very high lot sizes from several thousands (e.g. in medical technology) to literally billions (e.g. resistor end caps in the electronics industry). Moreover, such micro parts are often produced under severe cost pressure, with the market price for an individual part often being only a fraction of a € cent.

From an economic point of view, this means that the whole process chain has to be designed such that the output or production rate (e.g. measured in parts produced per minute) is maximized, whilst the defect rate (measured in ppm (parts per million) rejected) has to be minimized. From a technical point of view, these economic constraints are often very challenging, especially due to the specific characteristics of micro production with regard to materials, processes and products. Typical examples—in contrast to macro production—are a higher fragility of parts, scatter in and anisotropy of properties of the materials and parts produced, the highest tolerance requirements, and size effects.

In this context, and in order to deal with such challenges specific to micro production, an integrated "Micro-Process Planning and Analysis" (μ-ProPlAn) methodology and procedure with the required level of detail and complexity are suggested and implemented in software (Sect. 3.3). With this prototype software, the integrated planning of manufacturing, handling, and quality inspection at different levels of detail shall be enabled, reducing the overall effort and time for designing micro production processes, also pursuing a Simultaneous Engineering approach. Thus, three views with increasing levels of detail are provided: the process chain view, the material flow view, and the configuration view. The latter view is the closest to the part produced and the manufacturing process as such, and incorporates both qualitative and quantitative cause-effect networks to combine expert knowledge with statistical and learning-based data analysis methods. The challenge is that, between and within all views, complex interrelations occur, which are, for example, due to micro specific features such as high variances in properties. The μ-ProPlAn software prototype shall support process designers during all stages of planning from manufacturing process design to process configuration and the evaluation of process chain alternatives by analyzing and optimizing the underlying models, e.g. with respect to time- or cost-related parameters.

On a level closer to the actual micro manufacturing technology, in both bulk and sheet metal forming, the micro specific challenges illustrated above have to be dealt with, too (Sect. 3.2). In order to improve the planning especially of handling and assembly operations and systems, a micro specific enhancement of data exchange models on the basis of ISO 10303 is suggested. As handling of individual micro parts is fairly difficult, it is proposed and discussed to produce, convey and store these parts in the form of linked parts (either line-linked, ladder-linked or comb-linked). This reduces the effort for handling and storage (both long-term storage and highly dynamic buffering), but increases the effort for referencing parts with respect to their position and orientation at high production rates. Moreover, the force transfer especially during forming between neighboring parts due to low stiffness has to be considered, e.g. in the design of conveyors or (adaptive) guides, and can have an effect on part quality and accuracy. Understanding and modeling such scatter in accuracy can potentially allow us to widen the tolerance fields by methods such as selective assembly, thus minimizing waste. However, a prerequisite there is to master appropriate strategies for synchronization.

3.2 Linked Parts for Micro Cold Forming Process Chains

Philipp Wilhelmi*, Ann-Kathrin Onken, Christian Schenck, Bernd Kuhfuss and Kirsten Tracht

Abstract The method of production as linked parts allows a significant increase of output rates compared to the production of loose micro parts. In the macro range, it is already applied to sheet metal forming, for example, in the production of structural parts in the automobile or aviation industry, but a simple downscaling is not appropriate. The goal of producing high volumes of high-quality parts requires micro-specific methods. A holistic concept is presented considering planning and production. The main aspects of planning are the design of linked parts and the production planning. Furthermore, an approach is introduced to widen the tolerance field for assemblies to the production of linked micro parts. This approach requires the consideration of all affected process chains during planning. Linking of the parts enables them to be conveyed as a string and thereby simplifies handling, but also causes new challenges during production. Specific handling technologies are studied to overcome these challenges. Finally, a concept for the physical synchronization with regard to the investigation of tolerance field widening is presented.

Keywords Handling • Assembly • Production Planning

3.2.1 Introduction

Achieving high production rates is the key factor for a profitable industrial production of micro parts because of the typically low unit prices. Due to high tooling costs, this is especially relevant for cold forming. In addition to the small tolerances, fragility and sticking of the parts to each other or machine components make handling a major bottleneck in this context. This is mainly related to the surface-to-volume ratio, which increases with decreasing part size [Qin15]. The strength and weight of the parts depend on the volume, while the adhesion forces are related to the surface. Fantoni et al. have carried out extensive research in this field related to grasping [Fan14]. In the literature many different concepts for the handling of micro parts can be found, where grasping is involved at least for orientation or fixation, as presented in [Fle11], for example. However, most of these systems do not achieve the desired clock rates. Vibratory bowl feeders are the most common systems applied in the industrial environment so far [Kru09]. They are robust in their function and comparatively inexpensive, but only applicable for a limited range of types of micro parts and do not allow the production in fixed cycle times. The concept of production as linked parts, presented in [Kuh11], eliminates or strongly simplifies most of the speed-limiting handling operations like separation, grasping, releasing, orientation and fixation of individual parts. A linking structure keeps the parts interconnected and allows feeding of the parts as a string throughout the whole of the production stages. Examples of the industrial application of production as linked macro parts are mainly found in the manufacturing of sheet metal parts by punching, bending and deep drawing. The sheet is fed stepwise as a strip and processed by progressive dies. Lead frames for surface-mounted devices (SMD) or electrical connectors are produced as linked parts with features in the micro range. The sheet thickness for these applications is typically $s \geq 100$ µm. In the MASMICRO project, the feeding and processing of thinner sheet material with progressive dies has been investigated [Qin08]. Merklein et al. considered a concept for massive forming as linked parts. Before the actual forming process, billets are formed from a metal strip (thickness: $1 \text{ mm} \leq s \leq 3 \text{ mm}$) and remain therein [Mer12].

Until now, only single conceptual ideas were addressed. In the following, a holistic concept for the production planning and production of micro parts as linked parts is presented. This is done on the basis of two exemplary process chains illustrated in Fig. 3.1. The involved processes of deep drawing (see Sect. 4.2), laser rod melting (see Sect. 2.2) and rotary swaging (see Sect. 2.3) are considered in detail in the mentioned sections. Process chain 1 deals with ladder-linked parts, which are produced based on sheet material. Process chain 2 deals with the wire-based line-linked parts and involves massive forming. In addition to the single process chains, the synchronization of linked parts is included in the investigations.

In order to enable the mass production of different types of linked micro parts, two major fields of research are addressed. The first one is the planning of linked micro parts. In combination with the linking of the parts also micro-specific planning methods are required. The buffering of processes as well as the arrangement of the processes changes significantly, as shown in Sect. 3.2.2. The main focus of the second field of research is on the production of linked parts. Due to the linking of the parts, specific handling concepts are required. These are presented in Sect. 3.2.3 together with the synchronization of parts for building pre-assemblies.

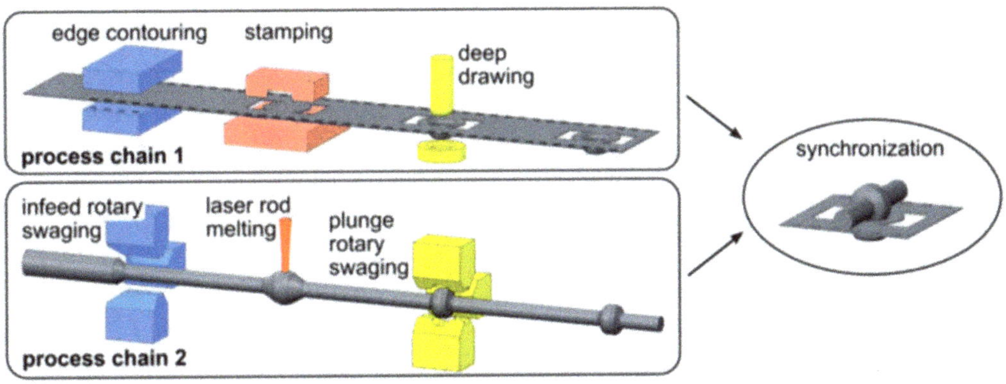

Fig. 3.1 Considered process chains

3.2.2 Design and Production Planning of Linked Parts

The planning and controlling of processes is required to ensure an economic production. Considering the production of linked micro parts, the manufacturing and assembly of parts in high quantities and quality is influenced by the interconnection of the parts, the small dimensions, and low tolerances. Therefore, three subject areas are addressed in the following. In the first area, the design of the linked parts as well as a product model for deriving the data throughout the production stages are presented (see Sect. 3.2.2.1). These results are required for deciding which kinds of linked parts are applicable, and how to define the linking. Furthermore, the product model and the connection of processes are important for the second field of research. Due to the fact that every manufacturing stage influences the dimensions of workpieces in a different way and that the statistical distributions of the macro range are not applicable, micro-specific distributions are presented for the simulation of process chains, as well as recommendations for the arrangement of process steps for increasing the throughput (see Sect. 3.2.2.2). The third field involves the control of the production stages. In contrast to conventional approaches, the increase of assembled parts is reached by a design adjustment. The linking of the parts enables the consideration of trends occurring, for example, as caused by wear. The deviation of the geometrical properties is used for the identification of assemblies and for playing back this information to the design, as shown in Sect. 3.2.2.3.

3.2.2.1 Design and Product Model of Linked Parts

The design of linked parts is influenced by the production processes, and further functions that can be derived by the linking of parts. In general, the handling of parts in bulk requires functions like separation, orientation, positioning, and transfer [Kuh09]. In the case of micro parts, furthermore, scaling effects such as the dominance of adhesion forces increase the handling effort. The manufacturing and assembly as linked parts overcomes these size effects [Kuh11] and eliminates handling functions like the orientation of parts [Wek14a]. The transportation of each type of linked part differs in the flux of forces. The three basic types are line-linked parts, ladder-linked parts, and comb-linked parts. Line-linked parts are, for example, interconnected by the wire they are made of. An example of line-linked parts is shown in Fig. 3.2. Relevant design parameters for line-linked

Fig. 3.2 Example of line-linked spheres

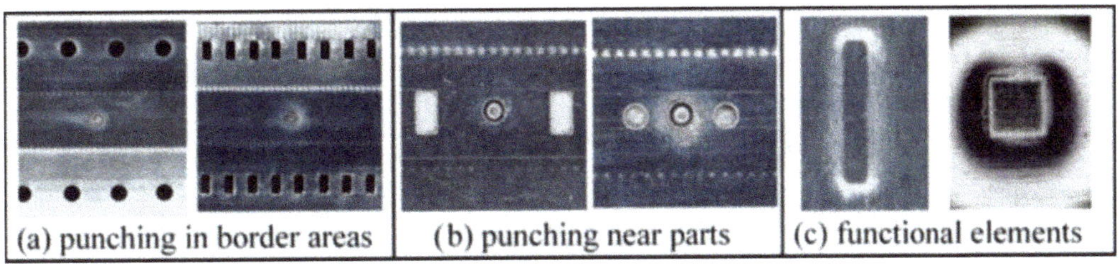

Fig. 3.3 Examples of the design of the border area of ladder-linked parts for absorbing forces

parts are the diameter of the spheres, the diameter of the wire, and the distance between parts.

The flux of forces during transport of the linked parts runs through the parts themselves. The ladder-linked parts have a border area that is used for conveying the linked parts. The flux of forces does not run directly through the parts in this case. An example of ladder-linked parts are cups, which are left in the metal foil they are made of, as depicted in Fig. 3.3. Comb-linked parts are a hybrid type, where the flux of forces does not run through the parts themselves [Tra13].

Besides the transfer of forces, such as the accelerating forces, the design of linked parts could be used for providing further functions and assisting while manufacturing parts. Especially the border areas of the ladder-linked parts could be utilized for this [Wek14]. Examples of the utilization of the border areas are lateral guide elements, pleats to increase the local stiffness, reference points, elements to modify the feeding force transmission in a belt conveyor, and spacers to protect the parts when they are coiled for storage [Kuh14]. Functional punching and elements in the rim of ladder-linked parts are depicted in Fig. 3.3.

The linked parts are produced in multistage processes. For this reason, product data models for a standard exchange of data are required. One possibility is the application of ISO 10303 STEP, which is the abbreviation of "Standard for the Exchange of Product Model Data". Therefore, different methods, guidelines and application protocols are used. ISO 10303 is divided into classified series [ISO04]. Special areas are defined in the 200-series like the AP 207, where a product data model for sheet metal die planning and design is defined [ISO11]. Ladder-linked parts are made of sheet metal. For this reason, they are manufactured using sheet treating processes and could be integrated into AP 207 [Tra12].

Line-linked parts, on the other hand, have several requirements that are not covered by the AP 207. For this reason, a new application model, which covers the scope of micro forming is required. Therefore, the four steps of building an application activity model (AAM), developing an application reference model (ARM), mapping of the ARM, and developing an application interpreted model (AIM) have to be made [Wek16]. Regarding line-linked parts, interviews with researchers of the CRC 747 as process experts, and analysis of technical datasheets are used for identifying the required application objects [Tra12a]. An extract of this information is shown in Fig. 3.4 based on the example of line-linked parts.

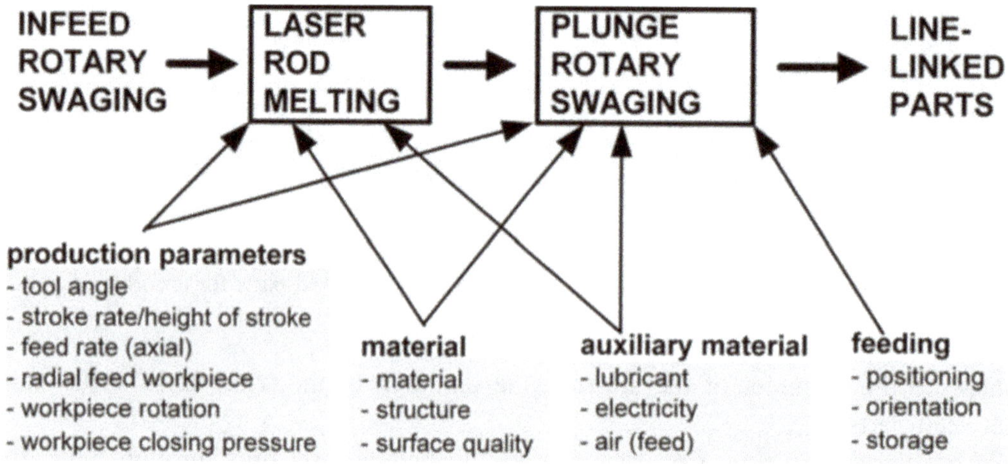

Fig. 3.4 Information and data for plunge rotary swaging and laser melting

The identified application objects and their relation are the basis for the development of the AIM of the production process for linked parts. For completing the AIM, the geometry of the linked micro parts needs to be modularized. Therefore, it is especially important to summarize the piece part instances and the linkage parts instances [Wek16].

3.2.2.2 Production Planning

For achieving high output rates while guaranteeing an adequate part quality, it is important to profoundly understand the whole production process with its interfaces, technologies, and process conditions [Vol04]. In the following, important aspects for planning the mass production of linked micro parts are introduced, based on the example of manufacturing and assembling linked spherical parts and cups.

The occurrence of disturbances or scrap parts requires recommendations for actions that can improve multi-stage processes. Due to the fact that the impact of defects in shape, dimensions, and surfaces is much higher than in macro production [Wek14a], micro-specific process properties have to be taken into account. Simulation studies are an important tool in the planning of manufacturing lines even in the micro range. For reliable simulation results, it is important to use micro-specific distributions during the simulation of micro manufacturing processes. Therefore, micro-specific statistical distributions of ladder-linked parts [Wek13] and line-linked parts [Wek13a] are considered.

Looking at micro cups, the deviation of the rib thickness and the variability of the rib altitude, for example, are used for measuring the quality of the cups. Under consideration of the basic material, and the coating of the punching tool, the distributions for ladder-linked parts are investigated. In many cases, the burr distribution is applicable [Wek13].

For determining the statistical distribution of defects and changes in the geometry of line-linked parts, three different properties of line-linked parts are investigated as examples for distributions of micro parts. The first one is the diameter of the spherical preforms before processing by rotary swaging. The results show that, in total, 12% of the diameters of the preforms are not in the tolerated field. The second defect is the distance between the preforms. In total, 18% of the preforms are not in the tolerated distance. As the statistical distribution for the distance of preforms manufactured by laser rod melting, a Cauchy distribution is appropriate. The third quality criterion, which is especially important for further steps such as rotary swaging, is the volume of the parts. In this case, the Rayleigh distribution is appropriate for simulation of the parts properties. Furthermore, the results show that the volume of the produced parts more often exceeds the targeted volume [Wek13a]. Due to the fact that deviations of the shape, dimensions, and surfaces have a high impact on the production of micro parts, knowledge about these changing distributions is crucial for simulating processes. The simulation offers the possibility of analyzing different scenarios for the identification of appropriate places and capacities for buffering and discharging stations, as depicted in Fig. 3.5.

Due to the rigid linking of the parts, the processing steps are also rigidly linked. Therefore, the configuration of buffers has a major influence on the throughput of the system and balances the varying process times. Furthermore, scrap parts are also present in the linked parts. Therefore, it is important to evaluate if the extraction of only one part or even a whole charge is to be recommended. Due to the fact that the unit prices are comparatively lower as in the macro range, it could be more profitable to extract charges [Wek14a].

The distribution and the amount of scrap parts influence the way in which the processes have to be arranged to improve the production of linked parts. In the following, recommendations for process chains are presented that are investigated by simulating the process chains of linked parts. In the case of a large amount of widespread scrap parts, it is best to extract only single parts instead of batches, due to the fact that there is a minor possibility of batches without defective parts. The implementation of buffers has a positive impact on the output, due to the balancing of the processes. Especially when scrap parts are extracted, it is important to buffer these processes. The decision whether defective parts should be extracted or manufactured in further steps is also dependent on the number of defective parts. When there is a large number of scrap parts, the extraction of the scrap parts has a positive impact on the throughput [Wek14a]. For a further increase of the number

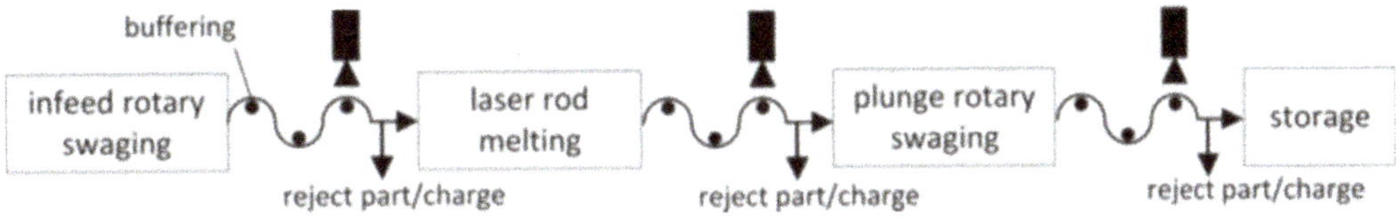

Fig. 3.5 Simulation scenario of line-linked parts with and buffering and rejecting

of assembled good parts, the geometrical deviations, which are caused by the whole manufacturing line, need to be considered. An approach to the widening of the tolerance field is presented in the following section.

3.2.2.3 Tolerance Field Widening

The linking of the parts maintains the production order. This facilitates an investigation of the deviation of geometrical dimensions over the production time. The knowledge of these deviations enables a widening of the tolerance field by identifying sections that could be assembled according to the fit [Onk17]. This approach is similar to selective assembly, where parts are sorted and matched for assembly according to their geometric characteristics. Several approaches to selective assembly, like the identification of assembly batches [Raj11], and reducing surplus parts using genetic algorithms [Kum07], are known. The adaptation of the production based on selective assembly has also been focused on in research. Approaches for selective and adaptive production systems offer the possibility of adjusting the process parameters directly during production. Kayasa et al. offer a simulation-based evaluation of adaptive and selective production systems [Kay12]. Lanza et al. provide a method for real-time optimization in selective and adaptive production systems [Lan15].

Due to the fact that the adaptation of processes is especially used in hard-to-control processes, the named approaches are not relevant for micro cold forming processes. A further contrast to these approaches is that the widening of the tolerance field enables a consideration of long sections of linked parts by using trends of geometrical deviations to identify matchable sections [Onk17]. This is applicable for two reasons. The first one is the sorted feeding by the interconnection of parts. Small tolerances for assembly are often caused by the required interchangeability of loose micro parts. Due to the retention of the production order, precise matching is feasible. The second reason is that in micro production, assemblies like valves, in the case of a defective component, are exchanged as a whole. Consequently, a widening of the tolerance field does not influence the interchangeability for maintenance or repair processes. For widening the tolerances, sequences of parts are matched under consideration of their fit. The methodology of widening the tolerance field is divided into three steps. The steps of clustering, matching and adjusting of the nominal value are performed after the production of the first batch, as shown in Fig. 3.6 on the basis of the example of cups and spheres [Onk17]. The results are used for the manufacturing and assembly of further batches. Within the first step, the trends are in focus. The consideration of the trend enables long sections that are matched in the second step [Onk17]. This is important for ensuring the mass production of further batches, due to the fact that a large number of interruptions for the changing of sections will reduce the productivity [Onk17a].

Linked parts clustering (LPC) is able to identify sections that are homogeneous according to the trends occurring. Essential for clustering trends is that the

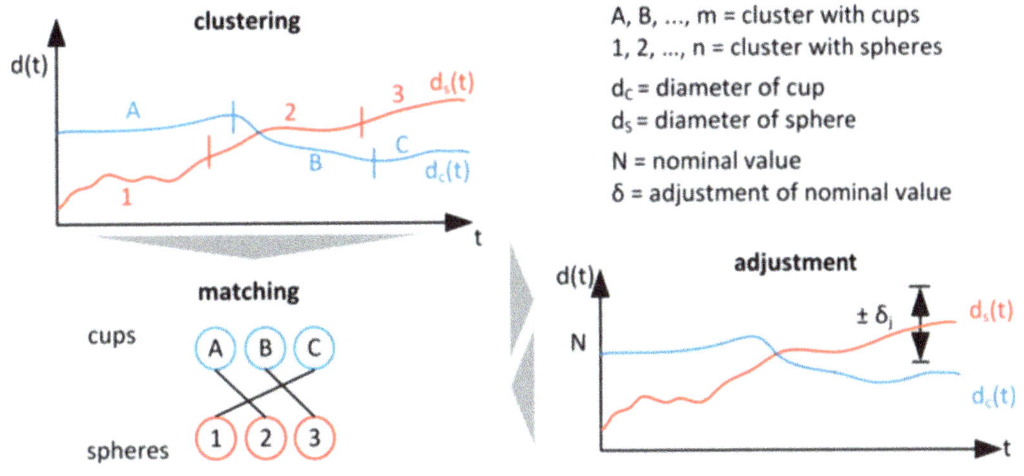

Fig. 3.6 Schemata of the tolerance field widening

algorithm has to consider the linking of the parts. This could be performed by using cluster algorithms for networks. For building the clusters, similarity measures, which compare the characteristics of the parts, are used. This is important for considering the trends. The logic of the LPC is a basic configuration that could be adapted to the trend changes occurring using the assigning Factor (F). The algorithm is able to consider the linking of the parts and decides, on the basis of the Euclidian distance and F, whether the parts should be assigned to the present cluster or to a new one, as depicted in Fig. 3.7. The diameter of the parts is used for the calculation of the Euclidian distance [Tra18].

In the second step, the parts are matched under consideration of the fit. Due to the fact that the matching of these two parts is similar to bipartite graphs, matching algorithms for bipartite graphs like the Munkres algorithm could be used [Onk17]. For the matching of clusters with varying lot sizes, these algorithms must be enhanced by adding additional rows due to the fact that most of them require N × N matrices.

As the third step of the tolerance field widening, the nominal value is adjusted. Due to the known development of the trends, an ideal nominal value can be

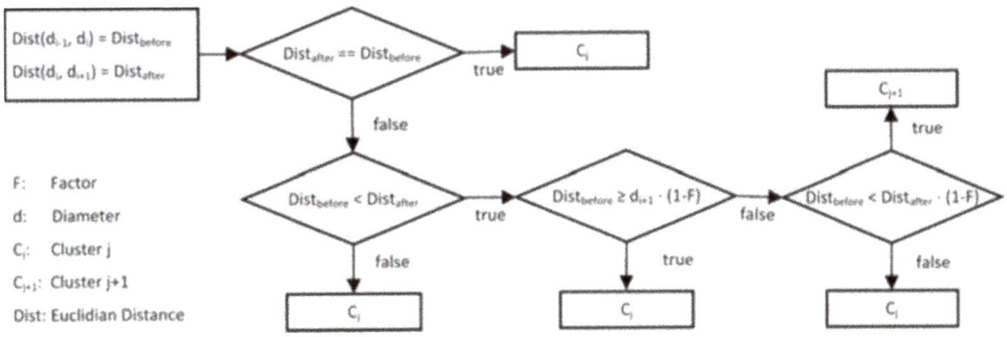

Fig. 3.7 Linked parts clustering [Tra18]

identified. This means that the ratio of the two trends enables a large number of parts to be assembled. Therefore, after every adjustment of the nominal value, the number of pre-assemblies achieved is verified by another matching [Onk17]. The results of the tolerance field widening are influenced by the number of identified sections for matching and the trends occurring. Simulation results showed that with a larger number of sections also the number of mountable parts increases. The results emphasize that there is a point at which an explicit increase of sections only initializes a small increase of the number of pre-assemblies. Due to the fact that micro parts have a comparatively low unit price, such a small increase of assemblies is insignificant. Every interruption for changing sections leads to an increase of handling operations and thus decreases the throughput. The clock rates of the manufacturing and assembly of the parts are increased [Onk17a].

3.2.3 Automated Production of Linked Micro Parts

By applying the method of production as linked parts, on the one hand, time-critical sub-functions of the handling task like the grasping of individual parts and part orientation are avoided. On the other hand, by linking the parts, new challenges arise that must be dealt with to achieve the desired high output rates. The positioning in the processing zone cannot necessarily be considered independent of the transport. Processing may cause interactions and affect neighboring parts. Further, when applying the linked parts method to the production of assemblies, different process chains need to be synchronized. To ensure a stable material flow at high cycle rates and at the same time adequate precision for the positioning in the single production stages, specific equipment is necessary.

3.2.3.1 Handling Concept and Equipment

An approach for realizing a precise positioning in the macro range is to punch reference holes in the first production stage. In the following stages, pilot pins, which are integrated in the tools, realize a fine positioning. This approach, which is typically applied in the aforementioned example of production with progressive dies, hypothesizes that the general structure of the linked parts has a certain inherent stiffness and that it does not change during production. However, for the manufacturing of micro parts on the basis of thin sheet (h $\leq$ 100 μm) and for massive forming of line-linked parts, these requirements cannot be guaranteed. A further measure to simplify positioning, which is applied in the progressive dies, is the integration of all production stages into a compound tool. This allows low tolerances to be realized for the pitch between the single stages, which corresponds to the part distance *a*. In this way, a certain number of neighboring parts can be positioned and processed simultaneously. This is possible under the condition that all production stages have a similar character and can be actuated by a single press.

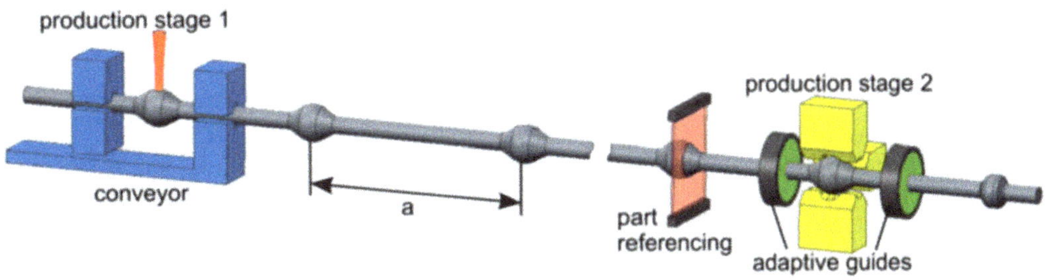

Fig. 3.8 Multi-stage production of linked parts

The proposed new approach references the parts with a sensor before positioning the processing zone. An example is illustrated in Fig. 3.8 by means of the investigated line-linked parts production. In production stage 1, preforms are generated by laser rod melting and consequently formed by rotary swaging in production stage 2. Without the referencing, a deviation from the desired distance *a* would lead to a positioning error in production stage 2.

When supplying parts in bulk, the referencing of individual parts is a multi-dimensional problem. Besides the part position, the orientation of the part must be detected. Because of limitations in measuring rates as well as the processing power of the data evaluation hardware, this may present a bottleneck with regard to output rates. The linked parts approach reduces the problem to one dimension, which is a linear shift in the feed direction. This is valid under the assumption that, except for the feed direction, guides block all other degrees of freedom. In this case, a stationary sensor can be applied for referencing the linked parts during transport, as illustrated in Fig. 3.8. For that purpose, a position-dependent trigger signal is generated on the basis of the position sensor information of the conveyor. Each time the linked parts have moved a certain distance $\Delta x_{trigger}$, the sensor is triggered and acquires data. This approach was tested for three different measuring systems: an optical micrometer, a laser profile scanner and a line camera setup. The optical micrometer [Kuh12a] and the laser profile scanner [Kuh13] are both measuring systems based on image sensors and optimized for standard measuring tasks. They show quite good performances with respect to accuracy, but do not meet the requirements for the aspect of feed velocity. The approach of building up a measuring system on the basis of a camera in combination with suitable optics and illumination provides the best results in terms of both the accuracy and the possible feed velocities. By introducing an interpolation into the evaluation algorithm, the referencing is suitable for applications with output rates of more than 400 parts/min. This is valid even under the conservative assumptions of a part distance of Δx_{part} = 40 mm and a non-continuous conveyor. Both increase the resulting maximum feed velocity, which is $v_{feed,max}$ = 1.6 m/s. Figure 3.9 illustrates the standard deviation of a measurement of the distance *a* between parts at different velocities compared to a reference measurement. The trigger distance was set to $\Delta x_{trigger}$ = 50 µm in order to reach the desired cycle rate. For each velocity, eight linked parts ($d_{part} \approx$ 700 µm, $l_{part} \approx$ 1100 µm) were

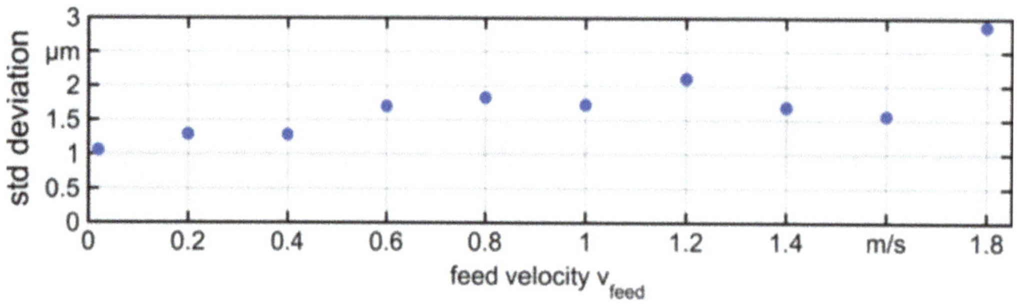

Fig. 3.9 Standard deviation in dependency on velocity for part distance measurement with part referencing system (line camera), $\Delta x_{trigger}$ = 50 µm

referenced and the seven distances between these parts were determined. The measurements were repeated ten times. The standard deviation is below 2.2 µm up to the targeted feed velocity of v_{feed} = 1.6 m/s. With higher feed velocity, the standard deviation increases noticeably [Wil17].

The sensor is ideally located close to the machining position, where the possible minimum distance depends on the actual machining process. Thereby, the applied conveyor must show an adequate positioning accuracy only for that limited feed distance. Two prototypes of conveyors were built and investigated during the experiments for the production of linked micro parts (see Fig. 3.10). One is a belt conveyor. The other conveyor works with two grippers, where one is stationary and the other is mounted on the slide of a linear axis, which is actuated by a linear motor.

Apart from the positioning accuracy of the conveyor, the linked parts influence the positioning. The low bending stiffness of the linked parts, which counts especially for thin sheet material, may lead to buckling and cause position deviations. In order to reduce this problem, the linked parts can be fed by pulling instead of pushing. Further, a tensioning of the linked parts with the help of a hysteresis brake or a second conveyor helps to solve this problem. Considering the accuracy of the feed axis with the grippers, small deviations are introduced as a result of the gripping. But, when assuming that after referencing no gripping is performed and the relevant part is directly positioned, the accuracy depends mainly on the feed axis and is in the sub-micrometer range. The belt conveyor has a large contact area between the linked parts and the coated belts. Thereby, high feed forces can be transmitted, and the local contact force may be reduced in order to avoid deformations. The rotational motion is converted by a friction power transmission in a linear feed. Consequently, different positional errors are introduced, for example, by conversion of the motion measurement from rotation to linear shift or the appearance of slip. Considering this, the feed axis shows a better performance in terms of positioning accuracy. Nevertheless, the belt conveyor can achieve a positioning accuracy in the lower micrometer range. One means of increasing the positioning accuracy is a direct measurement of the shift of the linked parts. Due to the fragility of the parts, contactless measurement is necessary. A sensor based on speckle

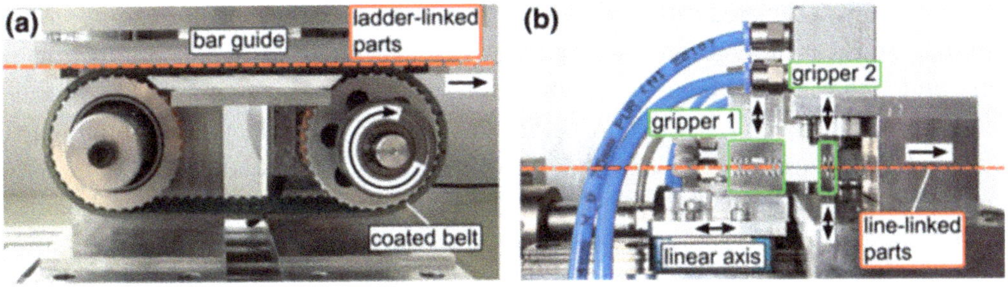

Fig. 3.10 Linked parts conveyors: **a** belt conveyor, **b** feed axis with grippers

correlation was investigated for that purpose and shows promising results, especially with regard to the achievable resolutions [Blo16].

Further, guides play an important role in the production as linked parts. Depending on the processes and their sequence, the guides along the process chain must fulfill different tasks. Besides the precise guiding of the linked parts transverse to the feed direction, they must isolate the individual processes from vibrations occurring in the other processes and damp the vibrations in the machining zones. Furthermore, compact size matters in order to enable the integration of the guides in the setups close to the machining zones. The guiding is particularly critical for line-linked parts, because of their longitudinally varying diameter. Hence, a concept for passive diameter adaptive guides for line-linked parts was developed [Wil17a]. The two variants illustrated in Fig. 3.11 were investigated under the aspects of guiding precision, suppression, and damping of radial vibrations as well as the influence on the feed force. Both variants were proven to enhance the guiding precision compared to non-adaptive guides. Further, they are suitable for suppressing the transmission of vibrations via the linked parts structure. Due to the material properties, the elastomer shows a better performance for damping vibrations in the machining zones. Nevertheless, form deviations of the linked parts from the desired geometry may result in inaccuracies for the guiding. The passive nature of the guides potentially introduces force peaks in the feed direction when a part passes through the guide. However, as they were experimentally determined to be below 0.1% of the applied feed forces, no negative effect is expected.

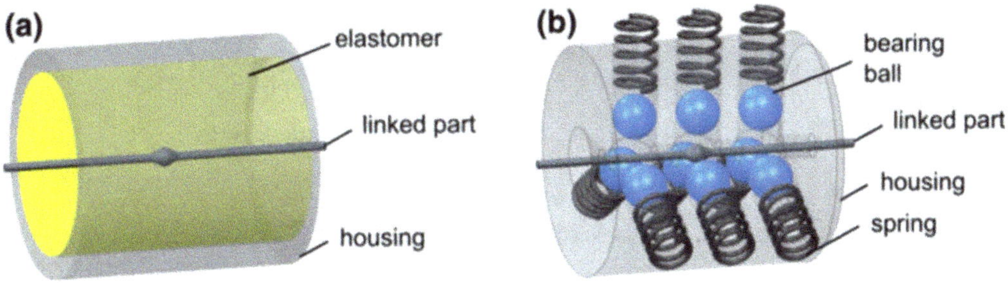

Fig. 3.11 Adaptive guide concepts: **a** elastomer; **b** spring load, bearing balls

Finally, the automated production as linked parts demands specific storage. Two different types of storage are required: long-term storage after the production of parts and buffering between the manufacturing stages. The two types of storage differ mainly in their logistical properties, which are capacity and dynamic capabilities. Apart from that, the storage design depends on the linked parts' geometrical properties (e.g. distance between parts, diameter and thickness) and mechanical properties (e.g. tensile strength, tolerable winding radius). Concepts for buffering are especially characterized by high process dynamics and low capacity. The linked parts could be buffered by extending the length of the conveyor line by using guide rollers as considered in [Rüc17]. Given that the required tensile stress is on the linked parts, guide rollers are arranged in line and the conveying line is minimal. The number of guide rollers defines the possible length of the buffer. For long-term storage, the well-known concepts of spooling and unspooling are suitable [Onk17]. These concepts are characterized by a high storage volume and a minor process flexibility. The line-linked parts are less sensitive to mechanical damage and can be stored in multiple paths and layers, and only the winding diameter is critical. Figure 3.12a illustrates line-linked parts on a spool. Spooling of ladder-linked parts is easier in one path only and with a protective layer in between, as depicted in Fig. 3.12b. This prohibits distortion of the ladder-linked parts caused by torsion. Furthermore, the insertion of the layer material prohibits the deformation and surface deformation of the sensitive parts [Rüc17].

3.2.3.2 *Effects Resulting from the Production as Linked Parts*

There are different challenges that result from the linking of the parts. One is related to the fact that the feeder is applied for the transport, positioning and optional feeding during processing. Further, the structural connection leads to a transfer of forces between neighboring parts or the structure is changed by the processes. The concrete form of the effects that occur depends strongly on the manufacturing process applied.

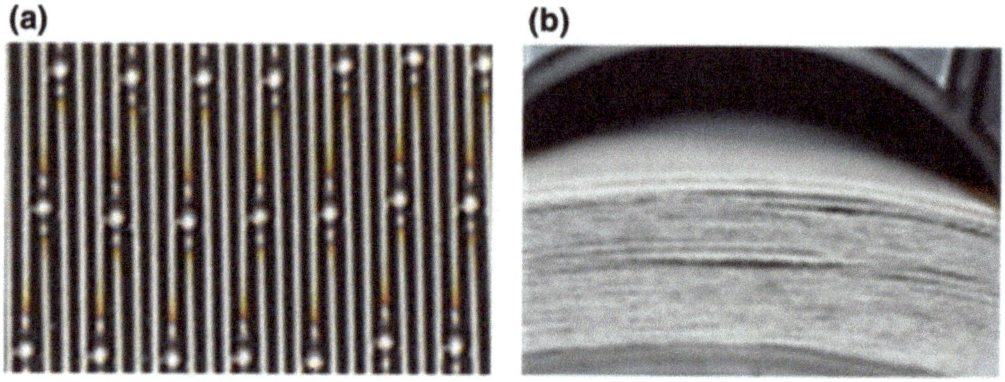

Fig. 3.12 Storage of linked parts: **a** spooled line-linked parts, **b** ladder-linked parts spooled with protective layer

For the production of line-linked preforms by laser rod melting (see Sect. 2.2), the applied conveyor is used both for the feeding during the process and for the transport afterwards. The requirements for these two tasks are different. Reaching the target position fast and a high stiffness against disturbing forces are the crucial points for the transport. For the laser rod melting, however, a good positioning behavior with respect to the desired feed velocity and a small overshoot is assumed to be favorable. The behavior of the conveyor and its effect on the production of preforms as linked parts was investigated experimentally [Wil18]. In the case of the control parameters, which are optimized for short transport times, the actual velocity of the investigated conveyor shows a significant deviation from the desired velocity especially for higher melting velocities. It could be shown that the process is relatively robust against such deviations. Further, an online adaptation of the control parameters can be applied as a countermeasure. The coupling of usually separate functions does not necessarily need to have an adverse effect. Rather, it could be proven that a further increase in the clock rate can be achieved by parallelization. Figure 3.13 illustrates the adapted process scheme for laser rod melting.

The preheating and the closing of the grippers are parallelized to the forward stroke. The effect of that adaptation on the cycle time is illustrated in Fig. 3.14, a measured position curve of the parallelized process. The transport motion is merged with the feed during melting. Thereby, the melting process can start earlier. In this example, the cycle time could be reduced by about 24 ms. The maximum cycle rate for laser rod melting of linked parts achieved with the described system is up to 300 parts/min. It is strongly influenced by the part volume and distance.

A change of the structure of the linked parts is especially observable in the case of massive forming. It has been found that the material flow during plunge rotary swaging leads to a significant elongation of the linked parts, which has a retroactive effect on the process. Assuming that the processed linked part is clamped on one side by the gripper of the conveyor and on the other side preloaded with a defined force, the behavior illustrated in Fig. 3.15 is observed. Before the processing, the linking structure is straight and the part is positioned in the center of the tool cavity. During processing, the part is centered within the cavity and the material flow leads

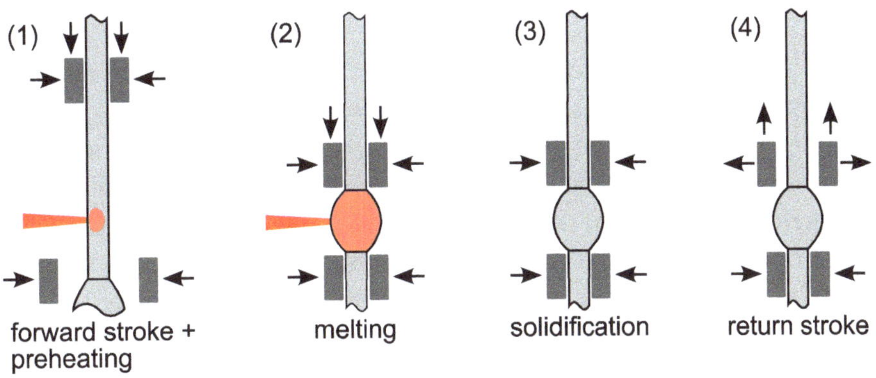

Fig. 3.13 Process scheme of parallelized laser rod melting

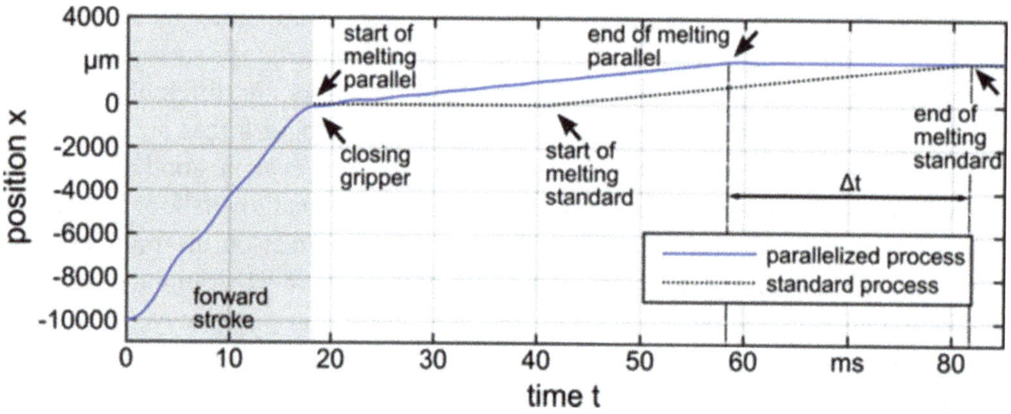

Fig. 3.14 Position measurement of parallelized process

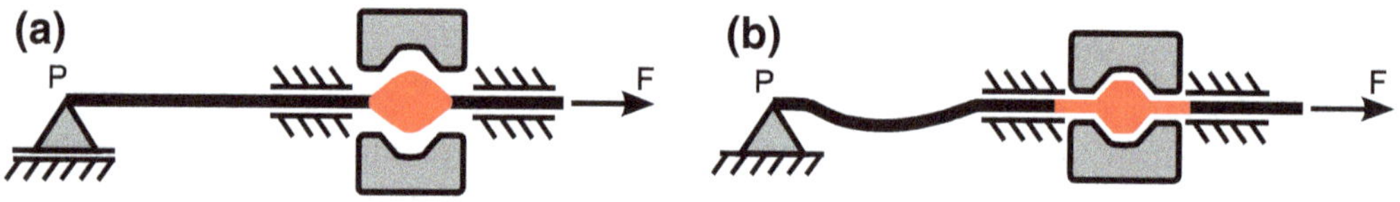

Fig. 3.15 Material flow during rotary swaging: **a** before processing, **b** during processing

to an elongation of the linking structure or wire on both sides of the part. Due to the low bending stiffness of the wire, this leads to buckling on the side of the conveyor and a spring force that presses the part against one side of the cavity within the tool. Thereby, inaccurate forming with asymmetric parts or even defects result [Wil15], as illustrated in Fig. 3.16. It could be shown that the effect on the part geometry can be more or less significant depending on the machine parameters. As shown in [Mou18], a convenient control of the applied forces or a compensation of the elongation improves the forming results and enables the production of symmetrical parts without defects.

3.2.3.3 Synchronization of Linked Parts

In the here presented considerations related to synchronization, the main focus is not on the aspect of cycle rates. The key point is that individual linked parts of one type, or rather groups of linked parts of one type, are assigned to certain linked parts of a different type. This is particularly important for the already presented method of tolerance field widening. At the production level, synchronization means that assignment of the linked parts is performed physically in a way that simplifies the further handling. Instead of a direct assembly, also new types of linked parts can be created and form a pre-assembly. In that case, there are the three options [Onk18]:

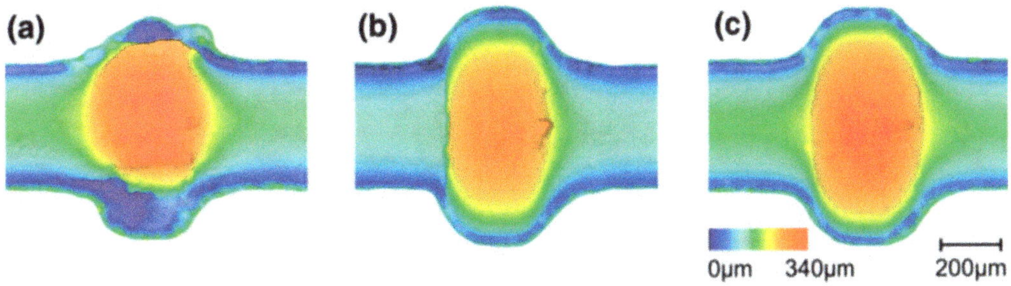

Fig. 3.16 Forming results: **a** bad part, **b** asymmetric part, **c** symmetric part

- directly joining both types of linked parts without separating them before;
- separating the linked parts of both types and creating new linkage;
- separating the linked parts of one type and joining them to the linked parts of the other type.

Option one is the direct joining of two types of linked parts without separation, which could cause problems in the further handling. An increased bending stiffness, for example, complicates winding. Comparing options two and three, the separation of only one type of linked part means less effort and therefore it is more convenient in most cases. For that reason, this option was investigated in an experimental setup. Figure 3.17 illustrates the basic function of the synchronization station. Both types of linked parts are referenced and positioned in the synchronization point. (The feed directions are indicated by arrows.) After that, they are welded via capacitor discharge welding in order to maintain the relative position and assignment between the synchronized parts for further handling. As the thin strip material is more sensitive to mechanical damage, the structure of the ladder-linked parts is kept and the line-linked parts are separated by mechanical shearing.

The complete setup is illustrated in Fig. 3.18. It requires all the previously presented handling technologies. The conveyor for positioning the ladder-linked parts is located on the left side of the central synchronization module. On the right side, another conveyor applies a constant force in order to tension the ladder-linked

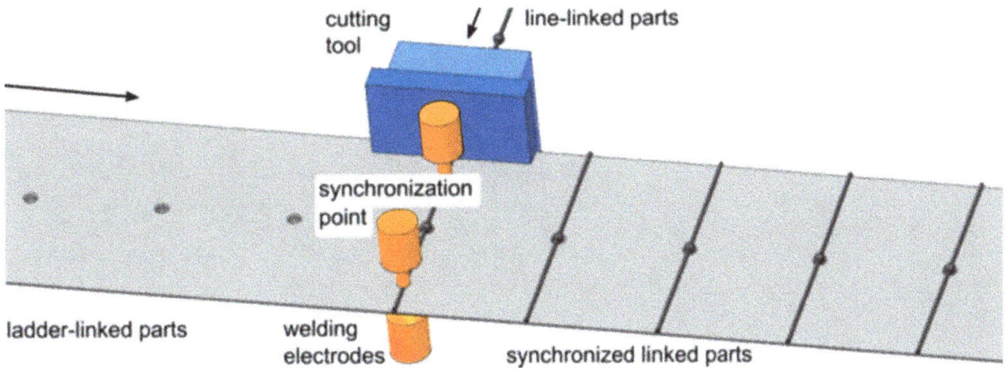

Fig. 3.17 Basic function of the central module of the synchronization station

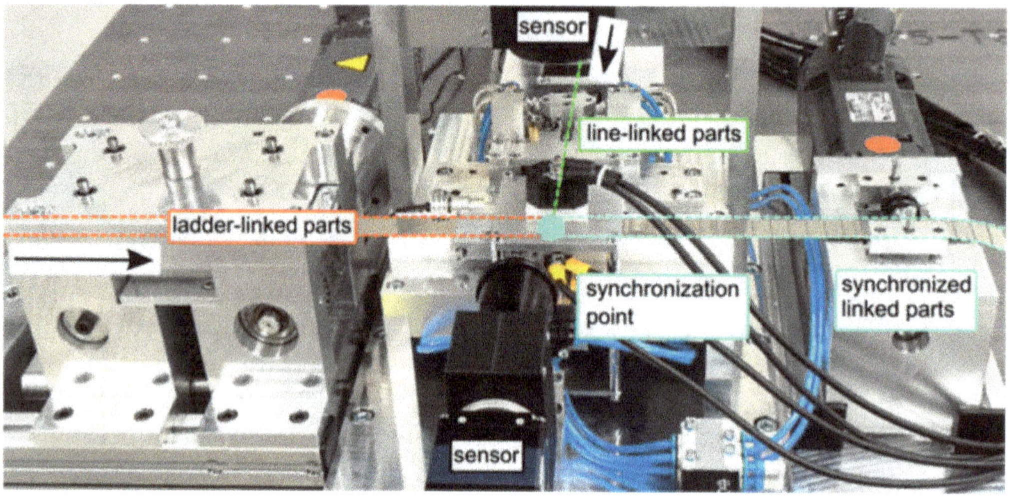

Fig. 3.18 Complete setup—synchronization station

parts. The line-linked parts are fed orthogonally to the feed direction of the ladder-linked parts, from the back to the front. The sensors for the referencing are located in such a way that the linked parts are referenced close to the synchronization point. The central synchronization module includes a pneumatically actuated down holder, which fixes the linked parts after positioning and moves down the welding electrodes and the upper part of the cutting tool. The synchronization station is not only a concept for a possible synchronization setup, but also demonstrates the collaboration of the developed technologies in a practical scenario. It shows that, even with the inclusion of further technologies such as joining and cutting, clock rates of more than 200 parts/min can be achieved.

3.3 A Simultaneous Engineering Method for the Development of Process Chains in Micro Manufacturing

Daniel Rippel*, **Michael Lütjen and Michael Freitag**

Abstract The occurrence of size-effects requires detailed planning and configuration of process chains in micro manufacturing to ensure an economic industrial production. Existing methods lack the required level of detail and complexity as well as the handling of partly missing data and information, which is required for the micro domain. Consequently, this chapter describes a methodology for the planning and configuration of processes in micro manufacturing that aims to circumvent the mentioned drawbacks of the existing approaches. The methodology itself provides tools and methods to model, configure and evaluate process chains in micro manufacturing. Whereas the process design allows the detailed specification of micro manufacturing processes, the configuration is performed by the application of so-called cause–effect networks. Such cause–effect networks depict the interrelationships between processes, machines, tools and workpieces by relevant parameters. Thereby expert knowledge is combined with methods from statistics and artificial intelligence to cope with the influences of size-effects and other uncertain effects.

Keywords Modeling • Production planning • Predictive Model

3.3.1 Introduction

During the last decades, the demand for metallic micro parts has continuously increased. On the one hand, these components have become increasingly smaller while on the other hand, their shape complexity and their functionality are constantly increasing. Major factors contributing to this development are the increasing number and complexity of applications for micro components as well as the increasing demand for such components in the growing markets of medical and consumer electronics [Mou13]. Besides the growing demand for Micro-Electro-Mechanical-Systems (MEMS), which are generally produced using methods from the semi-conductor industry, the demand for metallic micromechanical components is growing similarly. These micromechanical components are used as connectors for MEMS, casings, or as contacts. They cannot be manufactured using semi-conductor technology and are usually produced by applying processes from the areas of micro forming, micro injection, micro milling etc. [Han06]. In this context, cold forming processes constitute an option for the economic mass production of metallic micromechanical components, as these processes generally provide high throughput rates at comparably low energy and waste costs [DeG03].

The efficient industrial production of such components usually requires high throughput rates of up to several hundred parts per minute [Flo14], whereby very small tolerances have to be achieved. These tolerances result from the components' small geometrical dimensions, which, by definition, are below one millimeter in at least two dimensions [Gei01]. Additionally, so called size-effects can result in increasing uncertainties and unexpected process behaviors when processes, workpieces and tools originating from the macro domain are miniaturized [Vol08].

As a result, the planning and configuration of process chains is seen as a major factor of success for the industrial production of metallic micromechanical components [Afa12]. To cope with the occurrence of size-effects, companies require tools and methods for highly precise planning, covering the complete process chain. Thereby, they have to consider the interrelationships between processes, materials, tools and devices. In micro manufacturing, small variations in single parameters or characteristics can have a significant influence along the process chain and can finally impede the compliance with the respective tolerances [Rip14].

In order to facilitate the planning and configuration process, this chapter presents the "Micro-Processes Planning and Analysis" methodology which was designed to support process planners with a tool to plan and configure manufacturing process chains with the required level of detail. The methodology itself provides tools and methods to model, configure and evaluate process chains in micro manufacturing. Thereby, the configuration is performed by the application of so-called cause–effect networks, detailing the interrelationships between relevant process, machine, tool and workpiece parameters. These cause–effect networks combine expert knowledge with methods from the areas of statistics and artificial intelligence to cope with the influences of size-effects.

3.3.2 Process Planning in Micro Manufacturing

Within the literature, there are only very few approaches which allow the joint planning of process chains as well as the configuration of the processes involved. During the last few years, different articles have focused on the configuration of specific processes. Thereby, the configuration relies on detailed studies of the corresponding processes and is usually supported by very detailed physical models in the form of finite element simulations (e.g. [Afa12]). Another approach found in the literature focuses on the use of sample data (historical or experimental) as templates for the configuration of processes. Although these approaches enable a very precise configuration of a single process, the interrelations between different processes cannot be considered easily. Moreover, the construction of finite element simulations or the direct application of historical information requires a deep understanding of the processes, which is often unavailable due to size-effects or the novelty of processes.

Usually, classic methods like event-driven process chains, UML or simple flow charts are used in the context of process chain planning. Whereas these methods do not include the configuration of processes, Denkena et al. proposed an approach that indirectly addresses this topic [Den06]. This approach builds upon the modeling concept for process chains [Den11]. In this concept, a process chain consists of different process elements, again consisting of localized operations. These operations are interconnected by so-called technological interfaces that describe sets of pre- or post-conditions for each operation. While these interfaces can be configured manually, Denkena et al. also proposed the use of physical, numerical or empirical models to estimate the relationships between the pre- and post-conditions [Den06, Den14]. Although this approach enables configuration of the processes, the design of these models again requires very detailed insights into the processes. Moreover, within the micro domain, a single model can quickly be unsuitable to capture all relevant interrelations between process, machine, tool and workpiece parameters, particularly under the influence of size-effects, as its construction would require tremendous amounts of data.

3.3.3 Micro-Process Planning and Analysis (μ-ProPlAn)

The methodology "Micro-Process Planning and Analysis" (μ-ProPlAn) is being developed to support process designers in designing and configuring process chains (see [Rip17] for a more detailed description of the methodology). The modeling methodology μ-ProPlAn covers all phases from the process and material flow planning to the configuration and evaluation of these models [Rip14]. In this context, it enables the integrated planning of manufacturing, handling and quality inspection activities at different levels of detail. It consists of a modeling notation, a procedure model. and a set of methods and tools for the evaluation of the corresponding models. The evaluation covers methods to ensure the technical feasibility of the modeled process chains and production systems. Moreover, it includes a component to execute μ-ProPlAn models directly in terms of a material flow simulation to assess the modeled production system's logistic performance.

The modeling notation consists of three views as depicted in Fig. 3.19, which represents different levels of detail. The first view focuses on the top-level process chains. This view's notation closely follows the classic notation of process chains as described in [Den11]. Unlike the classical approach, process elements and operations are connected using process interfaces. In addition to the original technical parameters, these interfaces include logistic parameters. In an extension to the classical approach, operations act as interfaces to the next view. The material flow view further details operations by assigning those material flow objects used to conduct the operation (e.g. machines/devices, workpieces, tools, operating supplies or workers). This enables the modeling of specific production scenarios with specified resources and therefore allows an evaluation of the models regarding logistic aspects. For example, in the case of an existing production system, μ-ProPlAn can be used to assess the impact of new process chains (e.g. new products) on existing production plans or on the performance of the production system. Therefore, μ-ProPlAn offers the option to conduct material flow simulations based on the specified production system and the modeled process chains. This allows the early detection of e.g. bottlenecks or other undesired effects early during the planning stages. Consequently, it supports the selection of suitable resources, machines or devices for the new process chain. The third view focuses on the configuration of the processes and process chains using so-called cause–effect networks. These networks enable the evaluation of different process configurations (e.g. the use of different materials or different production speeds) and the assessment of the impacts of different choices on follow-up processes. As the cause–effect networks and material flow elements are closely connected, μ-ProPlAn reflects changes to the configuration of the material flow simulation and evaluates these e.g. regarding lead times, work-in-progress levels or the estimated product quality. Moreover, μ-ProPlAn provides several methods to evaluate the configured models. Besides the mentioned material flow simulation, the change propagation supports the process chain configuration and enables an estimation of a product's intermediate and final characteristics.

3.3.3.1 *Modeling View: Process Chains*

In general, process chains enable the structured modeling and planning of business processes, and depict the logic-temporal order of manufacturing, handling and quality inspection processes. Each process consists of one or more operations. Operations transform a set of input variables into output variables. These variables are called technological interfaces and describe the specific technical properties of

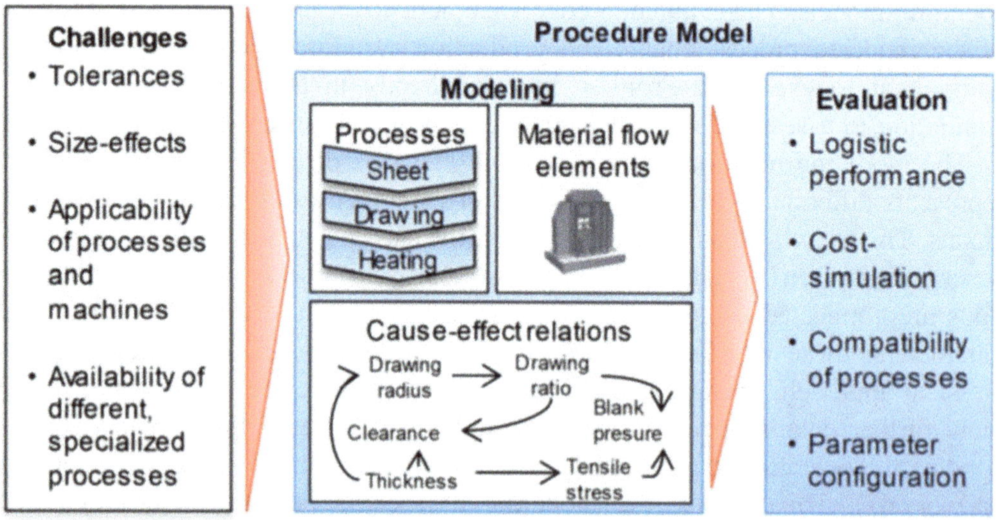

Fig. 3.19 Components of the μ-ProPlAn Methodology [Sch13]

the processed workpiece before and after each operation [Den11]. For example, a hardening operation modifies the variable *hardness*, which serves as an input and an output variable for the operation. On an aggregated level, processes provide and require similar technical interfaces with regard to other preceding or succeeding processes.

The notion of process chains provides an abstraction from material flow related resources and thus from specific manufacturing techniques. The hierarchical composition of the process chain enables the planning of the overall process chain based on processes and parameters relevant to the final product. In order to integrate the classic process chain models with the other views, the technological interfaces are extended to include logistic parameters, e.g. process durations or costs. Additionally, the operations themselves are extended to hold references of model elements from the material flow view, further detailing which devices and resources are involved in performing each single operation. At this, they directly provide the connection to the material flow modeling view.

3.3.3.2 Modeling View: Material Flow

The material flow modeling view consists of two parts. On the one hand, it is used to describe all the material flow elements within the considered system. On the other hand, it allows specific resources to be assigned to operations. Furthermore, it acts as an interface to the configuration or cause–effect view. Therefore, each material flow element is required to provide a minimal set of cause–effect parameters that differs depending on its type. During the evaluation, the material flow simulation uses these parameters to assess the logistic properties of the modeled system. Machines, tools, workers, operating supplies and workpieces constitute the major material flow elements. Table 3.1 summarizes the required parameters, which are necessary to estimate the logistic performance of the modeled system. Each operation requires at least one machine, a tool and a resulting workpiece. Additional resources can be assigned if applicable. During the machine definition, a setup matrix containing setup costs and durations has to be supplied.

Table 3.1 Mandatory parameters by resource (extended from [Rip14, Rip16])

Machines	Tools, workers, supplies	Operations
• Downtime on failure • Downtime on maintenance • Maintenance intervals • Setup times and costs • Mean time between failures • Costs while processing • Costs while idle	• Investment costs • (Tool) Service time in operations • (Worker) Costs while processing	• Duration • Probability of causing defect • Probability of detecting defect • Probability of falsely detecting defect

3.3.3.3 Modeling View: Configuration (Cause–Effect Networks)

To further detail each material flow element, μ-ProPlAn uses cause–effect networks, describing the interrelationships between relevant process parameters. Each network consists of a set of parameters and a set of cause–effect relationships, forming a directed graph. The set of parameters consists of all technical and logistic characteristics that are relevant to describe the object's influence on the production process and at least consists of the parameters given in Table 3.1. In the case of workpieces, these include material properties, costs per piece or geometrical characteristics. As for production devices, these parameters include production speeds, forces or other characteristics that can be set, calculated or measured (see Fig. 3.20 for an example). From a modeling perspective, the cause–effect networks are hierarchical. Each material flow object (workpieces, machines, tools, workers, etc.) has its own cause–effect network, or at least a set of describing parameters. When combining these single elements into operations, process elements or process chains, higher-level cause–effect networks are created by describing additional relationships between the parameters of the networks or by connecting them to previously specified process interfaces (see Fig. 3.20).

The design of cause–effect networks is divided into two steps: qualitative modeling and quantification. The qualitative model of the network is created by collecting all relevant parameters and denoting their influences among each other (see Fig. 3.20). Thereby, the process expert can facilitate the creation of the networks by pointing out the most relevant parameters and by indicating the relationships. Another option to attain a qualitative cause–effect network is the application of data mining techniques. For example, if experimental or production data are available, an analysis of this data can deliver preliminary insights into the cause–effect relations.

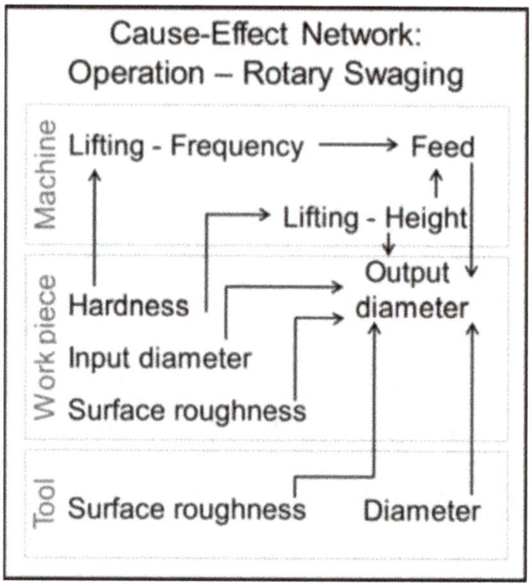

Fig. 3.20 Hierarchy of cause–effect networks—Example of composition of a rotary swaging operation from the networks of the corresponding machine, tool and workpiece (after [Rip14a])

The second step concerns the quantification of the cause–effect networks. The objective is to enable the propagation of different parametrizations throughout the network. For example, this propagation allows the estimation of the outcome of processes for different materials or machining strategies. In Fig. 3.20, the use of a different material for the forging tool will result in a different surface roughness, thus influencing the overall output diameter of the processes. Through quantifying the cause–effect relationships, it is possible to estimate the results of parameter changes for all connected parameters along complete process chains.

3.3.3.4 Basic Quantification of Cause–Effect Networks

In the case of simple or well-known relations, μ-ProPlAn allows the implementation of mathematical functions to calculate a parameter's value, based on its input parameters' values. For example, a continuous manufacturing process will have a *duration* according to the total *length* of the workpiece divided by the selected *feed velocity*. More complex but well-established cause–effect relations could be included from the literature. Nevertheless, in the area of micro manufacturing, different parameters can have a more significant impact than in the macro domain, resulting in the inclusion of parameters that can usually be neglected. In addition, size-effects may induce a different behavior than usually observed. Therefore, it is often impossible to describe all the parameters and cause–effect relations comprehensively. As a result, μ-ProPlAn provides the capabilities to quantify cause–effect relations from experimental or production data by applying methods from the areas of data mining and statistics. Therefore, the qualitative cause–effect network is divided into a set of sub-problems. Each of these consists of one single dependent parameter and its independent parameters. To achieve a quantification, μ-ProPlAn utilizes a set of statistical regression or learning methods further detailed in Table 3.2. While statistical regression methods usually calculate a mathematical function to describe the expected value of the considered dependent variable, learning methods usually provide a more abstract model for a mean estimator ($\mu(x)$). While these estimators can predict the value of the dependent variable, their models usually provide less insight into the mechanics behind the cause–effect relation and thus represent "black-box" models. Despite this disadvantage, they excel in their ability to learn arbitrary relations without prior knowledge of the relations shape (no prototype of the function has to be provided) and regardless of the problem's dimensionality. In practice, the application of locally weighted linear regression (or interpolation) models has yielded good results in estimating complex relations.

3.3.3.5 Characterization of Local Variances

While the application of the described methods enables the cause–effect networks to predict the estimated (mean) value of a parameter, they do not support the

Table 3.2 Regression and learning methods included in μ-ProPlAn [Rip14a]

Method	Description
Linear regression	As a fundamental multivariate regression function, μ-ProPlAn offers the option to perform a linear regression on the data provided. Additionally, μ-ProPlAn offers functionality to linearize e.g. exponential or logarithmic data
Polynomial regression	In the case of univariate cause–effect relations, a polynomial regression can be conducted
Tree/Rule regressions (e.g. [Hol99])	In contrast to the methods above, tree- or rule-based approaches do not result in a single analytic function. In general, they divide the search space into smaller segments for which a regression can be performed. Usually both methods use linear regressions for each segment
Locally weighted linear regression (LWL/LOESS) (e.g. [Fra03])	LWL constitutes an abstract prediction model, which performs a locally weighted linear regression each time a prediction is requested. Thereby, a kernel function is used to weight adjacent data points and finally a linear regression is performed. This method particularly excels in interpolating missing data in between available data points
Support vector regression(e.g. [Pla98])	Support vector machines are usually used as classifiers. Thereby, they learn models, which separate a set of data points into classes, maximizing the distance between each data point and the classification curve. The same method can be used for regression, particularly if the data provided contains strong variances
Artificial neural networks	A neural network usually consists of a number of layers, each containing a defined number of so-called perceptrons. The perceptrons of each layer are interconnected. During the supervised training of the network, these connections' weights are adapted to recreate the desired output on the last layer

characterization and use of variances. Nevertheless, coping with these variances poses a major task for the configuration of process chains in micro manufacturing, in order to achieve a stable and efficient production. Moreover, the variance of a process can change for different parametrizations, requiring a localized approach. For example, when using a laser-chemical ablation process, the variance can change according to the flow rate of the chemicals. Therefore, μ-ProPlAn provides an estimated distribution as a prediction for a parameter's value instead of a single mean value. Using these distributions, the process designer can e.g. try to select configurations with less variance, resulting in a more stable process/process chain, or assess the impact of less tolerated materials on the process chain.

Local calculation of the variance of each sampling point

In addition to the mean estimator, the local variance is computed for each point of the available sample data based on the corresponding standard deviation (σ). The method uses a set of experiments, each consisting of a number of independent variables (sampling points/configurations) $x = (x_1, x_2, \ldots)$ and a corresponding dependent variable (measurement). In the case of repeated measurements, each sampling point x corresponds to several values for the dependent variable, resulting in a vector y_x where the number of values $|y_x|$ equals the number of measurements.

In a first step, the mean estimator $\mu(x)$ over all experiments is acquired using regression or learning techniques as described before. As a second step, the approach first determines a localized value for the standard deviation at each given point of sampling data and then further interpolates this information to enable a prediction of the variance for arbitrary points/configurations ($\sigma(x)$). The localized standard deviation for each sampling point is thereby calculated as usual, whereby the target vector y and the matrix source $X^{m \times n}$, with $n = |y|$ and $m = |x|$, are specifically constructed using a newly introduced parameter t to control the degree of locality. In general, the algorithm only tries to incorporate measurements for the single sampling point/configuration x. In cases where the number of repeated measurements y_x for x are insufficient ($|y_x| < t$), the algorithm incorporates additional sampling points into the calculation. On the one hand, small values for t ensure a very high locality, focusing on single sampling points or at least on very small areas around them. While a strong locality reduces the window size, it can negatively influence the estimation precision, as a precise estimation of the standard deviation requires a sufficient amount of data. On the other hand, a broader window size can usually yield better estimation results, while smoothing local changes in the standard deviation.

In the case of normally distributed values, this approach provides an accurate estimation of the overall standard deviation for each sampling point. Nevertheless, several practical applications require the estimation of skewed distributions. In this case, μ-ProPlAn estimates distinct standard deviations for both sides of the estimated mean value. Therefore, each vector of repeated measurements y_x is separated into two vectors, one containing all values smaller than the estimated mean value and the other containing all values greater than the estimated mean value for that given sampling point. The only exception are values that are equal to the estimated mean. If such a measurement is available within the data, its value is added to both vectors. Afterwards, the local standard deviation is calculated for each of these vectors as $\sigma_{Ui}(x_i)$ and $\sigma_{Li}(x_i)$. This separation enables the estimation of different standard deviations for both sides of the curve at the cost of precision. Due to the reduction of the available data points, the window size will inevitably increase, resulting in a stronger smoothing for each sampling point. The only option to counteract this would be the inclusion of additional repeated measurements.

Interpolation for arbitrary points and example

By applying the described local approach, μ-ProPlAn can characterize each of the provided sampling points by its mean and its estimated local standard deviation. In order to achieve an estimation of the standard deviations for arbitrary points, an interpolation takes place. Figure 3.21 depicts an example of an application of the local approach to a dataset containing 230 measurements of an ablation process. The dependent variable y depends on two independent variables x_1 and x_2. Figure 3.21a depicts the mean estimation obtained by using a locally weighted linear regression using an Epanechnikov kernel spanning approximately eight neighboring points. Figure 3.21b additionally depicts the estimated 95% probability intervals as green and red grids respectively, assuming the measurements were normally distributed.

In general, μ-ProPlAn uses these standard deviations to describe the stability of a process in terms of its variance and its adherence to tolerances. In addition, it includes a technique to generate probability distributions based on these values. If only a single standard deviation was estimated, or if $\sigma_{Ui}(x_i)$ and $\sigma_{Li}(x_i)$ are close together, a normal distribution $\mathcal{N}(\mu(x), \sigma(x)^2)$ is generated. If they differ from each other, a Gumbel distribution is generated (also known as Extreme-Value Distribution). Therefore, a large number of sampling points is generated using two normal distributions derived as $\mathcal{N}_L(\mu(x), \sigma_L(x)^2)$ and $\mathcal{N}_U(\mu(x), \sigma_U(x)^2)$.

The resulting probability distributions can be used for further sampling, e.g. to calculate process capability indices c_p and c_{pk}. Therefore, μ-ProPlAn employs a Monte-Carlo simulation, whereby for a provided process configuration (e.g. materials, forces etc.) all dependent parameters along the process chain are sampled

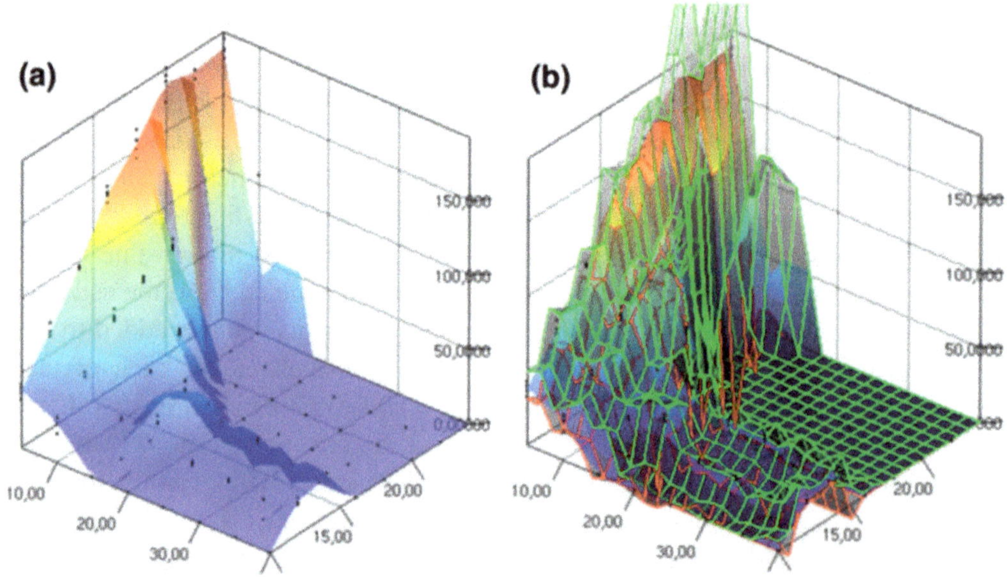

Fig. 3.21 Example of application on a three-dimensional problem. **a** The heat-map colored surface depicts the mean estimator. **b** Additional green and red grids depict the 95% confidence interval

and, if appropriate, used as input for the remainder of the cause–effect network. By repeating this several times, the behavior of the overall process chain can be estimated in terms of e.g. quality-relevant target parameters like diameters or surface properties. The overall process capability indices are then calculated using Eqs. 3.1 and 3.2, with USL and LSL being the upper and lower specification limits according to the specified tolerances.

$$C_p = \frac{USL - LSL}{6\sigma} \tag{3.1}$$

$$C_{pk} = \min\left[\frac{USL - \mu}{3\sigma}, \frac{\mu - LSL}{3\sigma}\right] \tag{3.2}$$

3.3.3.6 Simultaneous Engineering Procedure Model

To guide users through the process of model creation, a procedure model accompanies the methodology. This procedure model utilizes concepts of Simultaneous Engineering [Rip14]. In general, "Simultaneous Engineering describes an approach in which the different phases of new product development, from the first basic idea to the moment when the new product finally goes into production, are carried out in parallel." [Sch02]. By parallelizing the development of the product and of the manufacturing processes, all characteristics and demands regarding the product's complete life cycle can be taken into account early on. Ignoring such issues can lead to high costs in later stages of the product development or manufacturing [Pul11]. The application of Simultaneous Engineering techniques enables the early determination of these problems, which would usually only emerge during later stages of the production planning [Rei12].

The procedure model comprises three phases: process specification, configuration and analysis. The first two phases in particular should be conducted in parallel to the product development. Whenever the product design is extended or changed, these modifications should be integrated directly into the process chains.

The top-level process chain is designed during the process specification phase. First, the designer aligns all the required manufacturing processes with respect to the product design and defines the process interfaces. In a second step, the designer integrates quality inspection processes in between the manufacturing processes. As a last step of the specification, the designer integrates the required handling processes and operations. Some cases may require complex handling processes (e.g. the separation of single products or samples), while other cases can be satisfied by the integration of handling operations into existing processes. During the process configuration, suitable technologies for each process should be identified, as they determine or at least influence the process interfaces.

After defining the process chain and the according interfaces, the designer configures the processes. Based on the selection of technologies, he integrates relevant production parameters into the model and describes the parameters' interdependencies (cause–effect relations). For each parameter, the designer specifies tolerances, which describe the range of allowed parameter values, in order to guarantee the required output values. In a second step, the parameters are set, thus describing concrete values for all production-relevant parameters in the process chain. By first specifying the tolerances, the designer is able to select resources according to the requirements of the process. Consequently, the process is aligned to the product's quality requirements.

The last phase of the procedure is the analysis. Partially, it can occur in parallel to the specification of the material flow. Whenever a parameter is set to a specific value, the impact on the complete process chain can be analyzed according to the modeled cause–effect relations. As a result, the impact on later processes can be estimated and, if necessary, the designer can select other technologies or resources and adapt the model accordingly. In cases where no available technology is suitable to satisfy the product's requirements, modifications to the product design or the development of new technologies can be addressed. Due to the proposed parallel development of the product and the design of the manufacturing process, such incompatibilities can be detected early on in the development process and higher costs can be prevented. In the last step of the analysis, technically feasible process chains can be simulated, using a material flow simulation, to determine their logistic performance and their behavior. The designer can compare different setups or processes and select the best option. In particular, if certain tools or workstations have to be purchased, the simulation can facilitate the decision between options [Sch08].

3.3.3.7 Geometry-Oriented Modelling of Process Chains

The geometry-oriented process chain design aims to provide an alternative way of modeling and designing process chains in μ-ProPlAn. The procedure model described before assumes a manual modeling and evaluation of each alternative chain. The geometry-oriented approach focuses on additional annotations to already modeled operations, in order to select and combine suitable processes. Thereby, each operation is annotated with information on the processes' capabilities and limitations. In addition, workpieces can be specified by their geometrical features. Using a constraint-based search, suitable processes can be selected, combined and evaluated automatically. For these annotations, additional parameters are introduced to the cause–effect networks. These describe which geometries can be achieved by a process. Basically, μ-ProPlAn is adapted twofold. First, processes need to be annotated with their type (what they do). Second, each process requires the annotation of which geometrical features can be achieved under which circumstances.

The type of a process describes its function according to the standards DIN 8580 and VDI 2860. For manufacturing processes, the type describes if it is a (primary) forming process, a separating or joining process, if it is a coating process or if it just modifies the workpiece characteristics without changing the geometry. For handling processes, the type determines e.g. if it is a positioning or conveying process. Besides the definition of the type (optical, tactile, acoustic), quality inspection processes additionally require information on whether the process is in-line or destructive and about its resolution and measurement uncertainties.

The second adaptation describes the pre- and post-conditions of each process with respect to the geometrical features. Thereby, each process is annotated with a set of workpieces that can serve as input to the process, as well as a set of workpieces, which result from the process application to an input. In extension to the current model, each workpiece has a base geometry, describing e.g. if the workpiece is a sheet, a sphere, a wire etc. Each base geometry can be combined with additional base geometries to compose more complex workpieces. Figure 3.22, for example, shows a combination of a wire with a cone-shaped base geometry to represent a workpiece that could act as a plunger for a micro valve. Each base geometry can be assigned zero or more geometrical features like holes, steps, pockets or notches. These features and their parameters conform to the STEP-NC standard. Each of these features can again be assigned zero or more features on their own in a hierarchical manner.

By using these annotations, it becomes possible to apply constraint-based search algorithms to select suitable processes. During the modeling process, the designer specifies the hierarchical order of the base geometries and geometrical features required for the product. The software tool can search through all the stored processes within the model and select those that can achieve each feature. In the example in Fig. 3.22 there is only one alternative to join a sphere to a wire (melting). In contrast, there exist two alternatives to form this sphere into a cone. As the characteristics of each base geometry and each feature are expressed as parameters integrated into the processes' cause–effect networks, the configuration of each alternative process chain can be performed in the usual way.

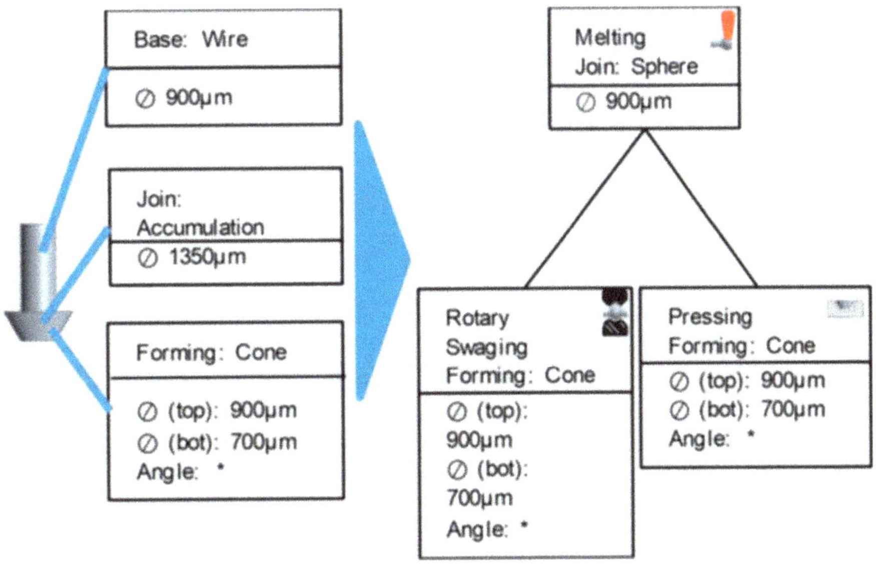

Fig. 3.22 Example of the geometry based annotations and generated process chain alternative

3.3.3.8 Analysis and Model Optimization

As mentioned before, μ-ProPlAn supports different methods to evaluate the process chain models using simulation-based and model-based evaluations. Technical characteristics (e.g. material or surface properties) or process capabilities can be estimated through the cause–effect networks, while logistic properties and the dynamic behavior of different alternatives can be estimated by transforming and simulating the different alternatives using the integrated discrete-event material flow simulation *jasima* [Hil12]. Using the respective results, the process designer can adapt the model and perform a manual optimization of different process chains, or compare different manufacturing scenarios. Nevertheless, a manual configuration of the processes can be time-consuming and prone to error. Therefore, μ-ProPlAn supports the configuration by providing methods to identify suitable values for input parameters, given a set of desired output values (e.g. for geometrical features) using meta-heuristics (i.e. a genetic search algorithm) and a pruned-depth first search algorithm. Both methods provide multiple settings, which are able to achieve the desired result within the specified tolerances. Moreover, Gralla et al. proposed a method to invert the cause–effect networks' prediction models [Gra17] using techniques from the mathematical field of inverse modeling. The results show more reliable results in shorter times using these inverted cause–effect networks in combination with mathematical optimization techniques [Gra18].

The proposed methodology supports process designers during all stages from the manufacturing process design, the process configuration until the evaluation of process chain alternatives. In combination with the software prototype, it can be applied to different topics, e.g. for cost assessments of different configurations [Rip14], for the evaluation of different machining strategies [Rip14a], for the process characterization and configuration [Rip17b], or even as an abstraction for physics-based finite-element simulations to enable fast and precise process planning [Rip18a]. In addition, the methods underlying the cause–effect networks can easily be extended to utilize time-related information, thus enabling a re-training of cause–effect networks to cope with a lack of initial training data [Rip18] or in the context of predictive maintenance [Rip17a].

References

[Afa12] Afazov, S.M., Becker, A.A., Hyde, T.H.: Development of a finite element data exchange system for chain simulation of manufacturing processes. Adv. Eng. Softw. **47**, 104–113 (2012)

[Blo16] Bloem, A., Wilhelmi, P., Schenck, C., Kuhfuss, B.: 2D position sensor based on speckle correlation—experimental setup for high-speed measurements and in field tests. In: Proceedings of 11th International Conference on Micro Manufacturing (ICOMM2016). Orange County, CA (USA), 29–31 March 2016, Paper# 17 (CD)

[DeG03] DeGarmo, E.P., Black, J.T., Kohser, R.A.: Materials and Processes in Manufacturing. Wiley (2003)

[Den06] Denkena, B., Rudzio, H., Brandes, A.: Methodology for dimensioning technological interfaces of manufacturing process chains. CIRP Ann. Manuf. Technol. **55**(1), 497–500 (2006)

[Den11] Denkena, B., Tönshoff, H.K.: Prozessauslegung und -integration in die Prozesskette. In: Denkena, B., Tönshoff, H.K. (eds.) Spanen – Grundlagen, pp. 339–362. Springer, Heidelberg (2011)

[Den14] Denkena, B., Schmidt, J., Krüger, M.: Data mining approach for knowledge-based process planning. Procedia Technol. **15**, 406–415 (2014)

[Fan14] Fantoni, G., Santochi, M., Dini, G., Tracht, K., Scholz-Reiter, B., Fleischer, J., Lien, T.K.: Grasping devices and methods in automated production processes. CIRP Ann. Manuf. Technol. **63**(2), 679–701 (2014)

[Fle11] Fleischer, J., Herder, S., Leberle, U.: Automated supply of micro parts based on the micro slide conveying principle. CIRP Ann. Manuf. Technol. **60**(1), 13–16 (2011)

[Flo14] Flosky, H., Vollertsen, F.: Wear behaviour in a combined micro blanking and deep drawing process. CIRP Ann. Manuf. Technol. **63**(1) 281–284 (2014)

[Fra03] Frank, E., Hall, M., Pfahringer, B.: Locally weighted naive Bayes. In: 19th Conference on Uncertainty in Artificial Intelligence, pp. 249–256. Morgan Kaufmann (2003)

[Gei01] Geiger, M., et al.: Microforming. CIRP Ann. Manuf. Technol. **50**(2), 445–462 (2001)

[Gra17] Gralla, P., et al.: Eine Methode zur Invertierung von Vorhersagemodellen in der Mikrofertigung. In: Vollertsen, F., et al. (eds.) 8. Kolloquium Mikroproduktion. BIAS Verlag, Bremen (2017)

[Gra18] Gralla, P., et al.: Inverting prediction models in micro production for process design. In: Vollertsen, F., et al. (eds.) MATEC Web of Conference, vol. 190. EDP Sciences, Les Ulis Cedex (2018). https://doi.org/10.1051/matecconf/201819015007

[Han06] Hansen, H.N., et al.: Dimensional micro and nano technology. CIRP Ann. **55**(2), 721–734 (2006)

[Hil12] Hildebrandt, T.: Jasima-An Efficient Java Simulator for Manufacturing and Logistics. https://www.simplan.de/software/jasima/. Accessed 27 Sep 2018

[Hol99] Holmes, G., Hall, M., Frank, E.: Generating rule sets from model trees. In: 12th Australian Joint Conference on Artificial Intelligence, pp. 1–12. Springer, Heidelberg (1999)

[ISO11] International Organization on Standardization: ISO 10303-207: standard for the exchange of product model data—Industrial automation systems and integration—Product data representation and exchange. Part 207: Application protocol: sheet metal die planning and design

[ISO04] International Organization on Standardization: ISO 10303-11: standard for the exchange of product model data—Industrial automation systems and integration—Product data representation and exchange. Part 11: Description methods: the EXPRESS language reference manual

[Kay12] Kayasa, M.J., Hermann, C.: A simulation-based evaluation of selective and adaptive production systems (SAPS) supported by quality strategy in production. Procedia CIRP **3**, 14–19 (2012)

[Kru09] Krüger, J., Lien, T.K., Verl, A.: Cooperation of human and machines in assembly lines. CIRP Ann. Manuf. Technol. **58**, 628–646 (2009)

[Kuh14] Kuhfuss, B., Schenck, C., Wilhelmi, P., Langstädtler, L.: Electromagnetic linked micro part processing. In: Proceedings of 11th International Conference on Technology of Plasticity, ICTP 2014, 19–24 Oct 2014. Nagoya Congress Center, Nagoya, Japan. Procedia Engineering, vol. 81, pp. 2135–2140 (2014). https://doi.org/10.1016/j.proeng.2014.10.298

[Kuh13] Kuhfuss, B., Schenck, C., Wilhelmi, P.: Advanced laser profile scanner application for micro part detection. Laser metrology and machine performance. In: Proceedings of 10th International Conference on Laser Metrology, CMM and Machine Tool Performance (LAMDAMAP 2013), pp. 365–372. Euspen, Buckinghamshire, UK (2013)

[Kuh12a] Kuhfuss, B., Schenck, C., Wilhelmi, P.: Laser profile scanner application for micro part detection. Appl. Mech. Mater. **275-277**, 2527–2530 (2013)

[Kuh11] Kuhfuss, B., Moumi, E., Tracht, K., Weikert, F., Vollertsen, F., Stephen, A.: Process chains in microforming technology using scaling effects. In: Menary, G. (ed.) International Conference on Material Forming (ESAFORM 2011), pp. 535–540. American Institute of Physics, Belfast, UK (2011) (online)

[Kuh09] Kuhfuss, B., Moumi, E.: Manufacturing of micro components by means of plunge rotary swaging. In: Proceedings of the Euspen International Conference, pp. 58–61 (2009)

[Kum07] Kumar, M.S., Kannan, S.M.: Optimum manufacturing tolerance to selective assembly technique for different assembly specifications by using genetic algorithm. Int. J. Adv. Manuf. Technol. **32**, 591–598 (2007)

[Lan15] Lanza, G., Haefner, B., Kraemer, A.: Optimization of selective assembly and adaptive manufacturing by means of cyber-physical system based matching. CIRP Ann. Manuf. Technol. **64**, 399–402 (2015)

[Mer12] Merklein, M., Stellin, T., Engel, T.: Experimental study of a full forward extrusion process from metal strip. Key Eng. Mater. **504-506**, 587–592 (2012)

[Mou18] Moumi, E., Wilhelmi, P., Schenck, C., Herrmann, M., Kuhfuss, B.: Material flow control in plunge micro rotary swaging. In: Vollertsen, F., Dean, T.A., Qin, Y., Yuan, S.J. (eds.) 5th International Conference on New Forming Technology (ICNFT 2018), vol. 190, p. 15014. MATEC Web Conference (2018). https://doi.org/10.1051/matecconf/201819015014

[Mou13] Mounier, E., Bonnabel, A.: Press Release Emerging MEMS (2013). http://www.yole.fr/iso_upload/News/2013/PR_EmergingMEMS_August2013.pdf. Accessed 27 Sep 2018

[Onk18] Onken, A.-K., Wilhelmi, P., Tracht, K., Kuhfuss, B.: Increased output in micro production by tolerance field widening and synchronisation. In: Vollertsen, F., Dean, T.A., Qin, Y., Yuan, S.J. (eds.) 5th International Conference on New Forming Technology (ICNFT 2018), vol. 190, p. 15006. MATEC Web Conference (2018). https://doi.org/10.1051/matecconf/201819015006

[Onk17] Onken, A.-K., Rückert, P., Perl, C., Tracht, K.: Joining linked micro formed parts through tolerance field widening and synchronization. In: Schüppstuhl, T., Franke, J., Tracht, K. (eds.) Tagungsband des 2. Kongresses Montage Handhabung Industrieroboter (MHI2017). Springer, Heidelberg

[Onk17a] Onken, A.-K., Wilhelmi, P., Rückert, P., Tracht, K., Kuhfuss, B.: Toleranzfeldaufweitung durch gezielte Kombination von Fügepartnern in der Mikroproduktion. In: Vollertsen, F., Hopmann, C., Schulze, V., Wulfsberg, J. (eds.) Fachbeiträge 8. Kolloquium Mikroproduktion, pp. 259–266, 27–28 Nov 2017. BIAS Verlag, Bremen (2017) (in German)

[Pla98] Platt, J.: Fast training of support vector machines using sequential minimal optimization. In: Schoelkopf, B., Burges, C., Smola, A. (eds.) Advances in Kernel Methods—Support Vector Learning. MIT Press, Cambridge, MA (1998)

[Pul11] Pullan, T.T., Bhasi, M., Madhu, G.: Application of object-oriented framework on manufacturing domain. J. Manuf. Technol. Manag. **22**(2), 906–928 (2011)

[Qin15] Qin, Y.: Micromanufacturing Engineering and Technology. William Andrew—Elsevier Science, Oxford (2015)

[Qin08] Qin, Y., Ma, Y., Harrison, C., Brockett, A., Zhou, M., Zhao, J., Law, F., Razali, A., Smith, R., Eguia, J.: Development of a new machine system for the forming of micro-sheet-products. Int. J. Mater. Form. **1** 475–478 (2008)

[Raj11] Raj M.V., Sankar S.S., Ponnambalam S.G.: Minimizing clearance variations and surplus parts in multiple characteristic radial assembly through batch selective assembly. Int. J. Adv. Manuf. Technol. **57**, 1199–1222 (2011)

[Rüc17] Rückert, P., Onken, A.-K., Tracht, K.: Beschädigungsfreies Speichern von Teileverbunden in der Mikroproduktion. In: Vollertsen, F., Hopmann, C., Schulze, V., Wulfsberg, J. (eds.) Fachbeiträge 8. Kolloquium Mikroproduktion, pp. 33–38, 27–28 Nov 2017. BIAS Verlag, Bremen (2017) (in German)

[Rei12] Reinhart, G., Meis, J.F.: Requirements management as a success factor for simultaneous engineering—enabling manufacturing competitiveness and economic sustainability. In: ElMaraghy, H.A., (ed.) Enabling Manufacturing Competitiveness and Economic Sustainability, pp. 221–226. Springer, Heidelberg (2012)

[Rip14] Rippel, D., Lütjen, M., Scholz-Reiter, B.: A framework for the quality-oriented design of micro manufacturing process chains. J. Manuf. Technol. Manag. **25**(7), 1028–1049 (2014)

[Rip14a] Rippel, D., et al.: Application of stochastic regression for the configuration of a micro rotary swaging processes. Math. Probl. Eng. (2014) (article ID 360862)

[Rip16] Rippel, D., Lütjen, M., Freitag, M.: Geometrieorientierter Prozesskettenentwurf für die Mikrofertigung. Ind. 4.0 Manag. **32**, 50–53 (2016)

[Rip17] Rippel, D., Lütjen, M., Freitag, M.: Local characterisation of variances for the planning and configuration of process chains in micro manufacturing. J. Manuf. Syst. **43**(1), 79–87 (2017)

[Rip17a] Rippel, D., Lütjen, M., Freitag, M.: Simulation of maintenance activities for micro-manufacturing systems by use of predictive quality control charts. In: Chan, W.K.V., et al. (eds.) Proceedings of the 2017 Winter Simulation Conference, pp. 3780–3791 (2017)

[Rip17b] Rippel, D., et al.: Charakterisierung der Einflüsse einzelner Prozessparameter auf die Werkstückgeometrie beim Einstechrundkneten im Mikrobereich. In: Vollertsen, F., et al. (eds.) 8. Kolloquium Mikroproduktion, pp. 219–226. BIAS Verlag, Bremen (2017)

[Rip18] Rippel, D., Lütjen, M., Freitag, M.: Enhancing expert knowledge based cause-effect networks using continuous production data. Procedia Manuf. **24**, 128–137 (2018)

[Rip18a] Rippel, D., et al.: Application of cause-effect-networks for the process planning in laser rod end melting. In: Vollertsen, F., et al. (eds.) MATEC Web of Conference, vol. 190. EDP Sciences, Les Ulis Cedex (2018). https://doi.org/10.1051/matecconf/201819015005

[Sch02] Schroeder, R.G., Flynn, B.B.: High Performance Manufacturing. Global Perspectives. Wiley, New York (2002)

[Sch08] Scholz-Reiter, B., Lütjen, M., Heger, J.: Integrated simulation method for investment decisions of micro production systems. Microsyst. Technol. **14**(12), 2001–2005 (2008)

[Sch13] Scholz-Reiter, B., Rippel, D.: Eine Methode zum Design von Mikroprozessketten. Ind. 4.0 Manag. **29**(2), 15–19 (2013)

[Tra18] Tracht, K., Onken, A.-K., Gralla, P., Emad, J.H., Kipry, N., Maaß, P.: Trend-specific clustering for micro mass production of linked parts. CIRP Ann. **67**(1), 9–12 (2018)

[Tra13] Tracht, K., Weikert, F.: Handling of microparts. In: Vollertsen, F. (ed.) Micro Metal Forming, pp. 331–342. Springer, Heidelberg (2013)

[Tra12] Tracht, K., Weikert, F., Hanke, T.: Suitability of the ISO 10303-207 standard for product modeling of line-linked micro parts. In: Chryssolouris, G., Mourtzis, D. (eds.) 45th International Conference on Manufacturing Systems (CMS'2012), pp. 406–413. Athens (2012)

[Tra12a] Tracht, K., Weikert, F., Hanke, T.: Suitability of the ISO 10303-207 standard for product modeling of line-linked micro parts. Procedia CIRP **3**, 358–363 (2012)

[Vol04] Vollertsen, F., Hu, Z., Schulze Niehoff H., Theiler, C.: State of the art in micro forming and investigations in micro deep drawing. J. Mater. Process. Technol. **151**(1–3), 70–79 (2004)

[Vol08] Vollertsen, F.: Categories of size effects. Prod. Eng. **2**(4), 377–383 (2008)

[Wek16] Weikert, F., Tracht, K.: STEP product model for micro formed linked parts. Prod. Eng. Res. Devel. **10**, 293–303 (2016)

[Wek14] Weikert, F., Tracht, K.: Konstruktionsbaukasten für Leiterverbunde des Mikroumformens. Konstruktion 3 70–74 (2014) (in German)

[Wek14a] Weikert, F., Kröger, A., Tracht, K.: Auslegung von Prozessketten der Mikroumformung für die Verwendung von Teileverbunden. wt Werkstattstechnik online 104, 11/12, 722–727 (2014) (in German)

[Wek13] Weikert, F., Weyhausen, J., Tracht, K.: Statistical study for micro forming technologies used for linked parts production. In: Zaeh, M.F. (ed.) 5th International Conference on Changeable, Agile, Reconfigurable and Virtual Production (CARV'13), pp. 41–46. Munich, Germany (2013)

[Wek13a] Weikert, F., Weyhausen, J., Tracht, K., Moumi, E., Kuhfuss, B., Brüning, H., Vollertsen, F.: Beurteilung von Linienverbunden der Mikroumformung Tagungsband. In: Tutsch, R., (ed.) 6. Kolloquium Mikroproduktion. Shaker Verlag, Braunschweig (2013) (in German)

[Wil18] Wilhelmi, P., Schattmann, C., Schenck, C., Kuhfuss, B.: Interactions between feed system and process in production of preforms as linked micro parts. In: Vollertsen, F., Dean, T.A., Qin, Y., Yuan, S.J. (eds.) 5th International Conference on New Forming Technology (ICNFT 2018), vol. 190, p. 15014. MATEC Web Conference (2018). https://doi.org/10.1051/matecconf/201819015004

[Wil17] Wilhelmi, P., Schenck, C., Kuhfuss, B.: Linked micro parts referencing system. J. Mech. Eng. Autom. **7**, 44–49 (2017). https://doi.org/10.17265/2159-5275/2017.01.006

[Wil17a] Wilhelmi, P., Schenck, C., Kuhfuss, B.: Handling in the production of wire-based linked micro parts. Micromachines **8**, 169 (2017). https://doi.org/10.3390/mi8060169

[Wil15] Wilhelmi, P., Moumi, E., Schenck, C., Kuhfuss, B.: Werkstofffluss beim Mikrorundkneten im Linienverbund. In: Hopmann, Ch., Brecher, Ch., Dietzel, A., Drummer, D., Hanemann, T., Manske, E., Schomburg, W.K., Schulze, V., Ullrich, H., Vollertsen, F., Wulfsberg, J.-P. (eds.) Tagungsband 7. Kolloquium Mikroproduktion, pp. 31–38, Aachen, 16–17 Nov 2015. IKV, Aachen (2015). (CD) ISBN: 978-3-00-050-755-7

Chapter 4
Tooling

Frank Vollertsen, Joseph Seven, Hamza Messaoudi, Merlin Mikulewitsch, Andreas Fischer, Gert Goch, Salar Mehrafsun, Oltmann Riemer, Peter Maaß, Florian Böhmermann, Iwona Piotrowska-Kurczewski, Phil Gralla, Frederik Elsner-Dörge, Jost Vehmeyer, Melanie Willert, Axel Meier, Igor Zahn, Ekkard Brinksmeier and Christian Robert

F. Vollertsen (✉) · J. Seven (✉) · H. Messaoudi (✉) · S. Mehrafsun
BIAS—Bremer Institut für angewandte Strahltechnik GmbH, Bremen, Germany
e-mail: info-mbs@bias.de

J. Seven
e-mail: seven@bias.de

H. Messaoudi
e-mail: messaoudi@bias.de

F. Vollertsen
Faculty of Production Engineering-Production Engineering GmbH, University of Bremen, Bremen, Germany

M. Mikulewitsch · A. Fischer · G. Goch
BIMAQ—Bremen Institute for Metrology, Automation and Quality Science, University of Bremen, Bremen, Germany

G. Goch
University of North Carolina, Charlotte, USA

O. Riemer (✉) · M. Willert
Leibniz-Institut für Werkstofforientierte Technologien—IWT, University of Bremen, Bremen, Germany
e-mail: riemer@iwt.uni-bremen.de; oriemer@lfm.uni-bremen.de

O. Riemer
LFM—Laboratory for Precision Machining, Leibniz Institute for Materials Engineering—IWT, Bremen, Germany

P. Maaß · I. Piotrowska-Kurczewski · P. Gralla · J. Vehmeyer
Center for Industrial Mathematics, University of Bremen, Bremen, Germany

F. Vollertsen et al. (eds.), *Cold Micro Metal Forming*, Lecture Notes in Production Engineering, https://doi.org/10.1007/978-3-030-11280-6_4

4.1 Introduction to Tooling

Frank Vollertsen

Forming as a shaping process, in which the tool is used as an analog memory for the workpiece geometry, requires precise tools. Especially in micro forming, it is not only the geometry of the basic tool bodies that plays a role, but also their surface topology, since it can be used considerably to control the material flow. The determination of the permissible tolerances has so far been an only empirically solved problem, whereby the applied trial-and-error methods in the development of tool kits end up being costly in the tool development for new workpiece geometries. The method of tolerance engineering described in Sect. 2.5 (Influence of tool geometry on process stability in micro metal forming) provides a remedy for this with regard to the scatter occurring as a result of the forming processes. In addition, the wear of the tools has to be taken into account, which is the topic in Sect. 4.2 (Increase of tool life in micro deep drawing). In Sect. 4.2 the various issues of tool wear are discussed, including the measurement and the results of the wear, which determine the tool life.

Knowledge about tool wear and protection methods like hard coatings (see Sect. 4.2 for test results) are used to design forming tools. For the manufacture of tool elements, which have typical dimensions of some mm and feature sizes in the range of some 10 μm, both new and enhanced standard methods are applied. One new method is the 3D laser directed chemical etching (laser chemical etching, LCM). This method was initially used in a 2D variant for cutting thin sheet from delicate materials like nitinol. The control of the cutting depth is limited in these cases to ensure an appropriate cutting depth for complete material separation. 3D shaping makes proper control of the removal process in all three dimensions necessary. Section 4.3 (Controlled and scalable laser chemical removal for the manufacturing of micro forming tools) introduces the process development towards a fast process for 3D removal without disturbances and a path planning method for optimized workpiece (i.e. tool) quality. One of the features of the process is the ability to generate sharp edges. Section 4.4 (Process behavior in laser chemical machining and strategies for industrial use) addresses one problem with the introduction of the LCM method into

F. Böhmermann · F. Elsner-Dörge · A. Meier
Labor für Mikrozerspanung—LFM, Leibniz Institute for Materials Engineering—IWT, Bremen, Germany

I. Zahn
Bremer Goldschlaegerei (BEGO), Bremen, Germany

E. Brinksmeier · C. Robert
LFM Laboratory for Precision Machining, Leibniz-Institut für Werkstofforientierte Technologien—IWT, University of Bremen, Bremen, Germany

I. Piotrowska-Kurczewski
University of Bremen, Bremen, Germany

industrial use. While the process is very attractive to industry as it uses a low power cw laser instead of expensive ultra-short pulse lasers, there might be problems due to the usage of sulfuric acid or phosphorus acid as etchants. Therefore, alternative etchants like citric acid and others were tested and are presented in Sect. 4.4. Section 4.4 also shows the feasibility of optimizing the cell layout for the process by FEM simulation. Computational fluid dynamics (CFD) is used to design the cells for a laminar and homogeneous flow across the workpiece.

While it is shown in Sect. 4.3 that sound control of the overall geometry of the tool is possible using laser chemical etching, the control of the surface microstructure appears to be difficult with this method. Due to the fact that the surface structure plays an important role in the material flow in (micro) forming, two sections present results on that. The standard method of tool-making by milling was investigated in respect to the development of surface structures and their prediction and control by optimization methods. The results in Sect. 4.5 (Flexible manufacture of tribologically optimized forming tools) show the structures that develop and how to predict the processing parameters, starting with the desired structure in an inverse optimization. One problem which had to be solved to master this inverse calculation was tool deflection by passive forces during cutting and tool wear during the milling process. Different tool wear models were tested and compared concerning accuracy and calculation time. Due to the small dimensions of the cutting tool, the deflection is much more pronounced than in conventional cutting. An additional size effect is the transition between the interaction modes of the tool and workpiece, i.e. ploughing or cutting appears, depending on the actual conditions. The related knowledge from Sect. 4.5 was used in Sect. 4.6 (Predictive compensation measures for the prevention of shape deviations of micromilled dental products) to optimize the milling of small free-form parts, i.e. dental prostheses. Like the work in Sect. 4.4, also the work in Sect. 4.6 was done together with industrial partners, having a discrete problem to be solved. This kind of knowledge transfer was also done in other work; see e.g. Sect. 2.4 (Conditioning of part properties) and Sect. 5.3 (Inspection of functional surfaces on micro components in the interior of cavities). The challenge in the work shown in Sect. 4.6 was to develop a method for optimized tool paths independent from the actual machine tool.

The second method for the control of the surface structure of tools was again a new method, which is elaborated in Sect. 4.7 (Thermo-chemical-mechanical shaping of diamond for micro forming dies). It is well known that diamond coatings and solid diamond tools (see Sect. 4.2) are suited for forming, especially for aluminum alloys, therefore the surface structure of such tools is of interest. On the other hand, diamond surfaces are very durable and therefore difficult to machine. Starting from the assumption that chemical wear of diamond cutting tools is dominant when cutting iron parts, experiments were conducted to use the diffusion of carbon into iron as a measure for structuring the diamond surface. This work led to a method based on a mechanism with thermal, mechanical and chemical aspects. The mechanisms and the achievable structures in diamond surfaces are explained in Sect. 4.7.

4.2 Increase of Tool Life in Micro Deep Drawing

Joseph Seven* and Frank Vollertsen

Abstract Micro metal parts are usually produced in large lot sizes at high production rates. In order to achieve sufficient product quality, excessive tool wear has to be avoided. In the micro range, so-called scaling effects and their influence on tool wear in micro forming have not been investigated so far. For the investigation, tests in micro deep drawing were carried out with metal sheets (s < 50 μm) of pure aluminum Al99.5, a copper alloy E-Cu58 and stainless steel 1.4301. The failure mechanisms of the micro cups produced were identified as bottom fracture and cup wall damage, and tool wear was measured and characterized with optical measurements and EDX analyses. Moreover, micro cups can be produced by DLC- and PVD-coated tools in a dry forming process. To investigate the wear behavior in a continuous process, a forming tool was developed with an integrated blanking and deep drawing die. The tool wear was measured optically, and the wear mechanisms could be identified in a combined micro and deep drawing process. As a result, the tool could be modified by a manufacturing in selective laser melting and the tool life increased by 290%. Furthermore, a method for tool wear examination in long-term tests was developed. In lateral micro upsetting, the tool wear history can be examined by analyzing the formed product. Several tool materials and coatings were tested with up to 500,000 strokes.

Keywords Deep drawing · Micro forming · Tool wear

4.2.1 Introduction

A long tool life is an important requirement for cost-efficient mass production. The tool life defines how long the tool can be used in production, which determines the downtime and productivity. Wear mechanisms during macro forming experiments have been studied. However, experiments in the micro range revealed that so-called scaling effects [Vol08] influence the tool wear behavior [Gei01]. Scaling effects comprise technical and physical properties that do not correlate linearly with a linear change of the tool geometry. In deep drawing experiments, the influence of scaling (1–8 mm punch diameter) was first analyzed by Justinger et al. [Jus07]. They determined that the tool size influences the friction. The results show that the friction increases with a decreasing punch diameter. In order to understand the wear mechanisms in this range, Manabe et al. [Man08] manufactured a combined blanking and deep drawing die. With this tool, circular blanks were cut in a first step and in the second step deep drawn to micro cups. These tests were performed as single tests as well as under static conditions. However, the dominant tool wear mechanisms and the influence on tool life in mass production have not been investigated so far.

Thus, the goal of this work is to increase the tool life for cost-efficient mass production in micro deep drawing. First, the tool life is defined in simple micro deep drawing experiments. The tool life is determined by micro cup failures. To further analyze the tool, a setup is constructed that allows a continuous mass production of micro cups. A forming tool was manufactured with an integrated blanking and deep drawing die. In these long-term tests, the tool life was increased by the characterization of tool wear and the subsequent modification of the dies. Furthermore, a simple method was developed to determine exactly the tool life. In lateral micro upsetting, a wire is penetrated by a punch in a long-term test. As the tool geometry is reproduced on the formed wire, the tool wear history of the tool can be examined. This method allows a simple investigation of different tool materials and serves as a simplified method for the selection of the tool material.

It is shown that the wear mechanisms in deep drawing depend on the microstructure, which also influences the tool life. Moreover, it is demonstrated that the tool can be modified according to the dominant wear mechanisms, which increases the tool life. The experiments also reveal that it is possible to substitute the deep drawing with a fast and economical lateral micro upsetting method to investigate tool wear mechanisms and the influence on tool life.

4.2.2 Definitions

4.2.2.1 Tool Life

In this section, the tool life of micro deep drawing dies is defined by how long micro cups can be produced without cup failures. In deep drawing, there are two typical cup failures – bottom tears and wrinkle formations. These failures occur when the tool geometry or blank holder pressure is not chosen properly. Furthermore, the process conditions can be changed during the production as a result of tool wear [Hu10]. If so, the friction between the tools increases, which leads to higher process forces and bottom tears. Another cup failure is cup wall damage, which can occur due to worn areas on the die radius [Hu10]. In this case, the circular blank cannot be uniformly deep drawn, which results in a cup with non-uniform cup walls.

4.2.2.2 Dry Forming

Dry forming comprises forming technologies without using lubricants. For economic and geo-ecological reasons, it is important to use lubricant-free forming processes in future [Vol18]. In the dry forming process, the friction is typically reduced by surface modifications like coatings and structures. Especially in the micro range, surface modifications have a great potential for efficient forming technologies [Hu11a]. Due to scaling effects, the influence of lubrication is reduced [Vol08]. Another advantage of lubricant-free forming is that an additional process step can be omitted, because cleaning of the lubricated products is not necessary. Cleaning of micro parts is especially difficult because the surface tension of the lubricant is relatively high compared to the macro range, which leads to the adherence of the micro parts.

4.2.3 Experimental Setups

4.2.3.1 Reciprocating Ball-on-Plate Test

A ball-on-plate test was applied to examine the friction coefficient of a sample [Hu11a]. The principle of the experimental setup is shown in Fig. 4.1.

A counter tool of X5CrNi18-10 mounted on a vertical spindle is moved to the workpiece and loaded with the force $F_z = 5$ N. The workpiece is mounted on a linear table, which is reciprocating at an oscillating distance of 5 mm and a velocity of 1 mm/s. The forces in x- and z-direction F_z and F_x are measured and recorded with a force sensor integrated into the spindle. The workpiece and the counter tool were cleaned before each experiment with 2-propanol. The experiments were carried out in dry conditions.

4.2.3.2 Micro Deep Drawing

Single tests in micro deep drawing were performed to identify the process parameters and characteristics of the wear behavior of micro deep drawing dies.

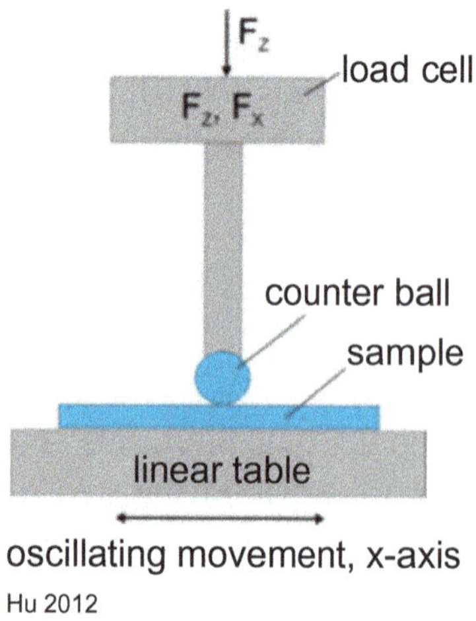

Process and geometry parameters	
Movement	Reciprocating
Counter ball	10 mm diameter
Counter ball material	X5CrNi18-10
Oscillation distance	5 mm
Load	5 N
Velocity	1 mm/s (sample)
Temperature	Room temperature
Lubricant	None

BIAS ID 180957

Fig. 4.1 Principle of the experimental setup for a ball-on-plate test

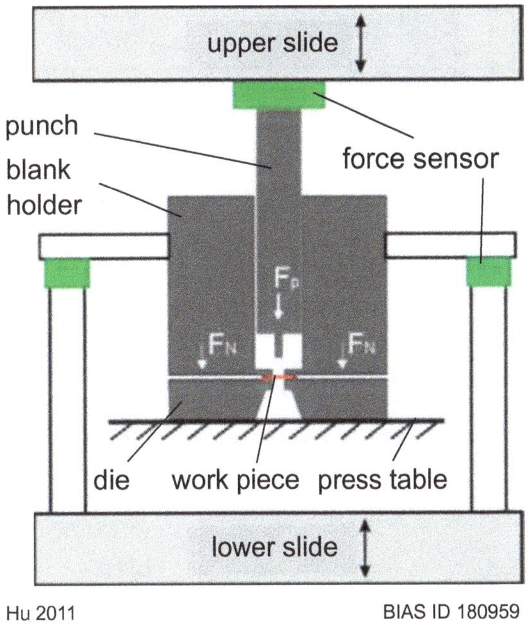

Fig. 4.2 Principle of experimental setup for micro deep drawing

Furthermore, the tool life and the failure mechanisms are determined. The experiments were carried out on a highly dynamic micro forming press [Sch08b] with two axes. The machine axes are driven by electrical linear motors with a maximum acceleration of 10 g and a maximum velocity of 3.2 m/s. The positioning error is below 3 μm at maximum acceleration and the positioning error is below 1 μm up to a stroke of 8 mm. The tool for the deep drawing process [Hu10] is shown in Fig. 4.2. Circular blanks are applied as workpieces for the deep drawing process. The blank holder, which is moved by the lower axis of the press, applies a pressure on the workpiece. The circular blank is deep drawn through the punch to a micro cup. The punch is moved by the upper axis. Both the forming punch and the blank holder are linked to a piezo force sensor, type Kistler 9311B, which has a measuring accuracy of 0.01 N.

4.2.3.3 Combined Blanking and Deep Drawing

The continuous mass production of micro cups requires a forming tool that can both cut and draw a foil. In a first step, the tool cuts the foil to a circular blank and in the second step the circular blank is deep drawn to a micro cup [Hu10]. The tool is built into a highly dynamic micro forming press [Sch08b]. This press with the integrated tool can produce micro cups in high quantities with 200 strokes per minute. For the progressive production, a feed system and sample reels are deployed for the feed of the sample foils. Figure 4.3 shows the steps inside the tool [Flo14a].

First, the blank holder is open and the blank (red line) can be transported by the feed system. Then the blank holder is closed and a pressure is applied on the blank (step 2). The combined blanking and deep drawing die cuts a circular blank (step 3)

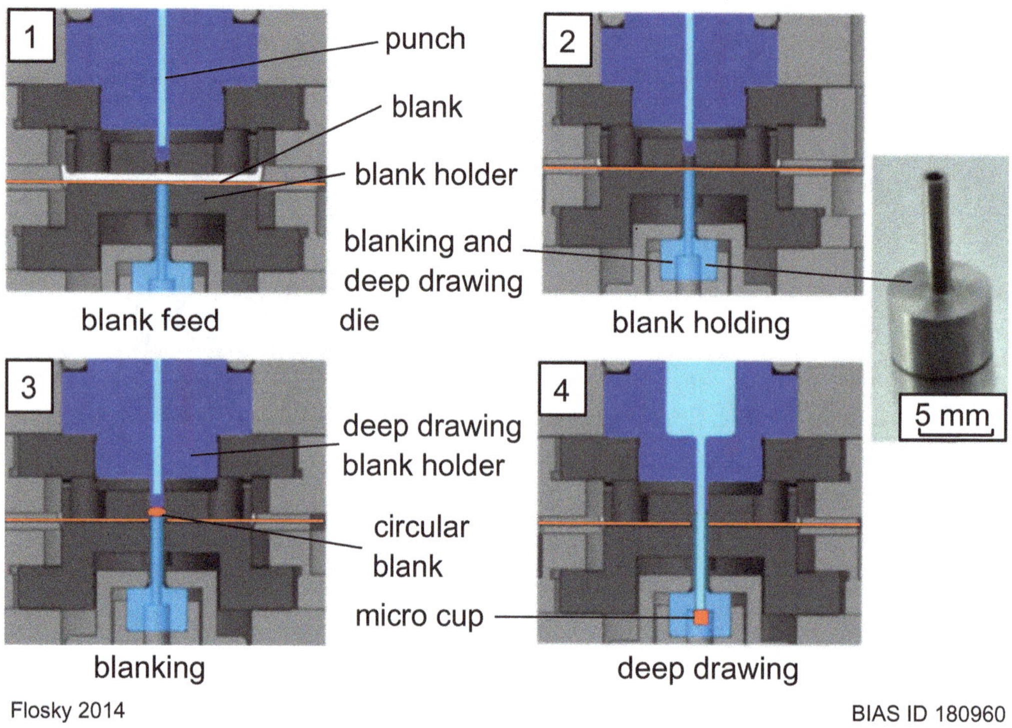

Fig. 4.3 Production steps in a combined blanking and deep drawing process

Table 4.1 Tool geometries for the combined blanking and deep drawing

Punch diameter	0.9 mm
Blanking diameter	1.7 mm
Inner diameter of deep drawing die	1.0 mm
Drawing gap	0.075 mm
Die radius	0.15 mm

and in step 4 the deep drawing punch deforms the circular blank. This production cycle is repeated until no micro cups can be produced any more. The tool geometries are listed in Table 4.1.

To investigate the running-in behavior at the beginning of the process, thermocouples are used and fixed in the tool [Flo14b]. The temperature is measured in a process with 400 strokes per minute without workpiece. Due to a wrong positioning of the tool, friction occurs through the contact between tool and tool guide. In the first 10 s, the temperature increases by 2 K and during the next 2 min there is a further increase of 1 K. Without positioning errors, the temperature remains almost constant and fluctuates with 0.5 K.

4.2.3.4 *Lateral Micro Upsetting*

Wear measurements in long-term tests requires a tool, which includes a long installation of the setup as well as expensive micro tools for only one test series. Moreover, the punch force cannot be attributed to the wear behavior, as the punch force is influenced by different parameters separately. Therefore, a method was developed to investigate the wear behavior [Sev18]. A lateral micro upsetting is used. The setup and process of the method is shown in Fig. 4.4. A punch penetrates a wire with a defined forming depth and a measured forming force. Then the punch reverses. As this is a progressive process, the wire is fed by a feed system so that the punch can deform the wire in a continuous process with different strokes per minute. To ensure that the forming process is reducible, the wire is fed through the wire guide. Compared to the combined blanking and deep drawing process, the lateral micro upsetting can be built simply and fast, the sample can be attributed for each punch force, and the micro tools per test series are cheaper. This method is used for the wear examination of different tool materials and coatings in a continuous dry forming process. Additionally, the tool geometry is measured by negative reproduction in polyvinyl siloxane without dismounting the punch.

4.2.4 Measurement Methods

4.2.4.1 *Confocal Microscope*

An optical and contact-free 3D laser scanning microscope Keyence VK-X210 is used for the recording of the tool state and surface measurements. The microscope has a maximum total magnification of 24,000 with a measurement uncertainty of 1 nm. The pictures can be analyzed with the software VK-Analyzer.

Fig. 4.4 Principle of setup and process of lateral micro upsetting

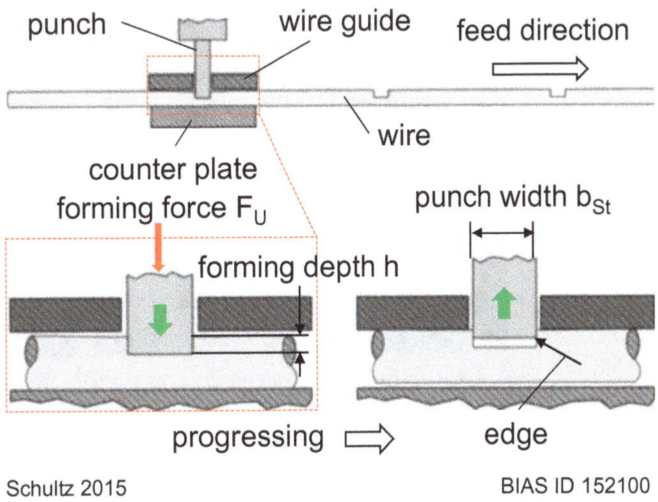

4.2.4.2 *Negative Reproduction of Tool Geometry with Silicone*

During the continuous micro cup production, the tool needs to be investigated for any tool wear. Dismounting and mounting the blanking and deep drawing die takes too much time and causes additional downtimes. Moreover, it could lead to unintentional defects in the process. The tool wear could be influenced additionally by wrong positioning when remounting the die. So a quality control without dismounting and remounting is necessary [Flo15]. The negative reproduction with polyvinyl siloxane offers a fast and simple investigation, of which an example is shown in Fig. 4.5 [Flo15]. Compared to a dismounting and mounting (duration of up to 1 h) this method takes only a few minutes.

On both microscopic pictures, there is the same tool pattern. Wear characteristics like grooves, spalling and deformations are all reproduced and can be measured in the negative reproduction with a confocal microscope. The red circle in the original picture shows residues of the cleaning material. The forming accuracy of the silicone reproduction is proved by a nominal-actual comparison of the geometry between the original tool and tool reproduction [Flo15]. Wear characteristics are measurable down to 1 μm, but the roughness cannot be reproduced with this measurement method.

4.2.5 Materials

4.2.5.1 *Workpieces*

Aluminum EN AW-1050A (Al99.5), copper alloy Cu-ETP (E-Cu58) and stainless steel 1.4301 were used as workpiece materials. For the reciprocating ball-on-plate

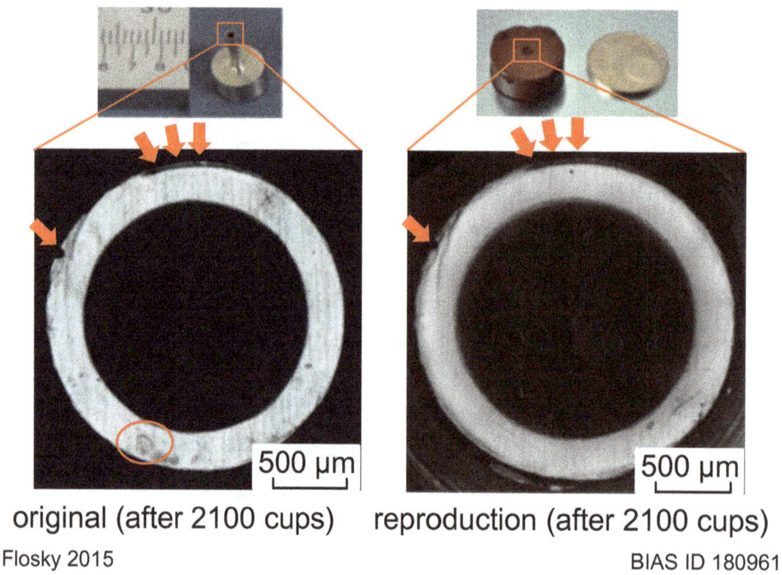

Fig. 4.5 Comparison between original tool (left) and tool reproduction (right)

test, the plates were made of 1.4301. In micro deep drawing, circular blanks of 1.4301 with a thickness of 25 μm were deep drawn. The circular blanks were cut by a nanosecond pulsed Nd:YAG laser with a wavelength of 1064 nm. In a combined blanking and deep drawing process, foils of Al99.5, E-Cu58 and 1.4301 (s = 50 μm) were cut to circular blanks, which were deep drawn to micro cups. In lateral micro upsetting, a 300 μm thick wire of 1.4301 was formed.

4.2.5.2 Tools

The tools in this work were made of conventional cold work steel 1.2379, Co–Cr alloy Stellite 21, high-speed steel 1.3343, powder metallurgical high-speed steel ASP23, hard metal MG30 and diamond. In the reciprocating ball-on-plate test, a ball of 1.2379 was on the plate. In micro deep drawing, the deep drawing die was made of 1.2379. Figure 4.6a shows a micro deep drawing die. Furthermore, a diamond deep drawing die is used for dry experiments.

In a combined blanking and deep drawing process, both 1.2379 and Stellite 21 were used for the manufacturing of a combined blanking and deep drawing die (see Fig. 4.6b). The blanking and deep drawing die of Stellite 21 was manufactured by selective laser melting (SLM) [Flo16]. A CAD model of the tool is transformed into a sliced model with a defined layer thickness and generated in a powder bed. The SLM technology is near-net-shape, but the tool must be micro milled and micro ground in a finishing process step. The advantage of the powder metallurgical materials used is the homogeneous microstructure with fine carbides. The relative density of Stellite 21 is above 99.95% and the hardness 38 HRC.

For the experiments in lateral micro upsetting, several materials can be used because of the simple geometry of the punch. Punches of 1.2379, ASP23, HSS and HM were manufactured.

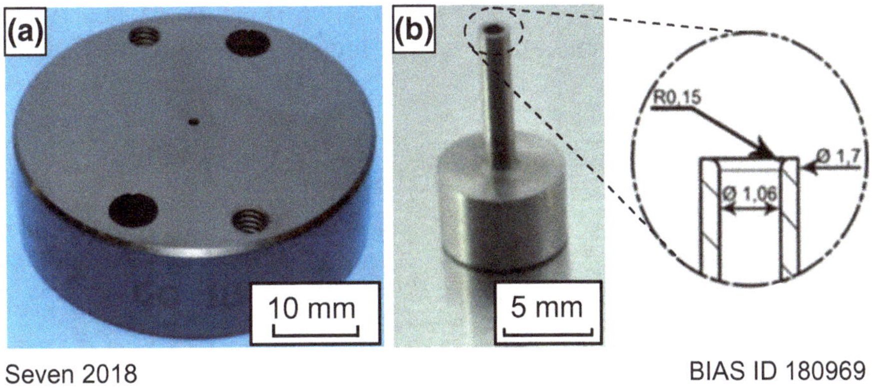

Fig. 4.6 Micro deep drawing die (**a**) and combined blanking and deep drawing die (**b**)

4.2.5.3 Coatings

Coatings are a potential method to reduce tool wear. In the ball-on-plate test, the plate of 1.2379 was modified with a DLC coating from Plasma-Consult (coating type PlascoDur). The hardness of the coating was measured before the tribotest with a Fischerscope H100VP XP from Fischer Types. The hardness of the DLC coating reached a maximum of about 2000 HV at the indentation depth of about 0.2 μm and decreased with further indentation down to 1400 HV at the indentation depth of 0.8 μm. This DLC coating was also used on a deep drawing die in micro deep drawing tests (see Fig. 4.7a). Furthermore, a PVD coating of TiN was applied on a deep drawing die and tested (Fig. 4.7b). For the lateral micro upsetting of the wire, hard metal punches were coated with the materials AlCrN, TiSiN, TiN and AlTiN (Fig. 4.7c).

4.2.6 Results

4.2.6.1 Characteristics of Tool Wear in Micro Deep Drawing

The tool life in the experiments is defined by the time micro cups within specifications are produced [Hu10]. For the experimental setup, the micro deep drawing is used. Figure 4.8 shows the maximal punch force and the failure rate of the micro cups depending on the number of experiments. The micro cups from the first 100 experiments have a failure rate of 0–20%, which might have resulted from the non-uniform properties of the thin foil used. After the first 100 experiments, there is a failure rate of 100%. The failures found are bottom tears and cup wall damage (see Fig. 4.8). Bottom tears are a result of a too high punch force [Hu10]. Tool wear causes higher friction, which leads to larger punch forces and bottom tears.

In the subsequent experiments, the worn area becomes smoother as a result of adhesion between the blank and tool, while some other worn areas are much worse [Hu10]. The result is a non-uniform distribution of friction along the forming

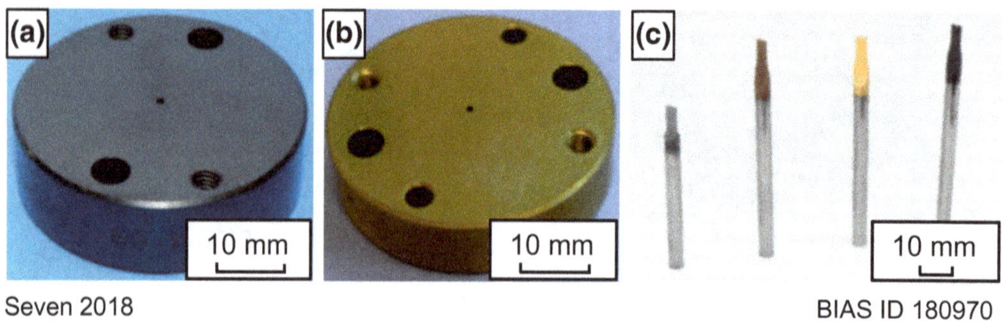

Fig. 4.7 DLC (**a**)- and PVD (**b**)-coated micro deep drawing dies and PVD-coated punches for lateral micro upsetting (**c**)

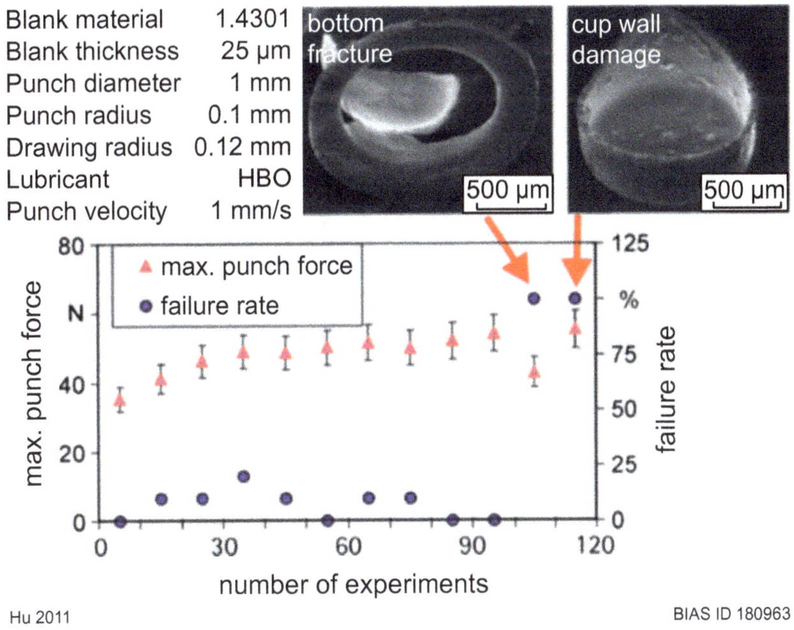

Fig. 4.8 Maximal punch force and failure rate of micro cups depending on number of experiments

zone. Thus, the circular blank is deep drawn non-uniformly into the die resulting in micro cups with cup wall damages. Figure 4.8 shows an example of a micro cup with cup wall damage. The maximum punch force of cups within specifications and with wall damage is almost equal. Therefore, tool wear cannot be detected by a change of punch forces and the tool life does not correlate with the maximum punch force.

The abrasion and adhesion between the tool and workpiece strongly influence the wear behavior. Therefore, the comparison of the tool wear is analyzed with a confocal microscope before and after several deep drawing cycles [Hu10]. The change of the roughness is visible in a before-and-after comparison and can be measured optically with the confocal microscope. Moreover, residues of the blank material on the draw radius caused by adhesion are measurable [Hu10]. Figure 4.9 shows the die radius of a micro deep drawing die with a residue of the circular blank. The residue is shown with a scanning electron microscope and characterized with an EDX analysis. The EDX analysis shows parts of Ni which are contained only in the blank material 1.4301 (X5CrNi18-10) and not in the tool material 1.2379 (X153CrMoV12). These residues cause cup wall damage as the circular blank is not properly deep drawn over the draw radius. Thus, the main wear mechanism occurring on the draw radius is adhesion.

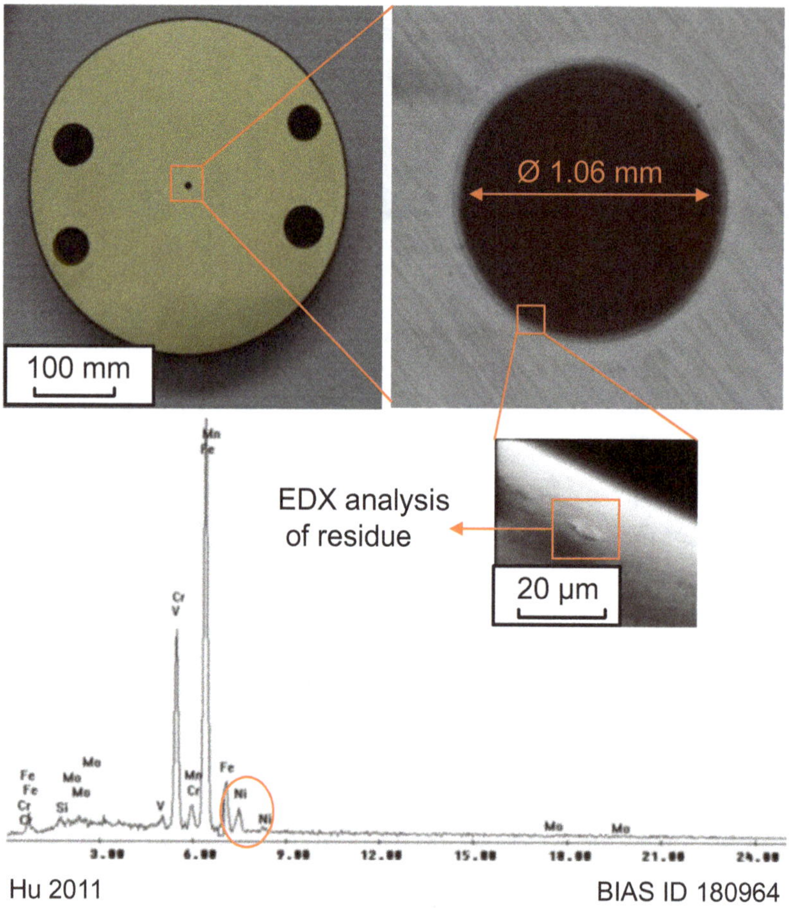

Fig. 4.9 EDX analysis of a residue on the draw radius of the die

4.2.6.2 Wear Behavior of Combined Blanking and Deep Drawing Dies

Compared to the single tests in micro deep drawing, there is an additional blanking of the foil in the combined blanking and deep drawing process. The blank is fed from the sample reel to the tool. With this setup, the combined blanking and deep drawing die produces micro cups with 200 strokes per minute.

In a continuous process with the lubricant Lubrimax Edel C from Steidle GmbH, 80,000 cups of Al99.5, 210,000 cups of 1.4301 and 300,000 cups of E-Cu58 could be produced [Flo14a]. After these strokes the tools failed. Figure 4.10 shows examples of the main characteristics of the wear behavior for the continuous blanking and deep drawing process. On the left, there is the surface of a new blanking and deep drawing tool. In the right picture, wear patterns are visible. The outer diameter is clearly smaller as a result of abrasive wear. Due to a wrong positioning of the tool, there is a bigger contact and stronger abrasion between the tool and blank holder. Other defects are edge disruptions caused through surface

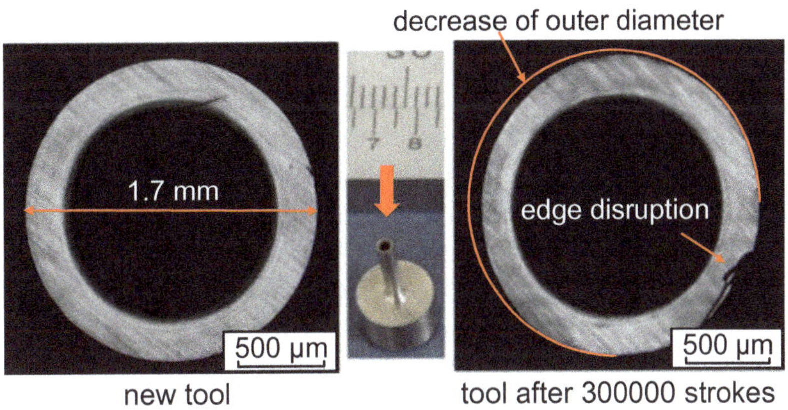

Fig. 4.10 New tool (left) and tool after 300,000 produced cups (right)

fatigue [Flo14a]. In cross-section, these tools show a high density of scattered pores and carbides with sizes up to 35 µm [Flo13]. Carbides in this range act as notches and can cause pitting and edge disruptions. For all tool material and workpiece material combinations, similar wear mechanisms were identified.

Furthermore, different drawing ratios were used for the production of Al99.5 cups [Flo13], one tool with a 1.6 mm outer diameter (drawing ratio 1.78) and one tool with an outer diameter of 1.7 mm (drawing ratio 1.89). The smaller tool (1.6 mm ≙ 1.78) failed after 20,000 strokes and the larger one (1.7 ≙ 1.89) after 80,000 strokes. So the wear of a smaller tool is larger. The smaller the tool surface the greater the influence of scattered carbides and pores [Flo13]. Moreover, a smaller tool surface leads to a higher contact pressure, which causes a higher dynamic stress. Related to the tool failure, it was observed that significant wear occurred just on the blanking edge and not on the draw radius.

The given results show that continuous micro cup production in a combined blanking and deep drawing is possible. The dominant wear mechanisms of the blanking and deep drawing die are identified with the decrease of the outer diameter through abrasion and edge disruptions on the blanking edge. The tool wear is mainly dependent on the microstructure, as carbides and pores can lead to pitting and edge disruptions. Compared to the failure mechanism in micro deep drawing, the tool life does not end through adhesion on the die radius. In a combined blanking and deep drawing process, the cup production is determined due to edge disruptions on the blanking edge.

4.2.6.3 SLM Tool in Combined Blanking and Deep Drawing

Carbides and pores in big sizes act as metallographic notches resulting in edge disruptions. The tool of the powder metallurgical material Stellite 21 generated by SLM [Flo16] has a homogeneous structure with fine carbides. The relative density of Stellite 21 is above 99.95% and the hardness is 38 HRC. In the following long-term test, the workpiece material Al99.5 (s = 50 μm) is tested with the lubricant Lubrimax Edel C and the same conditions as in Table 4.1. With the tool material 1.2379, 80,000 micro cups were produced. Compared to this, a total of 231,000 micro cups could be produced with the SLM tool, which means an increase of 290% [Flo16].

Figure 4.11 shows a comparison of the tool failures between the conventional (1.2379) and the powder metallurgical (Stellite 21) tool [Flo16]. The hardness of 1.2379 (63 HRC) is higher than the hardness of Stellite 21 (38 HRC). The harder material is more brittle, which is proved by the chippings on the blanking edge (red circle, 1.2379). Due to the broken blanking edge, the circular blank is cut tilted by the die. The tilted circular blank causes a micro cup failure as the tilted circular blank cannot be deep drawn properly [Flo16]. The surface and the edge of the SLM tool show some deformations. As Stellite 21 has a higher ductility and a fine microstructure, there are no edge chippings. The influence of edge chippings on

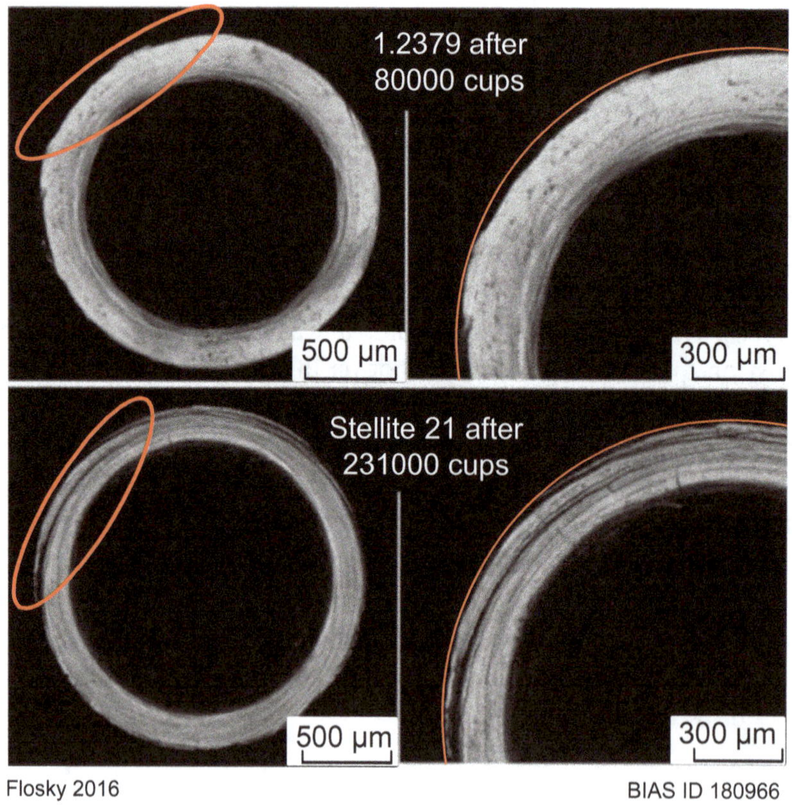

Fig. 4.11 Comparison of tool failures between 1.2379 and Stellite 21

tool life is greater than that of edge deformations because of the tilted cut of the circular blanks [Flo16]. Accordingly, tool materials with fine and homogeneous microstructures are required for the desired long tool life.

4.2.6.4 Dry Forming Processes

Coatings are a potential method to optimize the wear behavior. DLC and TiN coatings were used to reduce friction and wear. In a reciprocating ball-on-plate test, a sample of 1.2379 was tested with and without a DLC coating [Hu12] (see Fig. 4.12).

The tests were carried out in dry conditions. The balls are made of 1.4301 with a diameter of 10 mm. At a normal load of 5 N, the friction coefficient of the uncoated sample increases after 100 s steeply from 0.3 to more than 0.7. The friction coefficient for the DLC-coated sample is clearly lower with $\mu = 0.166$. Moreover, no damage due to the eggshell effect, cracks or spalling is observed on the DLC-coated sample [Hu12].

This DLC coating was applied on a deep drawing die to investigate the application in a dry micro deep drawing process. No damage on the produced micro cups was detected [Hu12]. The punch force of the DLC-coated die was measured and compared to a process with a lubricant and an uncoated die. In Fig. 4.13 the average punch force over the punch stroke is shown for both processes [Hu12].

The punch force of a lubricated process with an uncoated die is higher than a dry forming process with a DLC-coated die. Hence, the DLC coating can reduce the friction between the blank and die better than a lubricant. A similar result was achieved with a TiN-coated die, whereas the friction was higher than by a DLC-coated die [Hu11a]. Moreover, a dry and lubricated forming process was carried out with a TiN-coated die which reveals nearly no influence of the lubricant. According to the results, the DLC and TiN coatings can be used for application on micro deep drawing dies in dry forming processes. The punch force can be reduced

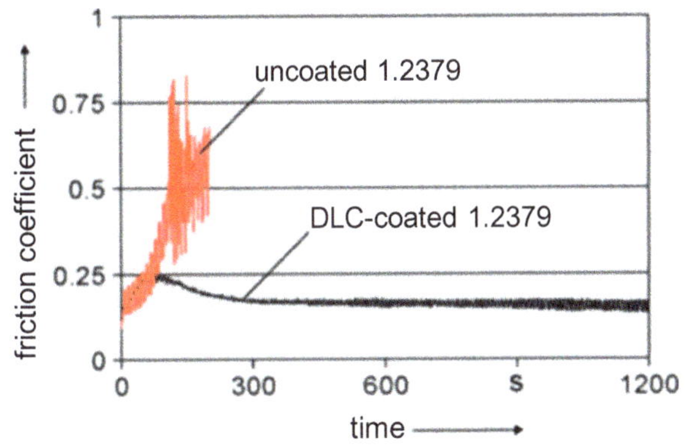

Process and geometry parameters	
Movement	Reciprocating
Counter ball	10 mm diameter
Counter ball material	X5CrNi18-10
Temperature	Room temperature
Load	5 N
Velocity	1 mm/s
Oscillation distance	5 mm
Lubricant	None

Hu 2012

BIAS ID 180958

Fig. 4.12 Ball on plate test of uncoated and DLC-coated 1.2379

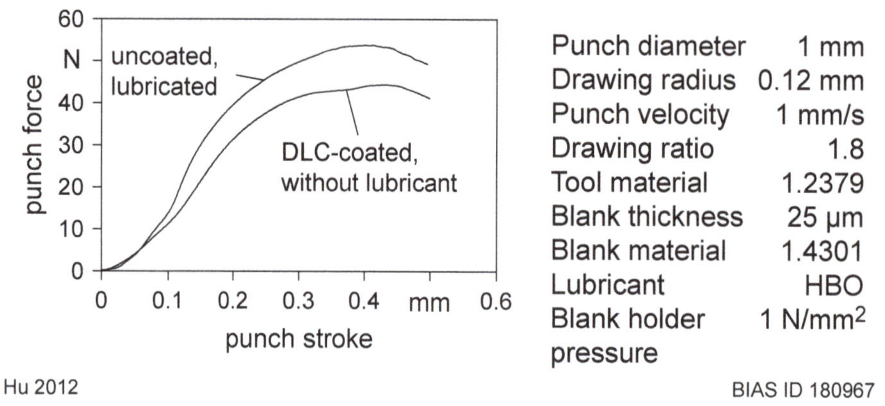

Fig. 4.13 Comparison of punch force versus punch stroke curves of micro deep drawing with DLC-coated and uncoated tools

in a dry process with a coated die better than in a lubricated process with an uncoated die.

Besides DLC- and TiN-coated deep drawing dies, a cubic diamond with the size of 3.5 × 3.5 × 2 mm^3 and a bore with 1 mm diameter (see Fig. 4.14b) are used as tool material for single experiments in deep drawing [Vol15]. Figure 4.14a shows the tool holder, in which the diamond is put and glued. The die radius is produced by an ultrashort pulse laser. In Fig. 4.14c, a cup of 1.4301 is shown with a diameter of 1 mm and a sheet thickness of 25 μm. The cup can be produced with the diamond die without any cup failures. The advantages of diamond dies are the material costs and the tribological properties.

Especially in the micro range, much less material is needed for the production of tools. The disadvantage lies in the difficult manufacturing of diamond tools with complex geometry due to the difficult processing of diamond.

Next to ball-on-plate and micro deep drawing tests, cups of Al99.5 were produced in a dry blanking and deep drawing process [Flo14]. The same conditions as in Table 4.1 were applied with the tool material 1.2379 and a blank thickness of

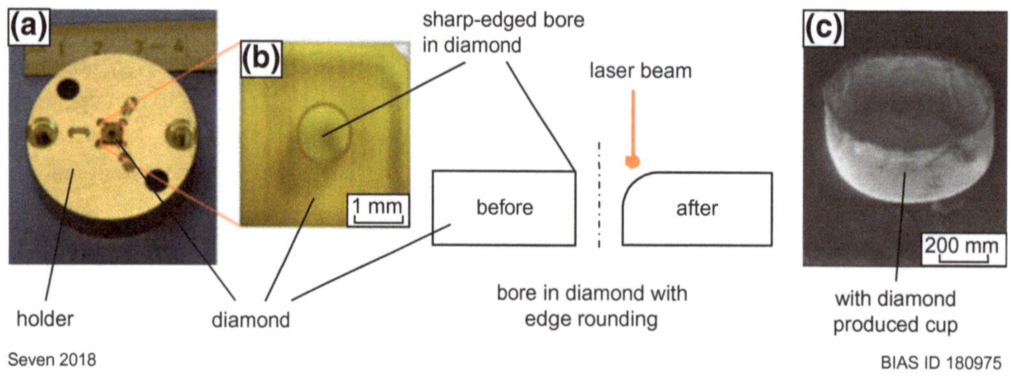

Fig. 4.14 Tool holder for diamond die (**a**), diamond die with laser-produced die radius (**b**) and with diamond die produced cup (**c**)

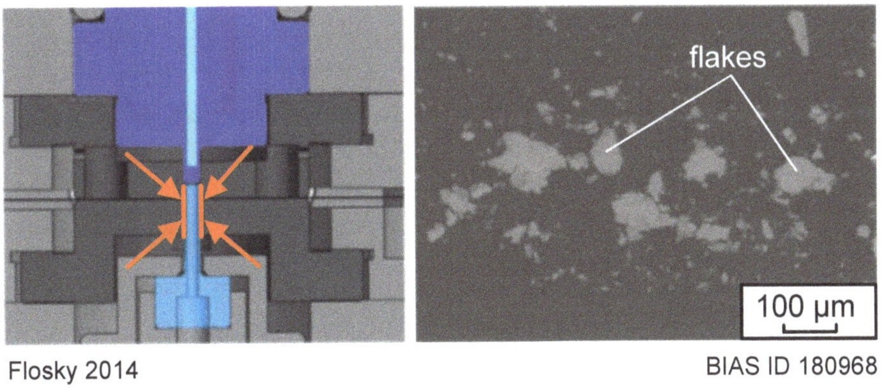

Fig. 4.15 Area influenced by flakes (left) and flakes on the surface of the blanking and deep drawing die (right)

50 μm. The result shows that micro cups can also be produced in a dry process, but the tool failed after 30 cups. Flakes are formed through the blanking process and get between the tool and blank holder (see Fig. 4.15, red line). Resulting from the higher friction, the blanking forces increase so that bottom tears appear and the tool fails. The flakes can be detected on the surface of the blanking and deep drawing die with an EDX analysis. To investigate the wear in a continuous test without the influence of flakes, the tool was cleaned with ethanol after every 10 micro cups. 4,300 micro cups were produced until tool failure [Flo14a]. Furthermore, observation of the quality of the dry formed micro cups indicates that there is no significant difference compared to micro cups produced with lubricant.

The results show that coatings can be used for reducing the friction in a ball-on-plate test and the punch force in micro deep drawing in dry conditions. Micro cups can be produced in a dry forming process with and without coatings. Due to the small tool size in the micro range, diamond tools are a cost-efficient method for producing micro cups. In a combined blanking and deep drawing process, a lubricant is necessary for the transport and removal of the flakes, as flakes cause higher friction and bottom tears. Thus, the tool life in a dry combined blanking and deep drawing process is mainly dependent on the removal of the flakes.

4.2.6.5 Wear Behavior in Lateral Micro Upsetting

In lateral micro upsetting, long-term tests were carried out to examine the wear behavior of several tool materials with and without coatings. The punches penetrate a wire with a forming velocity of 18 mm/s and 350 strokes per minute. The tool materials used are 1.2379, 1.3343, ASP23 and hard metal. Moreover, punches made of hard metal were coated with the materials AlTiN, TiSiN, TiN and AlCrN in a PVD process by Oerlikon Balzers. After 500,000 strokes, the tool wear was measured with a confocal microscope and compared in Fig. 4.16.

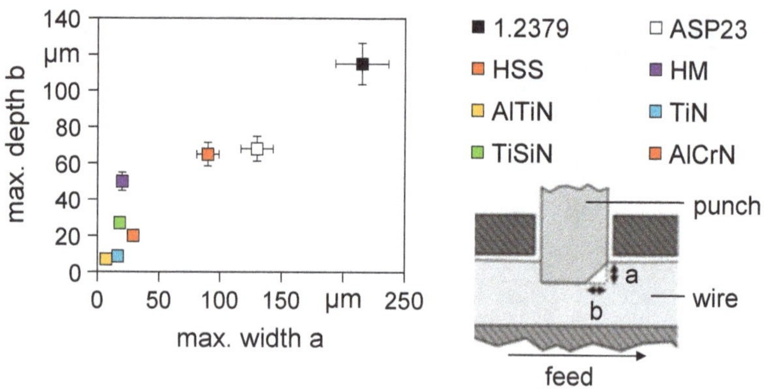

Wire material	1.4301	Forming speed	18 mm/s
Wire diameter	300 µm	Number of strokes	500000
Punch width	500 µm	Strokes per minute	350
Forming depth	150 µm	Substrate for PVD-coatings	HM

Seven 2018 BIAS ID 180972

Fig. 4.16 Tool wear of different tool materials with and without coating after 500,000 strokes

The quality of the tool materials correlates with the expected wear behavior. 1.2379, with a hardness of 780 HV and an inhomogeneous structure, shows the highest tool wear after 500,000 strokes. ASP23 has a hardness of 880 HV and the structure is more homogeneous and finer than the structure of 1.2379. In the micro range, the quality of the structure is more important with a highly dynamic, permanent load (see Sect. 4.2.6.2). Thus, the tool wear of ASP23 is lower. Hard metal has both a high hardness (1520 HV) and a fine structure, which results in lower tool wear. The PVD-coated hard metal punches, due to the improved hardness (2300–3500 HV), show the best wear behavior. In a comparison of the coated punches, it was found that a surface with a higher roughness leads to higher tool wear. Moreover, a high influence of the substrate is observed by testing a punch of HSS (800 HV) with the coating AlTiN (3300 HV). The coating is damaged due to eggshell effects, which leads to a strongly progressive wear and results in the highest tool wear of the coated punches.

Analysis of the wire surfaces assumes that punch features are formed into the wire while penetrating the wire. The hard metal punch after 500,000 strokes is measured with a confocal microscope and compared to the wire surface of the 500,000th stroke (see Fig. 4.17).

On both surfaces, similar features as disruptions are visible. The yellow marked line shows a disruption with a depth of 5 µm. The feature within the red square contains a depth of 1 µm. As a result, by a penetration depth of 150 µm in lateral micro upsetting, punch features up to 1 µm are reproduced on the formed wire. Furthermore, the forming accuracy of the reproduction was determined by a comparison between the height profiles of the tool and wire [Sev18]. The reproduction of the geometry allows the investigation of tool wear for each stroke. As a

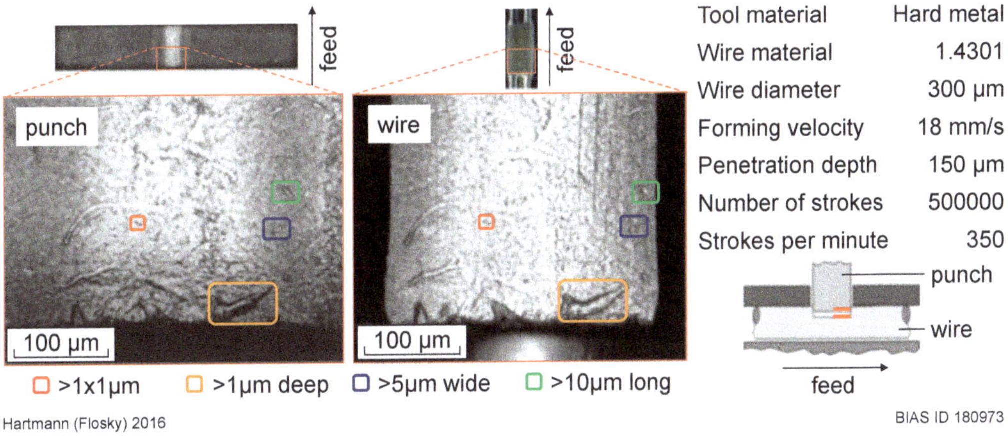

Fig. 4.17 Reproduction of punch features on wire surface in lateral upsetting

result, the tool wear can be tracked by the wire reproductions and the tool wear history determined in lateral micro upsetting. Through the tool wear history, it is possible to determine exactly when the wear occurred and exactly when the tool life is ended. In Fig. 4.18, two reproductions on the wire through the AlTiN-coated punch are shown in microscopic pictures.

The left wire reproduction belongs to the 200,000th stroke and the right one to the 250,000th stroke. After 200,000 strokes, no tool wear on the surface or edge can be measured, whereas after 250,000 strokes an edge disruption has occurred (see Fig. 4.18, white arrow). Hence, the punch is used in this experiment up to 200,000 strokes without any tool wear. To identify the exact stroke when wear first appears, the wire surface of each stroke has to be measured between the 200,000th and 250,000th stroke.

The investigations showed that lateral micro upsetting is a potential method to measure tool wear inline by using the analysis of wire reproductions. The tool geometry is reproduced on the wire with a forming accuracy below 1.5 µm. Thus,

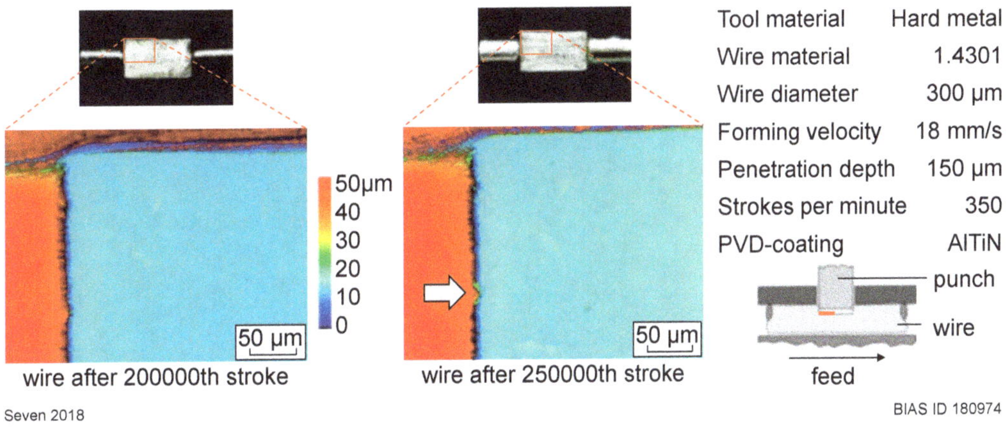

Fig. 4.18 Tool wear history of AlTiN-coated punch in lateral upsetting

the tool wear history can be analyzed by every wire reproduction of each stroke, which allows the exact determination of the tool life. Furthermore, there are several advantages of measuring tool wear by analyzing the formed product instead of analyzing the tool itself. No stopping of the production is necessary, which leads to reduced downtimes, the accessibility on the formed product is better, and no positioning errors occur from dismounting and remounting the tool.

4.3 Controlled and Scalable Laser Chemical Removal for the Manufacturing of Micro Forming Tools

Hamza Messaoudi*, Merlin Mikulewitsch, Andreas Fischer, Gert Goch and Frank Vollertsen

Abstract Laser chemical machining (LCM) is a promising micro processing method that is based on laser-induced thermal activation of anodic material dissolution and enables a gentle and selective micro structuring of metallic materials. However, profound understanding of the removal mechanisms as well as comparative studies with competing processes regarding quality and efficiency are still required to use this process, not only for 2D cutting but also for 3D removal and to further widen the industrial acceptance of the LCM process. For this reason, an analytical modeling to calculate the temperature distribution at the workpiece surface is developed. The spatial correlation with the resulting removal geometry makes it possible to determine the limit temperatures for a disturbance-free regime, as well as the boiling-related gas bubble formation and adhesion as the main cause for the occurring removal disturbances. Moreover, a closed-loop quality control based on inverse process models and an adaptive controller is designed to compensate the deviations of quality features of single or sequences of removals. Using post-production measurements, it is shown that the desired geometry can be achieved with shape deviations of 2.4 μm within only 3 pre-production steps. Furthermore, comparison of the machining quality with micro milling reveals that the LCM process is characterized by a high dimensional accuracy and sharp edge radii (11.2 ± 1.3 μm) especially for structure sizes <200 μm. Besides, additional examples of LCM-manufactured micro forming tools in different materials are presented to prove the process flexibility and diversity. Thereby, the achieved removal velocities in LCM of up to 1.25 mm/min in self-passivating metals are found to be comparable with those obtained in electrochemical machining.

Keywords Laser micro machining • Predictive model • Quality

4.3.1 Process Fundamentals

Laser irradiation is used in laser chemical machining as a localized and selective heat source that can induce a suitable thermal impact for the activation of a heterogeneous chemical reaction between a liquid environment and a metallic surface, and results in an anodic metal dissolution under the formation of hydrogen and water-soluble metallic salts following the chemical reaction [Bae11]:

$$Me + 2H^{+} \rightarrow Me^{2+} + H_2 \uparrow \tag{4.1}$$

In LCM, the laser-induced heat impact can induce or enhance reactions at the metal-liquid interface via changes in the electrochemical Nernst potential. The locally induced temperature gradients result in the generation of a thermobattery, allowing a current flow within the metal between the center of the incident laser light and its periphery. Despite the low generated electromotive forces (some 0.1 V for a temperature rise of 100 K), the electric field strengths are very high due to the small battery dimensions [Bae11]. Thus, self-passivating metals lose their natural passivation property, allowing then the dissolution of the base material [Ste10].

Within the LCM process the laser-induced temperatures define both the proton activity within the redox reaction and the electrochemical potential at the workpiece surface [But03]. It is therefore evident that certain threshold temperatures are required in order to realize a material removal above the background etching rate ($<10^{-8}$ µm/s) [Now96]. Thereby, the laser-induced temperatures depend on different factors that include the laser, material and electrolyte parameters, as illustrated in Fig. 4.19.

Beside the temperature, the electrochemical potential also depends on the chemical activity of the dissolved metal ions and on the mass transport limitation [Ste10]. The latter mainly determines the removal speed within the aqueous electrolytes. Further, the convective flow also represents an important factor by determining the transport of reaction products as well as the provision and exchange of reactants. Increased convection can enhance the reaction rates by several orders of magnitude. At high laser intensities, electrolyte boiling can occur, resulting in the

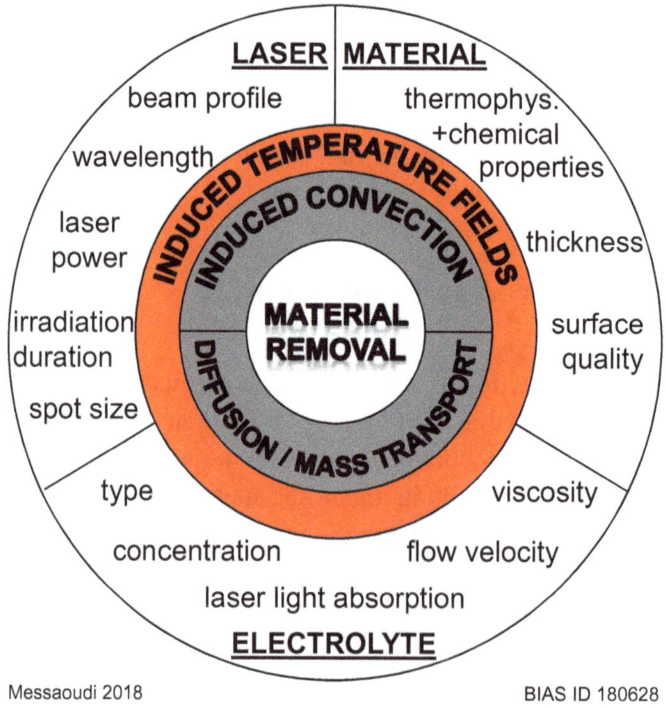

Fig. 4.19 Schematic illustration of the relevant parameters and the dominant induced factors in laser chemical machining

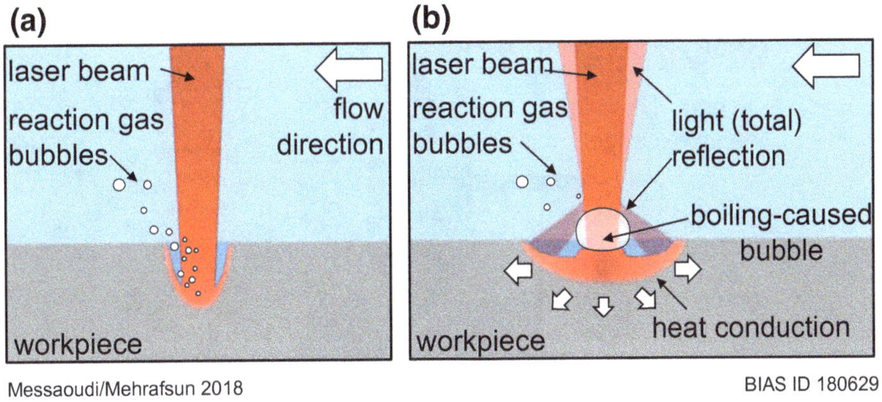

Fig. 4.20 LCM process with **a** disturbance-free removal and **b** disturbed removal caused by adhering boiling-related gas bubble

formation of vapor gas bubbles. In addition to Yavas et al. [Yav94], it has been reported that the bubbles formed result in removal disturbances that impede the controllability of the LCM process [Meh13]. Figure 4.20 shows a schematic illustration of the effect of adhering boiling gas bubbles on the LCM process. Depending on its size and dynamics, the bubble affects the amount and distribution of the deposited laser energy differently. It can shield the workpiece surface against the electrolyte, resulting in a reduced material dissolution reaction [Mes17a] and in the deposition of metallic salts and oxides [Eck17a]. However, the thermal impact due to the laser-workpiece interaction results in a lateral heat conduction towards the periphery, which explains partially the W-like removal profile. With increasing gas bubble size, the effect of light (total) reflection is assumed to increase too due to the transition from an optically dense medium (electrolyte) into an optically thinner medium (gaseous volume of gas bubble). The adhering bubble can act like a scattering center and deflect a part of the incoming laser beam to its periphery (light deflection effect). While the shielding effect results in a reduced removal depth and the deposition of metal salts and oxides, the light reflections lead to a broadening or a dislocation of the removal cavity [Meh18].

4.3.2 LCM Machines Concepts

Depending on the production requirements, two LCM process variants are used. For micro tool fabrication and deep structuring, the electrolyte-jet based LCM (JLCM) method using a coaxial electrolyte jet-stream with respect to the laser beam is suitable (Fig. 4.21). The etchant jet–stream provides a fast exchange of reactants which is assumed to enhance the removal rates. The feed velocity v_{feed} of the workpiece is controlled by an *xyz*-linear stage that allows the workpiece to be adjusted with respect to the focal position [Meh13].

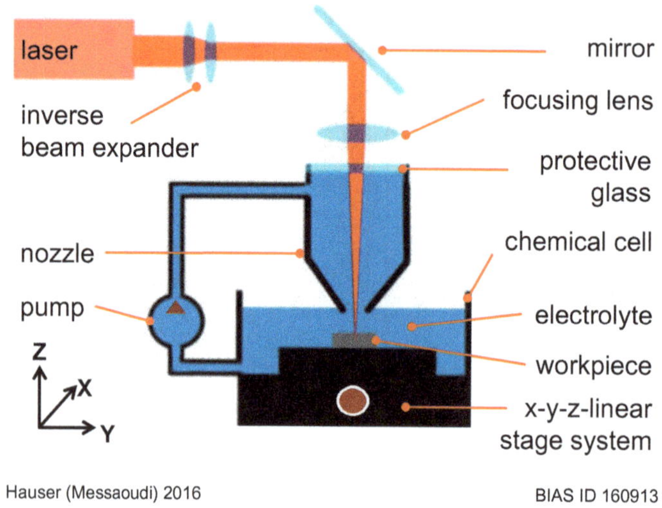

Fig. 4.21 Schematic illustration of the main components of the electrolyte-jet based LCM machine (JLCM)

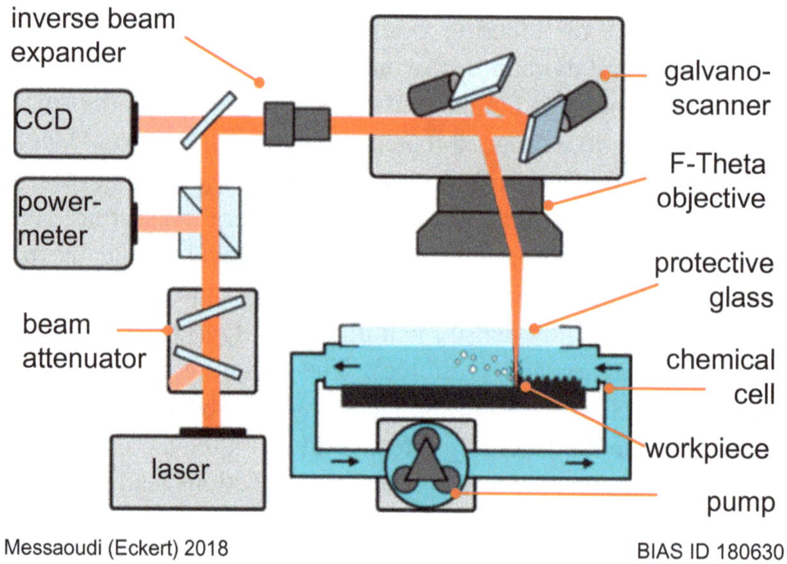

Fig. 4.22 Schematic illustration of the main components of the scanner-based LCM machine (SLCM)

For surface finishing and texturing, the scanner-based LCM (SLCM) can be applied, as depicted in Fig. 4.22 [Eck17a]. This is based on a closed process cell and a scanner system and allows flexible and safe machining of the metallic workpiece surface. In comparison to the JLCM setup, the workpiece is mounted in an etching chamber, where the electrolyte is pumped as a cross-jet through a 25 mm × 2 mm cross-section with a maximum velocity v_{flow} of 2 m/s.

One of the key challenges in laser chemical processing is to ensure a controllable and homogeneous laser energy deposition on the workpiece, considering the propagation throughout the aqueous etchant environment. In both concepts, fiber lasers emitting continuous-wave (cw) laser irradiation at a wavelength of $\lambda = 1080$ nm were used. The focus diameter d_{spot} of the incident laser beam can be varied between 24 and 74 μm in the JLCM machine and between 30.5 and 110 μm in SLCM using a flexible setting on an inversely positioned beam expander.

4.3.3 Influence of the Process Parameters on the Material Removal

4.3.3.1 Influence of the Electrolyte

The electrolyte is a decisive factor for the material dissolution reaction. Often, aqueous acid solutions such as phosphoric (H_3PO_4) and sulfuric (H_2SO_4) acids are used as electrolyte media. These ensure the highest efficiency, as was demonstrated by Nowak et al. [Now96] and Stephen et al. [Ste11]. However, more environmentally friendly electrolytes such as sodium chloride (NaCl) and sodium nitrate ($NaNO_3$) solutions were tested and compared with the acidic solutions [Hau15a]. The results will be discussed in more detail in Sect. 4.4.

Beside the nature and concentration of the electrolyte, the flow velocity v_{flow} and the wavelength-dependent transmission coefficient τ_E also represent important parameters in laser chemical removal. On the one hand, it was demonstrated that the increase of electrolyte feed rate from 0.5 to 20 m/s improves the aspect ratio (*depth/width*) in dependence on the laser power by factors up to 5 due to the ever narrowing cavities with the feed rate [Ste11]. On the other hand, the transmission coefficient of the electrolyte is mainly determined by the length of laser light propagation through the electrolyte [Mes14]. For example, the transmission coefficient τ_E decreases from 96 to 55% when the propagation path through a 5 molar phosphoric acid solution increases from 2 to 50 mm.

4.3.3.2 Influence of the Material

With respect to the electrolyte solution, the self-passivation of the material used is usually required in laser chemical machining. In many works, a wide range of materials has been determined to be well machinable. These included titanium [Hau15a], memory shape alloys (e.g. nickel-titanium alloys) [Ste10], stainless steel [Now96] and cobalt-chromium alloys (Stellite 21) [Mes17b].

Thereby, the amount and quality of the material dissolution is dependent on the chemical reactivity of the material elements. In [Hau15a] it was shown that the LCM of titanium in a 5 molar H_3PO_4 results in up to 39% higher removal rates and up to 30% deeper cavities in comparison with Stellite 21. This can be explained by the higher electronegativities of cobalt (63%) and chromium (25%), as the main elements in Stellite 21, in comparison with that of titanium. Higher electronegativity impedes the tendency to release electrons and thereby slows down the speed of the chemical reaction.

Moreover, the formation of gas bubbles, which could be related to the chemical reaction and to the electrolyte boiling, is on the one hand dependent on the material used and on the other hand exerts a significant influence on the machining quality [Meh16b]. For example, the machining of titanium in phosphoric acid solution was found to be much more vulnerable to removal disturbances compared with that of stainless steel [Meh18]. This confirms also the experimental observation in [Meh16b], in which records of the interaction zone in both materials revealed higher gas bubble activity in titanium than in stainless steel. This can be traced back in part to the smaller size of the gas bubbles formed in stainless steel due to the higher atomic number of iron ($z_{Fe} < z_{Ti}$) [Qia14]. In addition, the adhesive Van-der-Waals force considering equal bubble size is lower in stainless steel than in titanium due to the lower related atom radius ($r_{Fe} < r_{Ti}$) [Bob08]. Both the lower gas bubble sizes and Van-der-Waals force explain the reduced shielding effect in stainless steel in comparison with titanium.

The machining of tool steels without a self-passivation layer was investigated in [Mes18a]. There, the background etching rates in high-speed steels (HSS) were determined within phosphoric and sulfuric acid solutions in dependence on the cobalt contents (8–12%), interaction time and electrolyte temperature. As depicted in Fig. 4.23, the highest background etching rates amount to 0.135 μm/s at an electrolyte temperature of 60 °C after an interaction time of 1 h. Their continuous

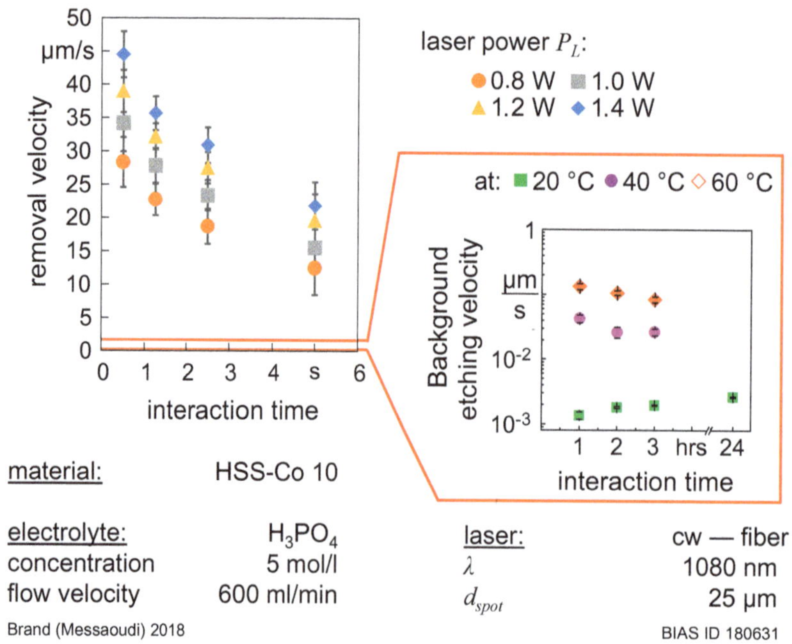

Fig. 4.23 Comparison between the LCM removal velocities and the background etching velocity of high-speed tool HS10-4-3-10 in 5 molar H_3PO_4 solution in dependence on the interaction time

increase at room temperature can be explained by an enhanced electrochemical potential due to the increased amount of metallic salts in the electrolyte solution and the availability of enough reactants. However, the background etching tends to get saturated at higher temperatures due to the fast consumption of reactants. In contrast, the laser-induced chemical machining results in a dependence on the parameters used in removal rates up to 50 μm/s. This provides evidence of the possible machinability of non-self-passivating metals without an external voltage supply and under consideration of the limited machining. Examples of laser chemically fabricated micro tools from HS10-4-3-10 are shown in Sect. 4.3.5.2.

4.3.3.3 Influence of the Laser Parameters

In previous works, the LCM of thin metallic foils was performed with an Ar^+-laser (wavelength of 514 nm) [Now96] and a Nd:YAG-laser (wavelength of 1064 nm) [Ste11] in several metallic materials. Although different absorption coefficients, laser spot sizes and laser powers, the removal results show a strong dependence on the effective absorbed laser power P_{abs} as well as a negligible influence of the laser wavelength. The material removal is found to be scalable with the laser power P_L and the spot size d_{spot} following the quotient P_L/d_{spot} [Mes17a]. Thereby, the disturbance-free removal in different materials such as titanium, stainless steel or cobalt-chromium alloys within a 5 molar phosphoric acid environment was ensured at P_L/d_{spot} between 20 and 50 mW/μm. The laser chemical machining of self-passivating metals takes place when a certain threshold laser power is applied. This was determined to be between 0.6 W and 0.7 W for LCM of titanium and stainless steel in a 5 molar phosphoric acid solution with a laser spot size of 30 μm [Mes17a]. Once the material dissolution is initiated, the material removal increases linearly in width and depth with the laser power. This is characteristic of a disturbance-free removal, which is labeled by a bell-like profile. A further increase of the laser power (over the electrolyte boiling point) results in removal disturbances [Meh13]. These are characterized by cross-sectional W-like or curved zig-zag 2D removal profiles, which are related first to the electrolyte boiling and second to the insufficient transport of reaction products, as well as the exchange of new reactants. Although reducing the processing quality, the removal rates in general continue to increase with the laser power [Ste11].

In addition, the feed velocity is also a key parameter in LCM process. Without consideration of the removal quality, it was shown that the removal depth decreases exponentially with the feed velocity [Ste09a]. Based on this, only feed velocities of <40 μm/s are used during the JLCM process [Ste09a]. In contrast, a single scan with feed velocities >100 μm/s leads, at suitable laser powers, to a non-visible material modification at the workpiece surface. However, the roughness peaks at the surface are chemically attacked and slightly removed. This is shown in [Eck17b], where a reduction of the surface roughness in titanium was achieved after some 10 scan repetitions with a feed velocity v_{feed} = 2 mm/s. Thus, surface finishing can be realized with the combination of higher feed velocities and multi-scans. It has to be mentioned that an areal mean roughness *Sa* of about 0.1 µm could be achieved when first the roughness peaks were removed and subsequently the remaining isolated surface valleys were leveled. This usually necessitates a removal depth of some 10 µm, depending on the initial surface roughness [Eck18].

Furthermore, the influence of lateral overlapping should be considered for large-area LCM. Hauser et al. [Hau15] have shown that constant absorption conditions can be presumed only for the first removal path. Due to the changed surface topography, the removal in the following path is different. Enhanced absorption due to multi-reflections on the inner walls of the first path as well as higher electrolyte dynamics result in deeper removal in the subsequent path. Moreover, the suitable lateral overlapping is found to be dependent on the removal width of a single path and on the material used [Mes18a]. Due to the Gaussian laser beam profile, the roughness *Sa* at the cavity ground is usually >2 µm [Mes17b]. This makes a subsequent finishing step necessary. Therefore, a two-step strategy with a roughing step based on single scans with velocities <40 µm/s and a finishing step based on multi-scans with feed velocities >50 µm/s is assumed to be the expedient approach [Mes18]. The results hereof are discussed in Sect. 4.3.5.

4.3.4 Strategies Towards a Controllable Laser Chemical Machining

4.3.4.1 Modeling of Laser-Induced Temperature Fields

Model Assumptions

As shown in Fig. 4.19, the laser-induced temperature field is dependent on different influence factors and is the subject of a complex interaction (chemically, physically and flow dynamics). To better understand the interrelations occurring, it is essential to precisely determine the induced surface temperature distribution. Therefore, the following assumptions were made in [Mes17a]:

- The laser beam is a TEM_{00}-mode with a Gaussian intensity distribution.
- The workpiece is moving with a constant speed v_{feed} in the x-direction.
- The material is assumed to be isotropic with temperature-independent properties. Moreover, phase changes, such as melting, are excluded.
- The effectively absorbed laser power $P_{abs} = P_L \cdot \tau_E \cdot \alpha_{abs}$, where P_L is the incident laser power, τ_E the electrolyte transmission coefficient and α_{abs} the absorption coefficient of the metallic material.
- Heat transfer into the etchant solution is considered through constant heat transfer coefficients H of the electrolyte.

Model Description

The thermal model developed is described in detail in [Mes17a]. It is based on the solution of the dimensionless heat equation *Ψ(X, Y, 0, τ)* using a Green-function approach and provides the following description of surface temperature rise *T(x, y, 0, t)*:

$$
\begin{aligned}
T = {} & \frac{P_L \cdot \alpha_{abs} \cdot \tau_E}{4\pi \cdot \kappa \cdot l} \cdot \int_0^{\tau} \frac{1}{1+4\phi} \cdot \frac{a}{(a-1)\sqrt{\pi\alpha^2\phi} + \sqrt{\pi\alpha^2\phi + a^2}} \cdot e^{-\left[\frac{(X-\mu\phi)^2+Y^2}{1+4\phi}\right]} d\phi \\
& - \frac{P_L \cdot \alpha_{abs} \cdot \tau_E}{4\pi \cdot \kappa \cdot l} \cdot \theta \int_0^{\tau} \frac{1}{1+4\phi} \cdot e^{-\left[\frac{(X-\mu\phi)^2+Y^2}{1+4\phi}\right]} \\
& \cdot \int_0^{\infty} \frac{a}{(a-1)\sqrt{\pi\left(\frac{Z'}{4\phi} + \Theta^2\phi + \Theta Z'\right)} + \sqrt{\pi\left(\frac{Z'}{4\phi} + \Theta^2\phi + \Theta Z'\right) + a^2}} \\
& \cdot e^{-\frac{Z'^2}{4\phi}} \cdot e^{-\alpha Z'} dZ' \, d\phi
\end{aligned}
\tag{4.2}
$$

Thereby, the first term in Eq. 4.2 represents the lossless heat equation, whereas the second one includes the heat losses when considering heat transfer into the electrolyte (i.e. $H > 0$). The list of the symbols used is illustrated in Table 4.4.

Modeling Results

On the example of machining titanium (3.7024) and stainless steel (AISI304) in a 5 molar phosphoric acid environment, the surface temperatures were calculated depending on laser power P_L and feed velocity v_{feed}. The main process parameters and the properties of the materials and electrolyte used are listed in Table 4.2.

Influence of Laser Parameters:

As described in Eq. 4.2, the induced surface temperature is proportional to the applied laser power. Thus, the surface temperatures increase linearly with the laser power. Moreover, it is found that the temperature rise becomes faster when the laser spot diameter is increased. At a power density of 50 kW/cm^2, the peak temperatures achieved in titanium amount to 51 °C, 88 °C and 130 °C with spot diameters of 30.5 µm, 68 µm and 109 µm, respectively [Mes17a]. This indicates that the laser intensity ($P_L/(0.25\pi \cdot d_{spot}^2)$) cannot be taken as a reference parameter. Instead, laser chemical machining is governed by a two-dimensional heating process that follows the quotient P_L/d_{spot}. This indicates that the surface temperature is one determinant factor in the LCM process. Regardless of the laser spot diameter, the thermal evolution of the LCM removal is unique for a defined electrolyte-metal combination [Mes17a].

Influence of Feed Velocity:

The thermal modeling reveals that the induced temperature fields with feed velocities up to some cm/s can be treated as quasi-static and quasi-symmetric.

Table 4.2 List of process parameters used for both modeling and experimental investigation

	Properties	Unit	Value
Titanium (3.7024)	Thermal diffusivity D	m²/s	$6.8 \cdot 10^{-6}$
	Thermal conductivity κ	W/(m·K)	21
	Absorption coeff. at NIR α_{abs}	–	0.4
Stainless steel (AISI304)	Thermal diffusivity D	m²/s	$4 \cdot 10^{-6}$
	Thermal conductivity κ	W/(m·K)	15
	Absorption coeff. at NIR α_{abs}	–	0.35
Laser beam	Wavelength λ	nm	1080
	Focus spot diameter d_{spot}	µm	30.5
	Feed velocity v_{feed}	µm/s	10 … 200
	Laser power P_L	W	0.5 … 2.5
Phosphoric acid (H_3PO_4)	Concentration C	mol/l	2.5, 5, 7.5
	Boiling temperature $T_{Boiling}$	°C	104
	Flow speed v_{flow}	m/s	2
	Transmission coefficient τ_E	–	0.96
	Heat transfer coefficient H	W/(m²·K)	$3 \cdot 10^3$ (T < 90 °C)
			$5 \cdot 10^4$ (T > 90 °C)

In this velocity range, no significant influence either on the temperature rise or on the shape could be observed. Starting at velocities >10 cm/s, the temperature distribution becomes asymmetric. Here, it is observed that the temperature rise (for $x > 0$) is faster than the temperature decay (for $x < 0$). Moreover, the peak temperature decreases with the feed velocity and is located farther and farther behind the center of the passing laser beam at $x < 0$.

Influence of Material:

The comparison of material-related energy conversion indicates that, despite similar absorption coefficients, the heat impact in titanium is lower than in stainless steel due to the higher thermal conductivity ($\kappa_{titanium} > \kappa_{steel}$). Moreover, the difference in the peak temperatures achieved augments with either increasing the laser power or reducing the feed velocity.

Influence of Electrolyte:

In the thermal modeling the influence of the electrolyte is considered through the transmission coefficient τ_E and the heat transfer coefficient H. For the definition of the heat transfer coefficients, the water flow in a micro pipe was chosen as a reference case due to the micro range of the interaction area (d_{spot} < 110 µm). Depending on the electrolyte temperature, two different heat transfer coefficients H were defined [Whi11]:

- H = 50,000 W/(m^2·K) for the case of water boiling. It was applied for temperature distributions with $T_{model,peak}$ > 90 °C. De Silva et al. [Sil11b] have shown that the risk of electrolyte boiling is extremely enhanced starting at 90 °C.

– H = 3,000 W/(m^2·K) representing an enforced convection within a moderate water flow. This value was applied for temperature distributions with $T_{model,peak}$ < 90 °C.

The modeling results in [Mes17a] show that the heat transfer into the electrolyte during the LCM can be neglected for surface temperatures between 20 and 200 °C. The calculated thermal losses in titanium were between 1.5 and 3 K with d_{spot} = 30.5 µm and between 4 and 10 K with d_{spot} = 109 µm.

Experimental Validation

Limit Temperatures for a Disturbance-Free Removal:

A spatial correlation of the temperature distribution with the removal profiles of the machined cavities, as depicted in Fig. 4.24, can be used to define the process characteristics. Thereby, the temperature at the distance $w_{rem}/2$ (half of the removal width) corresponds to the lower limit temperature $T_{R,min}$, at which the laser chemical machining is initiated. Moreover, the temperature at the maximum obtained depth ($T_{(x=depth,max)}$) represents the upper limit temperature $T_{R,max}$. In comparison with the temperature at the center position ($T_{(x=0)}$), it can be stated that the removal is disturbance-free as long as $T_{R,max} = T_{(x=0)}$. In contrast, a removal disturbance can be presumed if $T_{R,max} < T_{(x=0)}$, because the maximum removal depth is shifted from the center position x = 0 to the periphery [Mes17a].

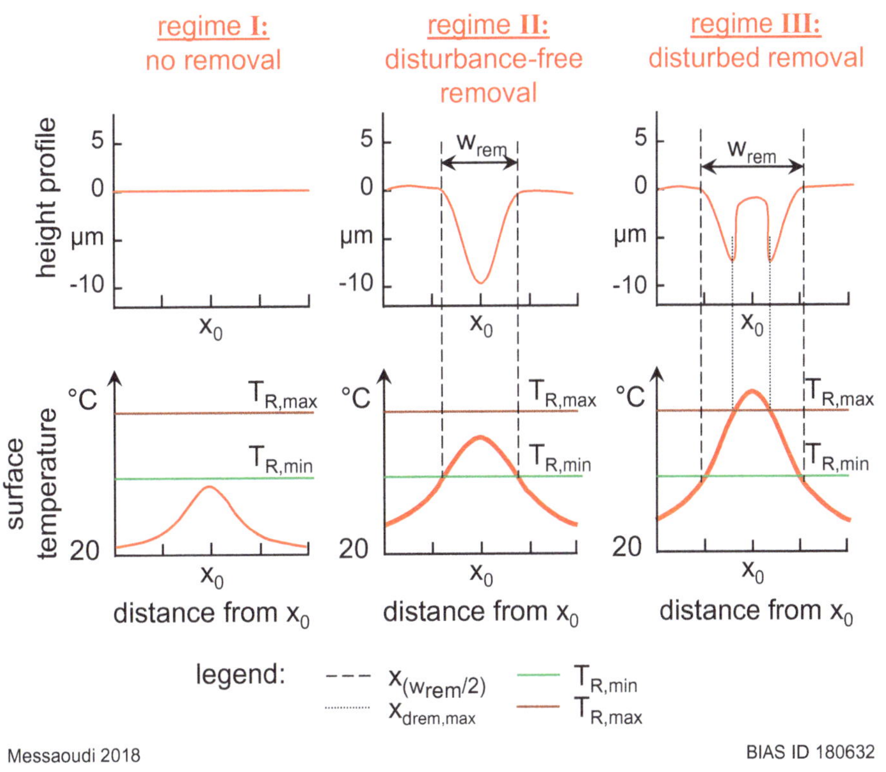

Fig. 4.24 Schematic illustration of the spatial correlation between removal profile and temperature distribution at the workpiece surface

Figure 4.25 shows the LCM limit temperatures $T_{R,min}$ and $T_{R,max}$ determined for titanium and stainless steel in a 5 molar H_3PO_4-environment. Within the identified disturbance-free regimes it is found that the lower limit temperature $T_{R,min}$ amounts on average to 65 °C ± 5 °C in titanium and 59 °C ± 4 °C in stainless steel. Further, it is observed that the removal width within the disturbed removal regime becomes broader and correlates with lower temperatures down to 47 °C, especially in titanium, which can be explained by light deflection occurring on adhering gas bubbles in titanium (see Fig. 4.20). This additional heat impact, which is not included in the thermal modeling, leads to a broadening of the removal cavity and to a shift of the limit temperature $T_{R,min}$.

The upper limit temperature $T_{R,max}$ that is defined to be located at the maximum depth is compared with the temperature $T_{(x=0)}$. There, it is observed that $T_{R,max}$ is equal to $T_{(x=0)}$ within the disturbance-free removal (between 0.75 W and 1 W for

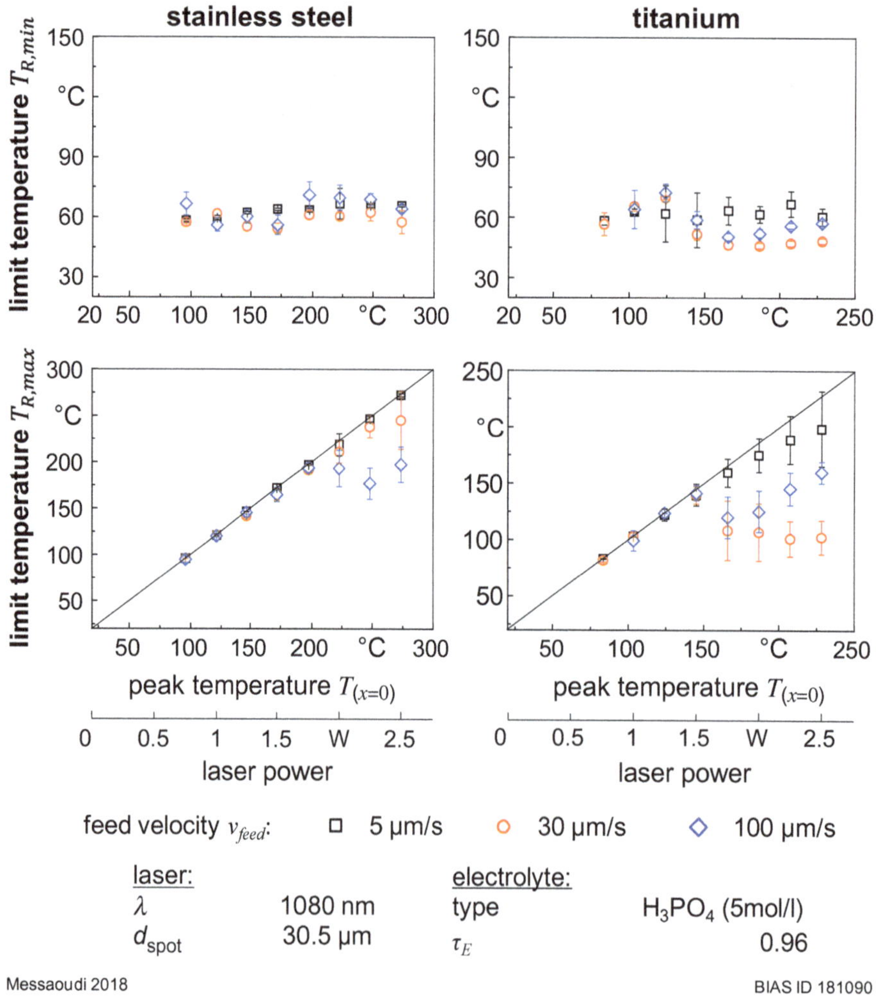

Fig. 4.25 Correlated limit temperatures $T_{R,min}$ and $T_{R,max}$ the laser chemical removal of titanium and stainless steel in a 5 molar phosphoric acid solution; for temperatures >150 °C the influence of adhering gas bubbles is not considered

stainless steel and 0.75 W and 1.25 W for titanium), which means that the maximum depth is located at the center of the incident laser beam. In contrast, the two materials reveal different behaviors when removal disturbances occur. In titanium it is observed that $T_{R,max} < T_{(x=0)}$ at induced temperatures $T_{(x=0)} > 145$ °C, which corresponds to the transition boiling regime and indicates a W-like removal profile with a shift of the maximum removal away from the center.

In contrast, the first deviations from $T_{(x=0)}$ are observed in stainless steel starting with temperatures >150 °C. However, $T_{R,max}$ remains located close to the center position $x = 0$ up to higher laser powers of 2 W. Thereby, $T_{R,max}$ decreases with the increased feed velocity. At a laser power of 2.5 W $T_{R,max}$-values of 272.5, 245.5 and 197.5 °C were determined at feed velocities of 5 µm/s, 30 µm/s and 100 µm/s, respectively. These high temperatures are assumed to be related to a lower influence of electrolyte boiling at the surface of stainless steel, as described in Sect. 4.3.3.2.

In general, it can be summarized that the induced temperature distribution determines the interaction area and thereby the removal zone, as seen from the fact that the removal widths remain constant. In contrast, the removal evolution into the material, represented by the removal depth, is mainly determined by the interaction time, which is inversely proportional to the feed velocity and can be described as:

$$interaction\,time\; t_{inter}[s] = \frac{spot\,size\; d_{spot}[\mu m]}{feed\,velocity\; v_{feed}[\mu m/s]} \tag{4.3}$$

As demonstrated in [Meh18], the removal depth increases following a nearly exponential function with the rising interaction time (i.e. it decreases exponentially with the feed velocity). However, the removal velocity shows a different behavior. For self-passivating materials, the highest removal velocities are obtained at interaction times around 1 s. This will be discussed in more detail in Sect. 4.3.6. Furthermore, it has to be noticed that the increase in removal depth slows down and tends to be saturated independent of the interaction time at induced surface temperatures >140 °C, which is characteristic of the beginning of film boiling e.g. gas bubbles adhering on the metal surface.

4.3.4.2 Quality Control System for Laser Chemical Machining

Factors such as laser-energy absorption dynamics, heat accumulation, chemical reactions, and hydrodynamic transport phenomena cause various disturbances during the material removal process of laser chemical machining (LCM) [Zha15]. In order to achieve competitive shape and surface quality in applications such as rectangular micro-forming dies, where the produced bottom surface requires a certain flatness and roughness, control of the workpiece manufacturing is imperative. The design and application of a high quality control scheme for the LCM application are presented in this section.

Control Specifications

The aim of LCM quality control is to govern the geometry of a single removal path as well as the workpiece geometry produced by multiple removals. Thus, a cross-path control is required for the entire process chain. Considering the geometry $\boldsymbol{G}$ of the workpiece as the superposition of multiple removal paths, it is difficult to achieve direct control of the workpiece quality features by adjusting all the process input parameters from every removal path, because of the many degrees of freedom and the non-linear process behavior.

In order to reduce the degrees of freedom, only the shape parameters $\boldsymbol{A}$ of every path $j = 1, \ldots, N_{path}$ are varied in the control loops. The shape parameters of N_{path} paths are gathered as the control variables $\boldsymbol{A} = (a_1, \ldots, a_{Npath})$ and the process input parameters of multiple paths as the actuating variables $\boldsymbol{U} = (u_1, \ldots, u_{Npath})$.

The general cross-process quality control concept shown in Fig. 4.26 has already been successfully applied in different process chains, such as a bearing ring production process chain [Zha12]. An inverse process model and a process chain prediction present the basis of the feed-forward control. For this purpose, the shape parameters of the single removal and the geometry of the workpiece are characterized and a mathematical model is derived to describe the relationship between the shape and the process input parameters. Both are combined with optimization algorithms to realize an inverse model and a process chain prediction based on artificial neural networks (ANN).

The feedback loop consists of a post-production measurement, an observer for rebuilding the control variables from the measurements, and an adaptive P-controller to achieve stability over the entire operating range of the LCM process. It should be noted that the quality control for LCM was implemented production-discrete, because the development of an in situ measurement technique is still the focus of

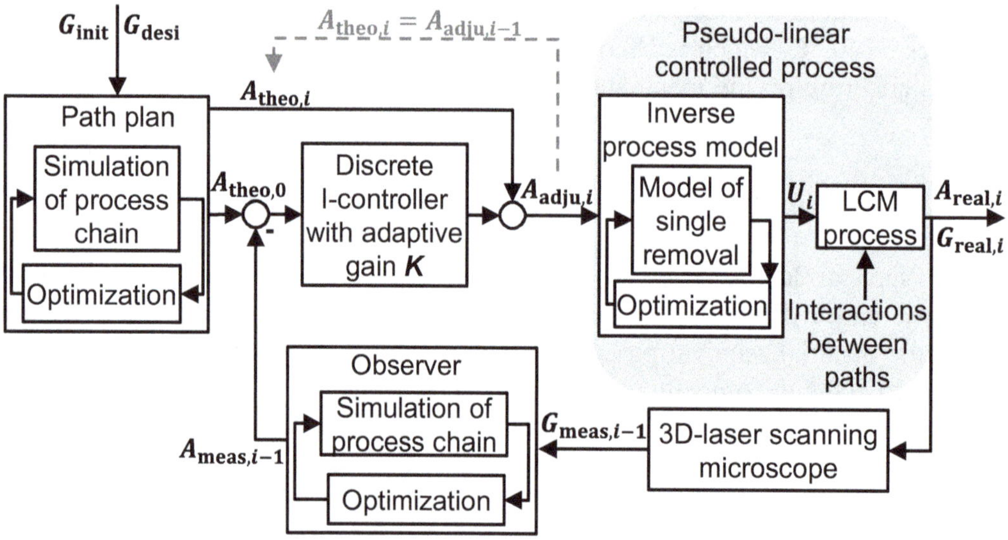

Fig. 4.26 LCM quality control concept with path plan, controller, inverse process model and observer

ongoing research [Mik17]. Production-discrete in this context means that the control loop is closed by post-production measurements with a confocal microscope after each production iteration $i = 1, \ldots, N_{prod}$. With this control loop, the process parameters are continuously optimized for the subsequent productions.

The challenges to be faced by this control concept are as follows:

- The LCM process is a multi-inputs-multi-outputs (MIMO) system.
- The interactions between overlapping removal paths are complex and partly unknown.
- The relationship between the process input parameters (actuating variables $\boldsymbol{U}$) and the shape parameters of one removal (control variables $\boldsymbol{A}$) are non-linear and can cover a broad range.

Cross-Path Interactions:

Due to the effective thermal energy distribution, the removal cross-section of the processed geometry is assumed as a superposition of single Gaussian curves. However, the angle-dependent laser energy absorption as well as the thermal and hydrodynamic condition can lead to a reduced removal rate at a sloped surface (flank), compared to a flat surface. Hence, a cavity produced by a simple overlapping of single paths does not necessarily meet the predicted geometry. Figure 4.27 shows a measured cavity cross-section (black curve) of two overlapping removal paths produced by identical process parameters with a center position distance x_d of 50 μm [Zha15]. The cavity cross-sections produced are measured by a confocal scanning microscope and show a distinctive deviation from the removal cross-section predicted by a process chain simulation with overlapping removal paths.

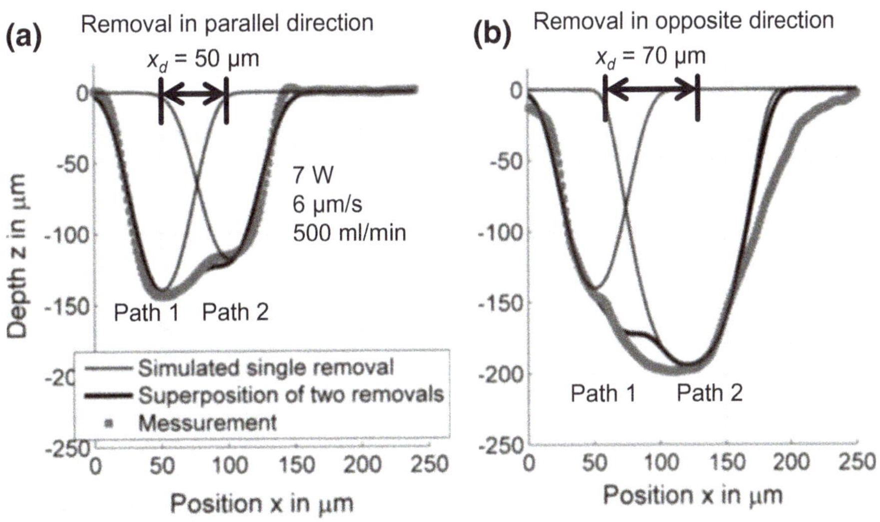

Fig. 4.27 Angle-dependent flank effect: Measured cross-section (black) and superposition (gray) of the Gaussian approximated removal paths with a distance x_d of 50 and 70 μm as well as the simulated prediction without consideration of the flank angle (dashed)

To produce a flexible geometry, an optimal combination of overlapping removal paths is calculated by minimizing the deviation in the z direction between the cross-section of the desired geometry $\boldsymbol{G_T}$ and the superposition $\boldsymbol{G_S}$ of the approximated individual Gaussian curves, considering the influence of the flank angle or the center position distance by a functional dependence. In this way, the path planning acts as a cross-process quality controller. Within the path planning, the flank effect is compensated by adjusting the quality features, the Gaussian form variables, in the removal depth and width.

Adaptive Gain Controller:

According to the calculated shape deviations of every path in the previous production, the process parameters for the subsequent production are adapted by the quality controller. A P-controller with proportional gain K_p is the simplest solution for this MIMO control system that is able to cope with the non-linear behavior of the LCM process. Thereby, an appropriate gain is based on a compromise between control speed and stability, as a low value of K_p normally results in a stable system, but a slow control speed. The use of a controller with a constant gain factor K_p for the LCM process has revealed that the stability and a satisfactory control speed could not be ensured simultaneously in the whole operating area. For this reason, an adaptive controller with the gain $K_p(i)$, was designed considering the previous set value of the shape parameters $\boldsymbol{A_{theo}}(i - 1)$ and the measured shape parameters $\boldsymbol{A_{meas}}(i - 1)$. This adaptive controller was verified by a simulated production shown in Fig. 4.28 and reveals a stable and fast control.

Experimental Validation

The quality control concept developed was validated by producing a micro forming tool having a rectangular shape $(500 \times 200 \times 100)\ \mu m^3$ that needs a processing

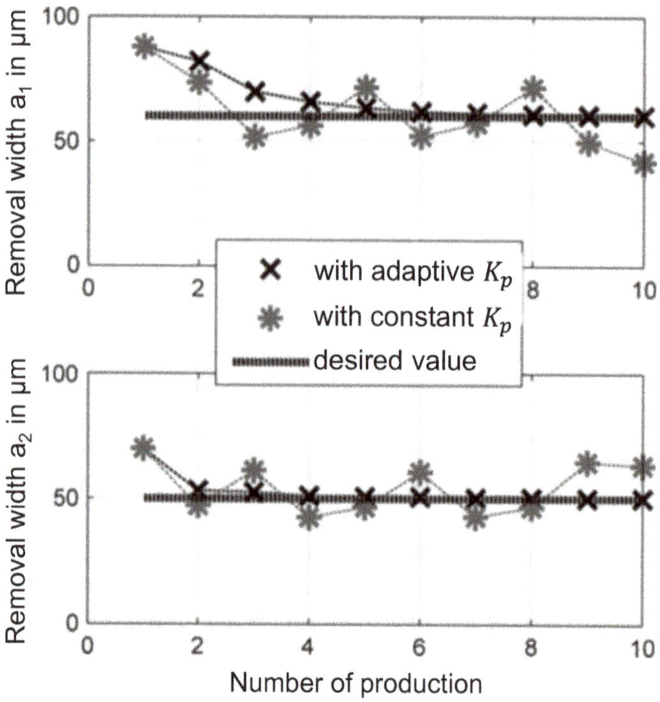

Fig. 4.28 Verification of the controller with a single path production. The set values of the shape parameters are $a_1 = 60$ µm and $a_2 = 50$ µm. The process chain simulation is applied

time of about 5 min. The quality features characterized were the removal depth d_{rem} (based on the arithmetic mean of the measured depths z_m at the positions x_m) as well as the cross-sectional straightness S of the produced bottom surface (with heuristically defined boundaries). The straightness is determined using a Chebyshev approximation [Goc08]:

$$S = \min_{c_1, c_2}\left\{ \lim_{n \to \infty} \sqrt[n]{\sum_{m=1}^{M} e_m^n} \right\} \tag{4.4}$$

where e_m describes the orthogonal distance of the points (x_m, z_m) from a linear function $z(x) = c_1\, x + c_2$. The manufacturing tolerances of these quality features were defined to be $\Delta d_{rem} = \pm\ 5$ µm and $S = \pm\ 2$ µm. The path plan and the mean cross-sections of the first three control iterations are shown in Fig. 4.29. The resulting mean depth $d_{rem} = 101.1$ µm and the straightness = 1.2 µm are determined as the average values obtained from different cross-sections. After two optimization iterations (i.e. two discrete productions), a workpiece with shape deviations that satisfy the manufacturing tolerances was produced. The maximum deviation of the mean depth, for instance, could be reduced from the initial 33.1 µm (without closed-loop control) to 2.4 µm (with closed-loop control) [Zha17]. The shape deviations of the produced micro dies from the desired geometry are thus shown to be distinctly reduced with the quality control developed.

This result demonstrates that the process-discrete control system, which is based on inverse models and a P-controller with adaptive gain, is able to cope with the non-linearity of the LCM process. It ensures a stable processing quality and exhibits satisfactory control speed and shape deviations. However, due to the reduced degrees of freedom, the control of single paths for complex geometries such as edge rounding is no longer possible. In addition, the post-process control system cannot

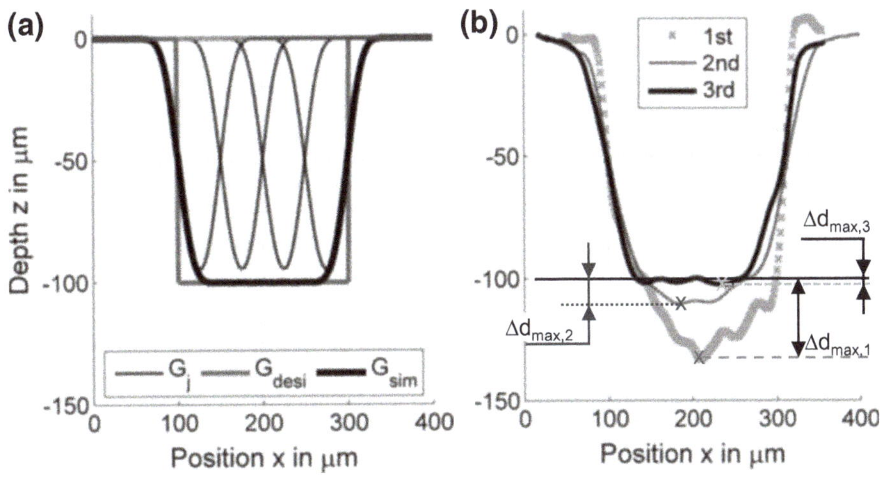

Fig. 4.29 Productions of a rectangular die using 4 paths: **a** path plan; **b** mean cross-sections of the produced workpieces for the first 3 optimization loops of the quality control [Zha17]. The maximum depth deviations for the 3 produced profiles are $\Delta d_{max,1} = 33.1$ µm, $\Delta d_{max,2} = 10.3$ µm and $\Delta d_{max,3} = 2.4$ µm

compensate random deviations along the material feed direction within one removal path. Therefore, a real-time quality control including in situ metrology techniques should be targeted in the future.

4.3.5 Tool Fabrication

4.3.5.1 Manufacturing of Stellite 21 Micro Forming Dies

Manufacturing Task

As demonstrated in the previous section, laser chemical machining can be controlled and its quality can be predicted precisely. The developed and validated temperature and control models represent two key aspects towards the high-quality manufacturing of metallic micro parts. Another important aspect is the experimental characterization of the quality achieved. Next, the tool material Stellite 21 (a cast cobalt-chrome alloy) is used to show the dimensional accuracy and the surface quality of laser chemically machined forming cavities, which are of interest for micro cold forging. Here, square micro cavities with side lengths of 150 and 300 µm and depths of 60 µm were targeted [Mes18].

To examine the manufacturing quality, the geometrical properties (side length L_K, removal depth d_{rem}) and the shape accuracy were recorded and characterized using a laser scanning confocal microscope (Keyence VHX970) and a scanning electron microscope (SEM, EVO M10-Zeiss). For the determination of the edge radius r_e, which describes the edge between the non-machined workpiece surface and the cavity wall, a 2D holistic approximation was used [Lue12], while the areal surface roughness *Sa* on the cavity ground was measured in accordance with ISO 25178.

Manufacturing Strategy

A detailed investigation was performed based on the results described in Sect. 4.3.4 in order to identify the influence of the relevant process parameters on the removal quality. With respect to the targeted dimensions, suitable laser and scan parameters were selected (see Table 4.3). Thereby, the cavity manufacture consists of two steps. As the first step, a roughing is applied to achieve the required removal depth.

Table 4.3 Summary of the selected parameters for the laser chemical manufacture of the micro cavities [Mes18]

	P_L x τ_E [W]	v_{feed} [µm/s]	$d_{overlap}$ [µm]	n_{scan} [–]	$t_{process}$ [min]
LCM roughing	0.6	10	6	2	12.5/50
LCM finishing	0.3	50	6	30	37.5/150

P_L x τ_E: laser power (after the propagation through the electrolyte), v_{feed}: feed velocity $d_{overlap}$: lateral overlapping, n_{scan}: scan repetitions, $t_{process}$: processing time

Table 4.4 List of the symbols used

Symbol	Description	Unit
a	Approximation constant ($a = \pi/(\pi - 2)$)	–
A	Control variable	–
A_{theo}	Target shape parameter	–
A_{meas}	Measured shape parameter	–
C_p	Specific heat capacity	kJ/(kg K)
$d_{overlap}$	Lateral overlapping	μm
D	Thermal diffusivity	m^2/s
d_{rem}	Removal depth	μm
d_{spot}	Laser spot diameter	μm
$G(X, Y, Z)$	Green function	–
G	Geometry parameter	–
H	Heat transfer coefficient	$W/(m^2 K)$
i	Production iteration	–
I	Laser power density	W/cm^2
K_p	Gain factor	–
l	Absorption depth	nm
l_{feed}	Feed length	μm
n_{scan}	Number of scan repetitions	–
P_L	Laser power	W
P_{abs}	Absorbed laser power	W
r	Laser beam radius	μm
r_i	Atom radius	10^{-12} m
r_e	Edge radius	μm
R_{rem}	Removal rate	mm^3/min
S_a	Areal arithmetic surface roughness	μm
S	Cross-sectional straightness	–
T	Temperature rise	°C
$T_{Boiling}$	Electrolyte boiling temperature	°C
$T_{model,peak}$	Peak surface temperature (model)	°C
$T_{R,max}$	Upper limit temperature for material removal	°C
$T_{R,min}$	Bottom limit temperature for material removal	°C
t	Time	s
t_L, t_{inter}	Irradiation time (interaction time)	s
$t_{process}$	Processing time	s
U	Actuating variable	–
v_{feed}	Feed velocity	μm/s
v_{flow}	Electrolyte flow velocity	m/s
v_{rem}	Removal velocity	μm/s
w_{rem}	Removal width	μm
$X\ Y\ Z$	Dimensionless coordinates	–

(continued)

Table 4.4 (continued)

Symbol	Description	Unit
$x\ y\ z$	Dimensional coordinates	m
z_i	Atomic number	–
α	Dimensionless quantity (r/l)	–
α_{abs}	Wavelength-related absorption coefficient of Ti	–
θ	Dimensionless quantity ($H{\cdot}r/\kappa$)	–
κ	Thermal conductivity	W/(m K)
λ	Laser wavelength	nm
ϕ	Integration variable/ substitution quantity (τ-τ')	–
ρ	Density	Kg/m^3
τ	Dimensionless time coordinate	–
τ_E	Transmission coefficient of the electrolyte	–
Ψ	Dimensionless temperature rise	–
Δd_{rem}	Removal difference/tolerance	µm

In the following step, laser chemical finishing is applied to improve the surface quality (Table 4.4).

Manufacturing Results

Figure 4.30 shows examples of the captured SEM images of laser chemically roughened and finished micro cavities with the targeted dimensions of (300 × 300 × 60) µm^3. The measured removal geometries have shown that the

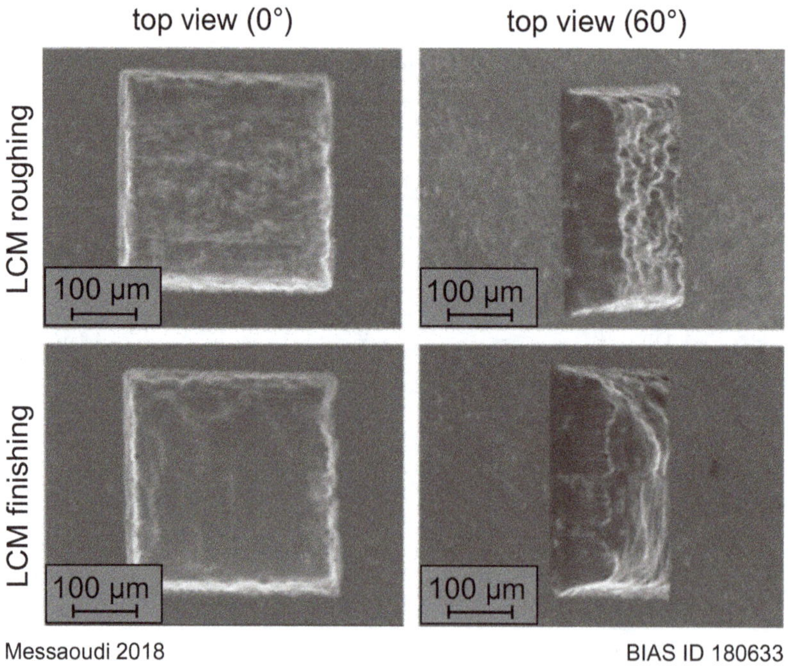

Fig. 4.30 SEM images of the laser chemically manufactured cavities (side length 300 µm), including the two steps roughing and finishing (top views under 0° and 60°)

LCM roughing results in mainly achieving the cavity dimensions, while the LCM finishing provides the final contouring and smoothing of the cavity surface. Indeed, sharp and accurate contours were successfully realized.

The quantitative characterization of the defined quality parameters revealed that the applied LCM strategy ensured the required depth of 60 μm, whereby the removal depth was increased by the LCM finishing step by an additional 10 μm. Moreover, the mean edge radii of the cavities amount to (10.5 ± 1.6) μm and (11.2 ± 1.3) μm for roughing and finishing, respectively. Depending on the machining task, an enhanced material removal, which is characterized by a poor surface quality, can be assured with slow scan velocities [Mes18]. Besides, controllable and low removal rates can be realized with increased scan velocities and reduced laser powers [Eck17b]. This combination was applied during the ensuing finishing step and led to an improved surface quality.

The mean roughness *Sa* was reduced from >2 μm down to 0.7 μm, while the peak roughness *Sz* was improved from 45 μm down to 8 μm. Moreover, the analysis of the chemical composition at different regions by energy dispersive X-ray spectroscopy indicates that the roughing step results in a noticeable deposition of metallic salts and oxides, whereas the finishing step results not only in a surface smoothing but also in removing the residues of oxygen and phosphor and thereby in resetting the conditions of the base material [Mes18].

4.3.5.2 Other Examples of Laser Chemically Machined Micro Tools

The square cavities shown in the previous section can be varied in size and replicated with high reproducibility. Figure 4.31a shows the logo of the CRC 747 patterned with the different square cavities in Stellite 21. Besides, hexagonal cavities with side lengths of 250 μm were also manufactured in Stellite 21, as reported in [Mes17b]. Following the aim of high removal rates and cavity depths of >200 μm, multiple scans with feed velocities <20 μm/s and relatively high quotient P_L/d_{spot} between 60 and 100 mW/μm were used. There, the occurrence of electrolyte boiling and related removal disturbances were not considered. The results show that cavity depths up to 450 μm can be achieved, while the edge radius increases with the removal depth up to 30 μm. Moreover, high shape accuracy at the surface is demonstrated with form deviations <3%.

In addition, as demonstrated in Sect. 4.3.4, the machining of non-self-passivating materials such as high-speed steel is possible considering the limited processing time (of some hrs). Based on this, micro cavities were machined in HS10-4-3-10, following the same strategy used to manufacture Stellite 21 [Mes18a]. Figure 4.31b shows square cavities with different side lengths from 150

Fig. 4.31 Examples of LCM fabricated micro forming tools: **a** CRC logo written with square micro dies in Stellite 21, **b** square micro dies in high-speed steel HS10-4-3-10 with different side lengths from 150 μm to 2 mm and **c** micro textured friction tool (stainless steel 1.4310)

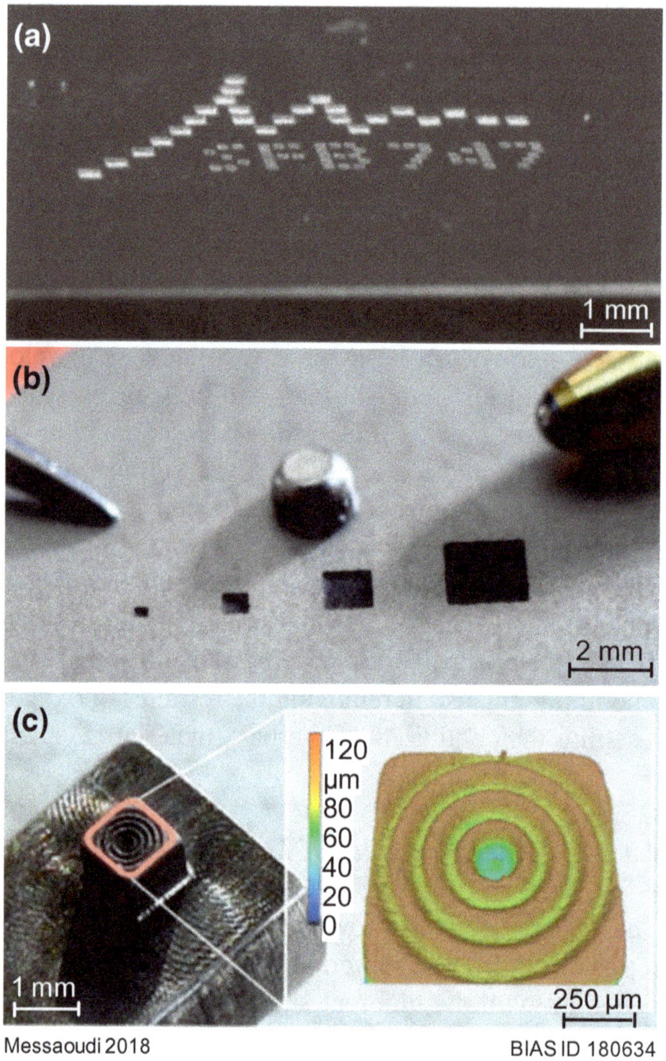

to 2 mm. Due to the immediately induced reaction at the metal-electrolyte interface, higher removal rates up to 50 μm/s were achieved. Moreover, similar edge radii of 15 μm on average were measured. However, due to the unequal chemical dissolution rates of the different alloy elements/phases with respect to the electrolyte solution used, an intensified grain boundary attack was observed during the finishing step, which led to an increase of the mean areal roughness Sa from 1.5 μm (after the roughing step) to 2.3 μm.

Among others, LCM was also used to fabricate molding micro tools (1 mm 1 mm) from stainless steel 1.4310, as shown in Fig. 4.31c. As reported in [Rob17], these tools were used for the structuring of monocrystalline diamond by ultrasonic-assisted friction polishing (see Sect. 4.7). Therefore, periodic circular and linear free-standing bridges <100 μm in width and >30 μm in height were realized. Thereby, it was found that the removal with a circular movement increases with a smaller radius and is more intense than with the linear one.

4.3.6 Comparison with Other Micro Machining Processes

Taking electrochemical machining (ECM), electrical discharge machining (EDM) and micro milling as competing methods, the maximum removal velocity and rates were defined as follows:

$$removal\,velocity\,v_{rem,max} = removal\,depth\,d_{rem,max} \cdot \frac{feed\,velocity\,v_{feed}}{spot\,size\,d_{spot}} \quad (4.5)$$

$$removal\,rate\,R_{rem} = removal\,volume\,V_{rem} \cdot \frac{feed\,velocity\,v_{feed}}{feed\,length\,l_{feed}} \quad (4.6)$$

Figure 4.32 shows examples of the maximum removal velocities $v_{rem,max}$ and rates R_{rem} determined within the disturbance-free regime ($T_{(x=0)} < 150$ °C) in titanium and stainless steel (AISI 304) in a 5 molar H_3PO_4 environment. In both

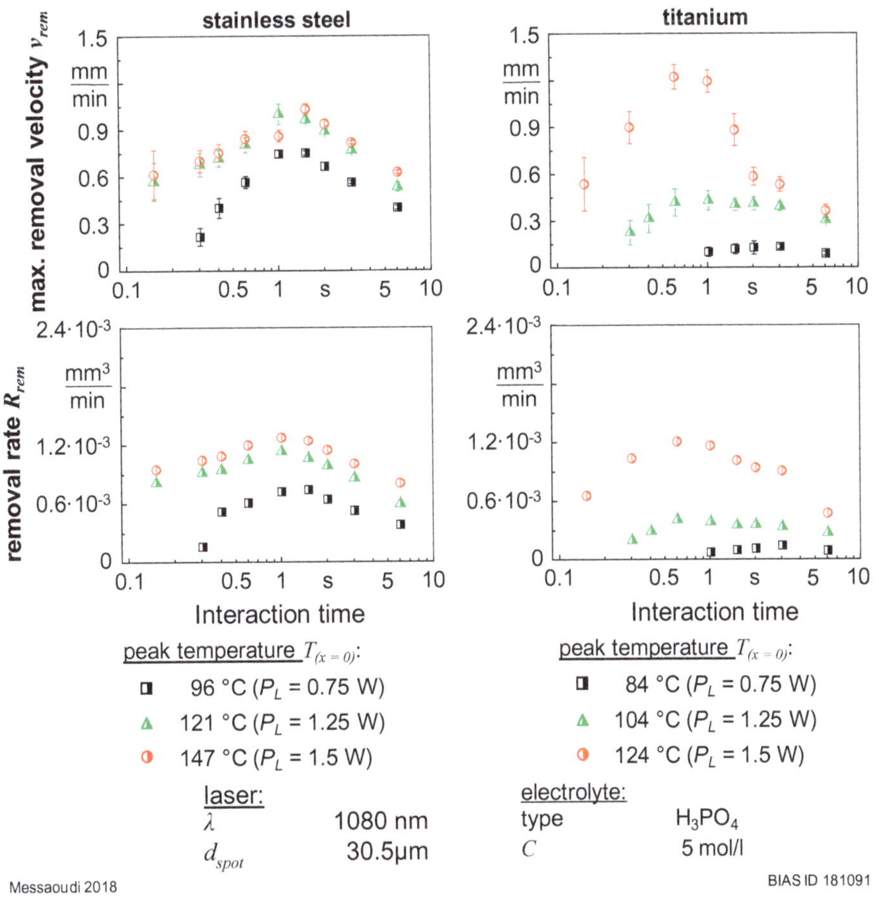

Fig. 4.32 Determined maximum removal velocities v_{rem} and removal rates R_{rem} during LCM of stainless steel (AISI 304) and titanium in a 5 molar phosphoric acid solution in dependence on the interaction time under different thermal loads

materials it is observed that the removal velocity/rate increases first with the thermal input and the interaction time and achieve a maximum at interaction times between 0.6 s and 1.5 s. In general, the highest removal is observed around $t_{inter} \approx 1$ s. In combination with a suitable thermal input, which is defined to be achieved with a temperature field having $T_{(x = 0)} = 125$ °C, removal velocities up to 1.25 mm/min in titanium and 1 mm/min in stainless steel can be realized. The related removal rates R_{rem} amount to about $1.25 \cdot 10^{-3}$ mm/min in both materials. Similar behavior is observed also in LCM of Stellite 21 [Mes18a], where maximum removal velocities of 0.75 mm/min were obtained at an interaction time $t_{inter} \approx 1.2$ s. In contrast, the LCM of the non-self-passivating alloy HS10-4-3-10 in a 5 mol/l H_3PO_4 electrolyte solution is characterized by a continuous decrease of v_{rem} with increasing interaction time [Mes18a]. In comparison with the self-passivating materials, the removal velocities obtained in HS10-4-3-10 including the background etching amount to 2 mm/min at $t_{inter} \approx 1$ s.

Moreover, these determined LCM removal velocities of some mm/min within the disturbance-free removal regime are comparable to the feed speeds usually applied in the ECM process. Despite the much higher removal rates (some mm^3/min), which are related to the size of the electrodes used, the ECM of titanium- and nickel-base for aero engine components is ensured with feed velocities between 0.5 mm/s and 1.5 mm/min, as reported by Klocke et al. [Klo13]. Taking similar interaction zones (some 10^4 µm^2), as is the case in micro-ECM, it can be stated that LCM shows higher removal efficiency. Han et al. [Han16] have determined feed velocities of 0.4 to 1 µm/s in micro-ECM of micro rods made of stainless steel (AISI304).

With the same target dimensions, as described in Sect. 4.3.5.1, micro square cavities in Stellite 21 were also manufactured by micro milling using hard coated tungsten carbide ball-end mills with diameters of 0.2 and 0.1 mm. A detailed description of the machining procedure comprising roughing and finishing steps can be found in [Mes18]. The cavities resulting here have undergone the same characterization as the laser chemically machined ones. Comparison of the two processes reveals that laser chemical machining (consisting of roughing and finishing steps) is more suitable for manufacturing cavities with dimensions <200 µm due to the higher shape accuracy with stable mean edge radii of (11.2 ± 1.3) µm, as can be seen in Fig. 4.33. However, the finish quality of micro milling with mean surface roughness *Sa* of 0.2 µm could not be achieved with laser chemical machining. Due to in-process induced waviness (at spatial wavelengths between 20 µm and 100 µm), the surface quality could only be improved from >2 µm down to 0.7 µm. Further, the metallographic analysis of the near-surface layers reveals that both manufacturing processes ensure gentle machining without any noticeable microstructural impact [Mes18].

In view of the machining time, micro milling appears much more efficient than laser chemical machining, e.g., 5 and 200 min are needed for the machining of one cavity of (300 × 300 × 60) µm^3 using micro milling and LCM, respectively. However, when considering the tool interaction area ($31.5 \cdot 10^3$ µm^2 for the 100 µm ball-end mill and $2 \cdot 10^3$ µm^2 for the laser beam), which is about 16 times larger for

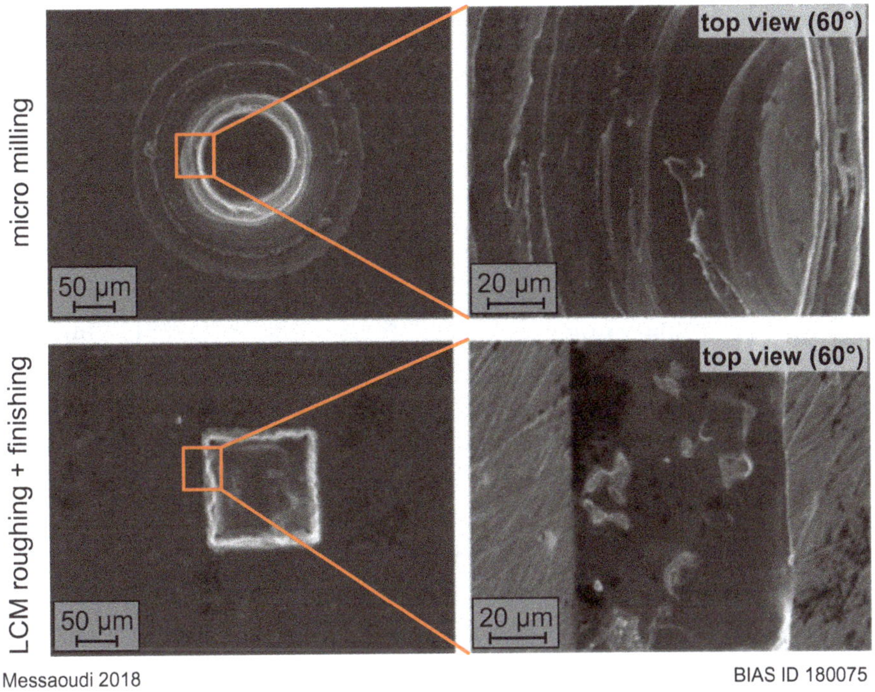

Fig. 4.33 Top view SEM images of a micro milled (above) and laser chemically machined (below) cavity with targeted dimensions of $(150 \times 150 \times 60)\ \mu m^3$ as well as under 60 ° magnified sections showing the cavity wall

micro milling, the LCM removal rates are close to those of micro milling. With regard to the tool area, the average removal rates within the applied machining conditions amount to $2.7 \cdot 10^{-5}$ mm^3/min for LCM and $6.75 \cdot 10^{-5}$ mm^3/min for micro milling.

In general, the presented examples show the diversity and flexibility of laser chemical machining as well as its efficiency, which is comparable to that of well-established processes such as micro milling and ECM.

4.4 Process Behavior in Laser Chemical Machining and Strategies for Industrial Use

Salar Mehrafsun, Hamza Messaoudi* and Frank Vollertsen

Abstract Laser chemical machining (LCM) is a competitive processing method in comparison with electrochemical machining (ECM) and micro milling for the manufacture of micro tools especially with structure dimensions <200 μm. However, the processing window for a disturbance-free removal is limited to only some 100 mW due to the process sensitivity to electrolyte boiling and related hydrodynamic effects. In this section, we focus on the environmental friendliness and the machine handling as two additional key aspects towards a transfer to near-production use. Therefore, alternative sustainable electrolytes, usually used in ECM, such as sodium-based solutions are applied. The resulting removal behavior and rates are compared and discussed with respect to the use of acidic etchants. Furthermore, monitoring concepts that ensure an automatic workpiece alignment and in-process removal control as well as the development of a fast and safe workpiece exchange are presented. Moreover, it is shown that fluid dynamic simulations can be used to improve the design of the chemical cells and to ensure a homogeneous electrolyte flow. The strategies presented enable an automated, user-friendly and safe operation.

Keywords Laser micro machining · Laser chemical machining · Concurrent engineering

4.4.1 Introduction

Electrochemical machining (ECM) processes are gaining in importance in the manufacturing of micro components because they avoid heat-affected zones and thermal stresses compared with conventional machining [Geo03]. To further increase the removal rates as well as the processing quality, the trend toward hybridized ECM is expected to increase [Gup16]. This includes the combination of ECM with other processes such as electro discharge machining (EDM) or laser machining (LM). In the case of laser-assisted ECM, a laser beam of 100 to 300 μm in diameter is coupled in a coaxial arrangement to the electrolyte stream through a nozzle electrode with a diameter of 1 mm [Paj06]. Due to the laser-assistance the removal rates of different metallic materials, such as stainless steel, aluminum, nickel and titanium alloys, have been increased up to 50% in comparison to those obtained with ECM [Sil11a]. Furthermore, Bäuerle has proved that an electrodeless ECM especially of metals within an electrolyte environment can be initiated by laser irradiation [Bae11]. Thereby, the laser-induced temperature field can be generated by a localized thermobattery that allows a current flow within the metallic surface and can thereby change the electrochemical potential to the range of an anodic material dissolution. Based on this knowledge, laser chemical machining (LCM) was developed especially for the micro machining of metallic surfaces. A similar concept is used by the Fraunhofer Institute for Solar Energy Systems (Fraunhofer ISE) to machine silicon-based photovoltaic elements. There, the laser beam is guided via total refection onto a liquid jet with a diameter of 30 to 80 μm generated in a nozzle with pressures up to 500 bar. Depending on the electrolyte solution, the laser heating or ablation can be combined with physical-chemical reactions such as etching, doping or deposition [Kra08]. A different method uses wet-chemical-assisted laser ablation, in which semiconductors such as gallium nitride (GaN) are irradiated with a fs-laser in hydrochloric acid (HCl) solution in order to realize micro channels. Nakashima et al. have demonstrated that the processing in a liquid environment results in smooth structures with sizes of some 100 nm [Nak09].

LCM and ECM have quite similar mechanisms of dissolution and surface modification. These processes have the common feature that material removal is based on the chemical dissolution on an atomic scale [Meh16a]. In LCM the material removal is accomplished by the laser-induced chemical reaction between an electrolyte and the metallic surface of the workpiece [Now94] and is mainly dependent on the laser-induced temperature distribution over the workpiece surface [Mes17a]. Besides, the thermophysical properties of the metallic material, the electrolyte characteristics (light absorption, concentration, flow direction and velocity, etc.) as well as the chemical reactivity of both are additional factors that influence the removal quality and rate. The influence of these factors is described in detail in Sect. 4.3. In general, laser chemical machining enables non-contact material processing using low laser power densities compared to the laser ablation process [Meh12]. Depending on the material-electrolyte combination, the material removal can be realized on the one hand by self-passivation or by means of an applied potential, resulting in a thin passivation layer. According to Nernst, the electrochemical potential is proportional to the temperature, and therefore to the laser power applied [Ste10].

In Sect. 4.3 the competitiveness of laser chemical machining (LCM) in comparison with electrochemical machining and micro milling for the manufacture of micro tools was demonstrated, especially for structures with dimensions <200 μm [Mes18]. Using thermal and closed-loop models, a profound understanding of the LCM removal mechanisms with the aim of avoiding removal disturbances has been developed. This opens up the opportunity for a predictable and controllable removal quality. However, the environmental aspects as well as the machine handling are still key aspects to consider prior to widespread industrial acceptance of the LCM process. For this reason, the use of alternative sustainable electrolytes frequently used in ECM, such as sodium-based solutions, is investigated and compared to the acidic etchants regarding the removal characteristics (width and depth) and quality. Furthermore, automation concepts for workpiece alignment, in-process removal control as well as fast and safe workpiece exchange are presented. In addition, fluid dynamic simulations are performed with the aim of improving the design of the chemical cells and to ensure a homogeneous electrolyte flow.

4.4.2 Materials and Methods

For the experimental investigations, self-passivating and technologically relevant (in the aerospace industry and medical technology) metallic materials, e.g. titanium (3.7024), stainless steel (1.4310), cobalt-chrome alloy Stellite 21 and the tool steel X100CrMoV8-2 (in hardened and tempered condition) were machined in different electrolyte solutions. Thereby, the passivation layers prevent corrosion or chemical dissolution of the material in direct contact with the electrolyte and enable the investigation of the laser-induced chemical reaction behavior without additional external currents. For greater environmental friendliness, the LCM process was performed in aqueous solutions usually used in electrochemical machining, such as sodium nitrate, sodium chloride and citric acid. For comparison, removal investigations were also done in phosphoric and sulfuric acid solutions. Table 4.5 provides an overview of the applied electrolytes and their concentrations.

The laser chemical machining was carried out in electrolyte-jet based (JLCM) and scanner-based (SLCM) LCM machines, which are described in detail in Sect. 4.3. To analyze the removal geometry, including width, depth and volume, laser scanning confocal microscopy (Keyence VK-9710) as well as a light microscopy (Keyence VHX-1000) were used. In addition, scanning electron microscopy (SEM) and energy dispersive X-ray spectroscopy (EDX) were used to analyze the chemical composition of the machined cavities.

4.4.3 Sustainable Electrolytes for LCM

The JLCM system (see Sect. 4.3.2) was used in machining titanium and Stellite 21 in different electrolytes [Hau15]. Thereby, it was found that titanium exhibits up to 30% deeper cavities and up to 40% higher removal rates than Stellite 21 when machining both in a 5 molar phosphoric acid solution. In addition, citric acid

Table 4.5 Overview of the materials and electrolytes used for the experimental work

Electrolyte	Concentration [mol/l]
Phosphoric acid (H_3PO_4)	2.5/5/7.5
Sulfuric acid (H_2SO_4)	1.9
Citric acid ($C_6H_8O_7$)	5
Sodium chloride (NaCl)	2/5
Sodium nitrate ($NaNO_3$)	2/5

(1 M $C_6H_8O_7$) was identified as a suitable environmentally friendly electrolyte, especially for the machining of the tool steel X110CrMoV8-2. However, the resulting etching reaction is significantly limited, which can be explained by the limited dissociation of citric acid within water and thereby by the lower number of reactants that can be provided for the chemical reaction. Compared to the LCM removal in phosphoric acid [Ste09b], the resulting cavity depths were about 80% lower [Hau15].

Using the SLCM setup (see Sect. 4.3.2), the threshold power required for laser-induced material removal was determined for different electrolyte-workpiece combinations. Phosphoric acid and sodium nitrate solution with concentrations of 5 mol/l as well as sodium chloride solution having a concentration of 2 mol/l were used as electrolyte environments during the machining of titanium (3.7024). Thereby, the resulting cavities were analyzed with respect to their removal volume V_{rem} after a constant irradiation time of 1 s and a gradual increase of laser power P_L. Depending on the electrolyte, the material removal starts at different laser powers, as can be seen in Fig. 4.34.

Using phosphoric acid, material removal is observed at threshold laser powers of 0.7 W. In contrast, material removal in sodium nitrate and sodium chloride solutions takes place at higher laser powers of 1 W and 1.7 W, respectively. To differentiate between the chemically driven and the thermally driven effects, the interaction between the titanium surface and laser irradiation was investigated additionally under an argon atmosphere. As mentioned in Sect. 4.3, the cooling effect of the liquid in LCM is rather low and can be neglected, making this comparison valid. In Fig. 4.34 it is shown that the first material removal, which is

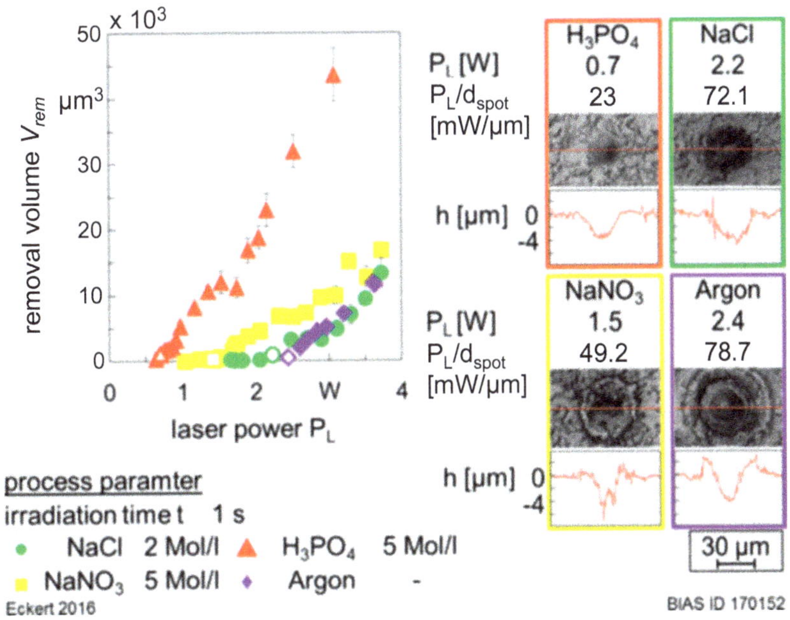

Fig. 4.34 Removal volume V_{rem} in titanium in dependence on the laser power and electrolyte processed with SLCM

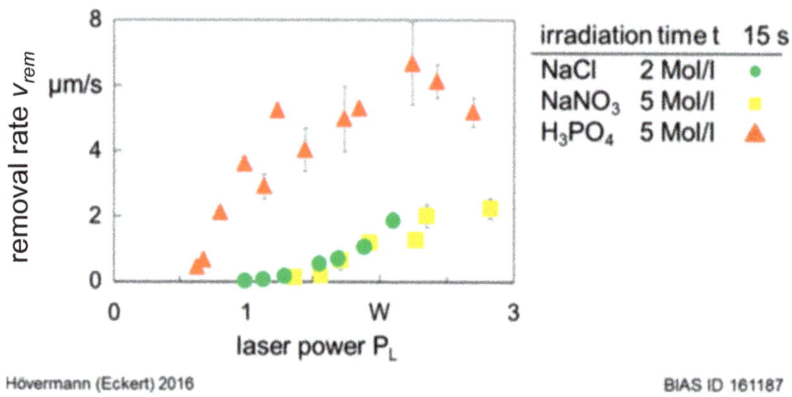

Fig. 4.35 Removal velocity v_{rem} in titanium in dependence to the laser power and electrolyte processed with SLCM

related to molten titanium, starts at laser powers of about 2.4 W. This removal is mainly the result of melt pool dynamics that throw out molten material from the center to the outer sides and is governed by temperatures that exceed the melting point of titanium (>1668 °C). This can be explained by a sudden jump in the absorption coefficient by temperatures >800 °C from 40% to >70% [Kwo12].

Figure 4.35 shows the resulting removal velocities v_{rem} in titanium in dependence on the electrolyte used. In the phosphoric acid environment, the removal velocity v_{rem} increases linearly between laser powers of 0.7 W and 1 W up to values of 4 µm/s. For $P_L > 1$ W, the removal rate decreases first before it continues to increase again with the laser power, with rising standard deviations showing a non-homogeneous removal. This is in accordance with the correlated surface temperatures (see Sect. 4.3.4) showing the first disturbances within the transition boiling regime (T between 104 and 140 °C) [Mes17a]. For the case of sodium nitrate and sodium chloride solutions, the removal rates indicate a similar behavior. At $P_L = 1$ W, etching rates <0.1 µm/s have been determined. These reach their maximum at P_L of 2.1 W, showing values of about 2 µm/s. These removal rates are significantly lower in comparison to those obtained in phosphoric acid. Thus, a higher irradiation time is required for an equally deep removal. With respect to the surface temperature, it becomes clear that the removal occurs first after the electrolyte boiling, which indicates the involvement of gas bubbles in the etching process.

A closer look at the SEM records of the resulting cavities in titanium (compare Figs. 4.36 and 4.37) shows the differences of the removal quality depending on the electrolyte used.

The removal of titanium in sodium nitrate $NaNO_3$ (Fig. 4.36, left) and sodium chloride NaCl (Fig. 4.36, right) is labeled by a pronounced intergranular corrosion, as already described by Wendler et al. [Wen12]. In accordance to [Uen08], the removal is assumed to comprise intergranular penetration and grain dropping occurring by turns. In contrast, the laser-induced chemical removal using acidic aqueous solution, e.g. phosphoric acid (H_3PO_4), results in much smoother surfaces

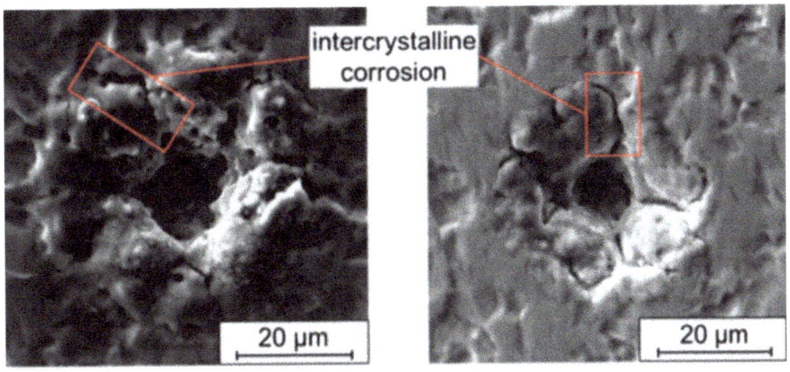

electrolyte	$NaNO_3$	electrolyte	NaCl
concentration c	5 mol/l	concentration c	2 mol/l
laser power P_L	1.7 W	laser power P_L	1.6 W
workpiece	titanium	workpiece	titanium
irradiation time t	15 s	irradiation time t	15 s

Hövermann (Eckert) 2016 BIAS ID 161186

Fig. 4.36 Removal cavities in titanium processed with SLCM and (left) $NaNO_3$ and (right) NaCl as electrolyte

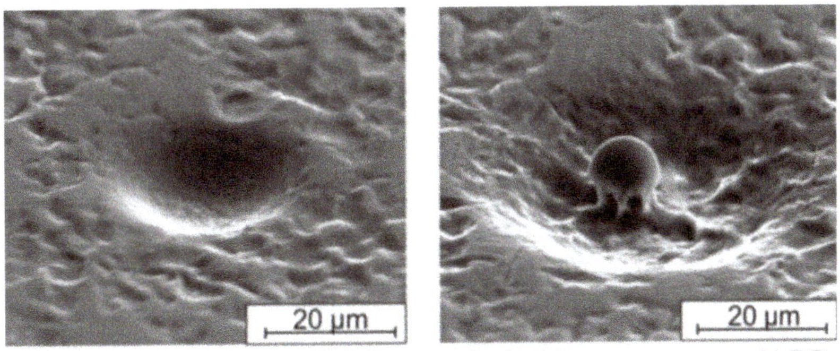

electrolyte	H_3PO_4	electrolyte	H_3PO_4
concentration c	5 mol/l	concentration c	5 mol/l
laser power P_L	0.9 W	laser power P_L	1.3 W
workpiece	titanium	workpiece	titanium
irradiation time t	1 s	irradiation time t	1 s

Hövermann (Eckert) 2016 BIAS ID 161180

Fig. 4.37 Removal cavities in titanium processed with SLCM and H_3PO_4 as electrolyte

without any significant intergranular corrosion effects (Fig. 4.37, left). However, the resulting cavity geometry depends on the processing parameters used. At higher laser powers, removal disturbances can take place in the cavity center (Fig. 4.37, right). The analysis of their chemical composition using energy dispersive X-ray spectroscopy revealed that within the disturbance-free removal only K_α and K_β titanium peaks can be detected, while the mushroom-like pile that fills the center of

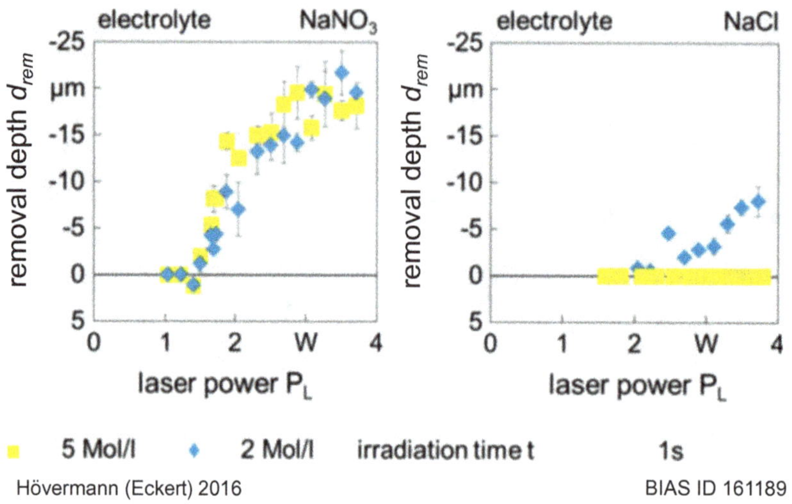

Fig. 4.38 Removal depth d_{rem} in titanium in dependence on the electrolyte and its concentration processed with SLCM

the disturbed cavities (Fig. 4.37, right) is additionally characterized by significant phosphor and oxygen peaks [Eck17a]. As reported by Gutfeld et al. [Gut86] and Nowak et al. [Now96], an etching in the periphery while plating in the laser beam center and vice versa can take place in dependence on the metal-electrolyte combination as well as on the laser and electrolyte parameters. Which process is dominant depends on the temperature gradient as well as on the type and concentration of the dissolved reaction products.

Figure 4.38 shows the influence of the electrolyte concentration on the resulting removal depth in titanium using NaCl and $NaNO_3$ solutions. Thereby, it can be clearly determined that material removal using NaCl is only possible with a concentration of 2 mol/l. This can be traced back to the enhanced salt crystallization due to local solution boiling at high concentration. Hence, this salt film formed on the surface can slow down or prevent the material dissolution at the workpiece surface [Dat87]. In comparison, similar removal depths are measured in $NaNO_3$ for the concentrations of 2 mol/l and 5 mol/l, which reveal no significant influence of the concentration for this etchant.

4.4.4 Strategies for Industrial Use of LCM

4.4.4.1 Automatic Workpiece Alignment for JLCM

For a further increase in the process stability and reproducibility of the LCM processing different strategies are pursued. One of these strategies is to increase the degree of automation with the aim of enabling more flexible and faster processing. Investigations have shown that the quality of laser chemical machined micro tools

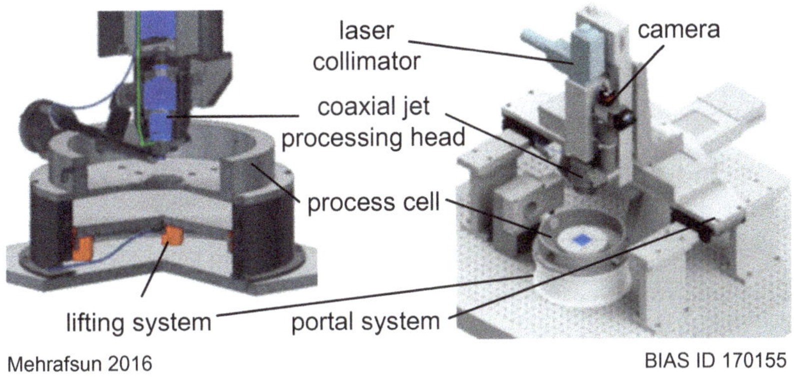

Fig. 4.39 JLCM system with integrated image processing for automatic workpiece alignment

is highly dependent on the positioning and alignment of the semi–finished products in the processing room (chemical cell). A first approach to improve the reproducibility is to implement automatic workpiece positioning using image processing. Therefore, the position of the workpiece within the processing cell is detected by means of reference points and compared with the predetermined position of the CAD data and the path planning algorithm. The system determines the positioning error by using image processing techniques and permits automatic alignment of the ideal starting and workpiece position with an accuracy of <1 μm with respect to the laser focal position. Thus, reproducible starting conditions for laser chemical machining can be ensured. This results in fewer parameter adjustments during processing and contributes to a reduced processing time. Figure 4.39 shows the design of the JLCM system with automatic workpiece alignment.

4.4.4.2 In-Process Monitoring and Fast Workpiece Exchange for SLCM

To provide a constant manufacturing quality, it is necessary to monitor and control all relevant process parameters, such as the pressure, workpiece potential, laser power and electrolyte flow rate. Otherwise, changes of these parameters over time could affect the machining behavior and result in losses in quality (see Fig. 4.40). Several factors limit the economic efficiency of the process. The conventional removal rates enable the use of only low process speeds (some μm/s), which means long processing times. Furthermore, as sample changing is performed manually, both the machining precision and speed are affected. To avoid this related downtime and to ensure user-friendly and safe operation, the system design must be trouble-free and low-maintenance. The development of an industry-compatible process cell was carried out in accordance with VDI 2221.

This cell, as depicted in Fig. 4.41, is based on a pneumatic locking system in combination with a modular workpiece holder that allows safe workpiece exchange with minimum contact with the electrolyte within a maximum of 30 s. Using a

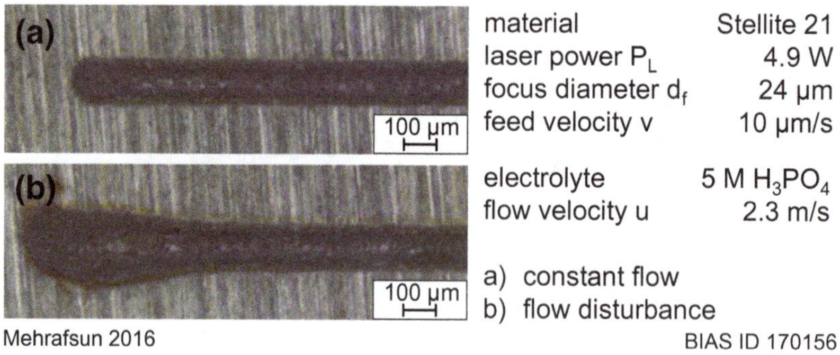

Fig. 4.40 Example of LCM removal path in Stellite 21 processed with phosphoric acid (5 mol/l) **a** without and **b** with disturbance in electrolyte flow velocity

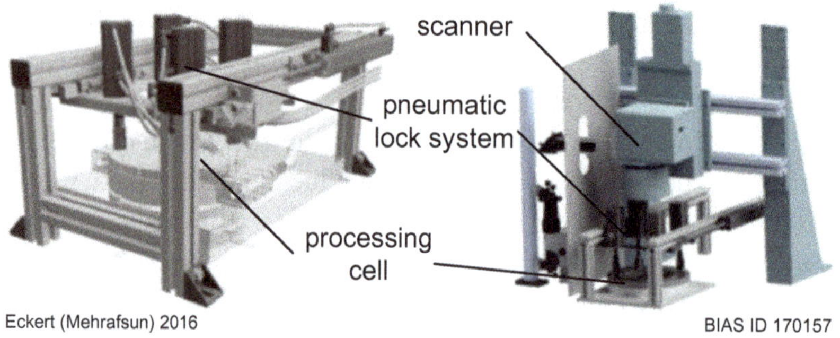

Fig. 4.41 SLCM system with pneumatic lock system for fast workpiece exchange

safety valve in the fluid circuit and locking units in the closure system, the process cell is protected from over-pressure and leakage of electrolyte during processing.

4.4.4.3 Demand-Oriented Multi-channel Flow in SLCM

As mentioned before, trouble-free and high-quality processing as well as reduced process times are essential for industrial implementation of LCM. Moreover, the removal rates must be enhanced to ensure that the machining is competitive. However, the increase of reaction rates is mainly accompanied by an increased formation of hydrogen [Meh13]. In addition, it necessitates the use of higher laser powers that can induce high surface temperatures and result in electrolyte boiling [Sil11a]. As a result, undesirable gas bubble formation and thereby removal disturbances can occur [Meh13].

Another source of disturbance could be the high saturation of the electrolyte during the LCM procedure. For industrial use of LCM, these factors should be taken into account, as well as the electrolyte flow behavior. Investigations have shown that the geometry of the process cell could affect the flow behavior, induces turbulence and gas bubbles, and results in a reduced processing quality (see

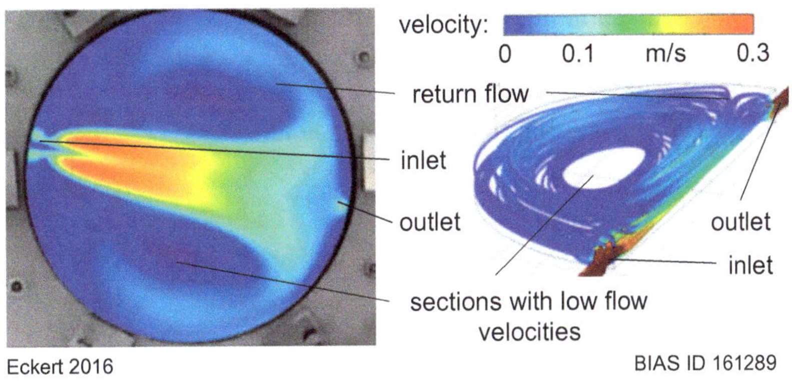

Fig. 4.42 Numerical fluid dynamic simulation of electrolyte flow within the process cell (half of the cell) before optimization

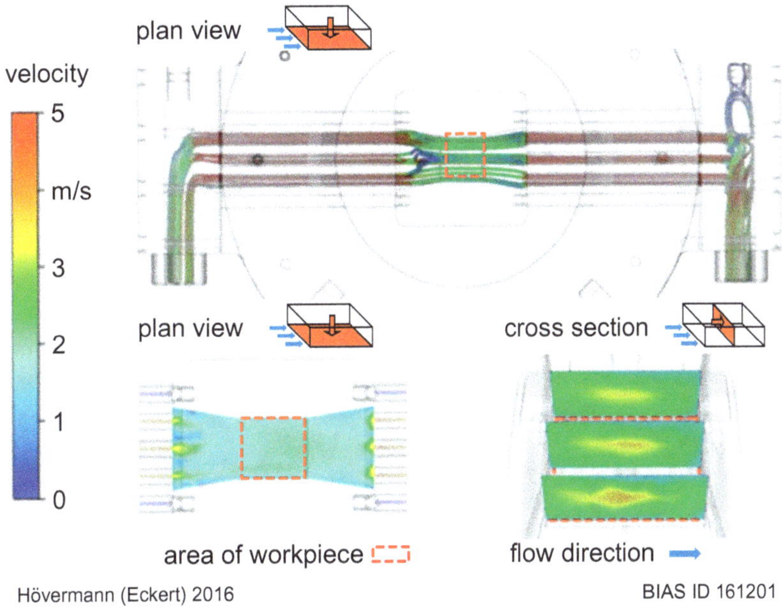

Fig. 4.43 Numerical fluid dynamic simulation of electrolyte flow within the process cell, (plan view above) electrolyte flow from inlet to outlet, (plan view below) flow distribution on workpiece surface, (cross-section) three cross-sections of flow distribution on workpiece surface

Fig. 4.42). Considering these factors, numerical fluid dynamic simulations (CFD) were performed to define a suitable cell geometry and electrolyte flow behavior.

The CFD simulation results are shown in Fig. 4.43. As can be clearly seen, the newly developed fluid system is able to ensure a constant and homogeneous stream profile at a flow velocity of 2.5 m/s. These flow conditions can guarantee a constant electrolyte exchange and continuous transportation of the gas bubbles that emerge, and thereby a stable machining operation. Further, it is possible to reduce or expand the interior of the process cell depending on the workpiece size using inserts, as well as to set the electrolyte flow velocity by switching on or off the inlet and outlet channels with respect to the pump setting parameters.

4.5 Flexible Manufacture of Tribologically Optimized Forming Tools

Oltmann Riemer*, Peter Maaß, Florian Böhmermann, Iwona Piotrowska-Kurczewski and Phil Gralla

Abstract In this work, mathematical and engineering methods are applied in union for the manufacture of efficient micro-forming dies exhibiting tribologically active, textured surfaces by micro-milling. This comprises the extension of analytical process models to wear characteristics of milling tool for die making and micro-tribological investigations. These models allow us to derive optimized process parameters for the manufacturing process of tribologically optimized micro-forming dies by using the non-linear inverse problems method. These optimized forming dies exhibit micro-textured surfaces and have a distinct impact on the micro-contact conditions between the die and the work piece, and thus leads to a reduction in dry friction during the actual forming process.

Keywords Surface • Cutting • Optimization

4.5.1 Introduction

Metal cutting is one of the most widely used methods to produce the final shape of manufactured products in macro dimensions [Alt00]. Therefore, its application for micro-parts is a logical step. However, due to size effects in the micro-regime, new challenges arise and need to be addressed for a successful application [Vol10]. Serendipitously, size effects in metal cutting can also offer new opportunities. The inconsistencies in material removal when undershooting the minimum uncut chip thickness in micro-milling leads to the generation of regular micro-textures. These micro-textured surfaces have been shown to reduce friction in dry tribological contact, which can be explained by changes in the micro-contact conditions determining the predominant friction mechanisms. The transfer of micro-textured surfaces generated by micro-milling is seen to be one key steps towards the development of robust dry deep drawing processes. The concept of micro-forming dies exhibiting textured surfaces (e.g. generated by micro-milling) can be seen in Fig. 4.44. However, identifying the most suitable textures offering minimum dry friction and finding the associated process parameter for their generation in micro-milling is not trivial.

Process modeling can help to overcome these issues. In this work, a process model based on sweep volumes incorporating the continuous change from chipping to plowing is presented. This addresses the unique mixture of material removal mechanisms that allow for the manufacture of regular surface textures in hardened tool steel where the texture's design is first of all determined by the process

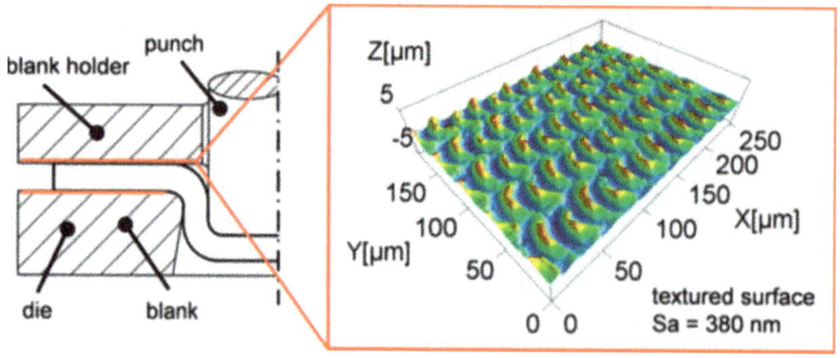

Fig. 4.44 Dry friction in a mold used for micro-cold forming

parameters feed per tooth and line pitch. Suitable process parameters for the manufacture of desired textured surfaces are identified by means of inverse simulation, while taking into concern inverse problems. Classic inverse problems identify the source of observed data. For example, in a computed tomography scan an image is reconstructed from a sinogram, which is the measured density via x-rays over multiple angles. One way to solve such an inverse problem is to use Phillips-Tikhonov-Regularization by minimizing a Tikhonov functional:

$$T^{p,q}_{\alpha,\delta}(u) := \frac{1}{p}\left\|F(u) - v^{\delta}\right\|^{p}_{Y} + \alpha\frac{1}{q}\mathscr{R}_q(u) \tag{4.7}$$

where F represents a forward model, u the parameter, v^{δ} a perturbed data, α the regularization parameter and $\mathscr{R}_q$ the regularization. Most current research deals with numerical implementation [Mor11] and aims to find suitable regularization [Jin12]. In the presented work, the discrepancy term is altered to account for tolerances in the data. The necessary forward model is a combination of analytical and statistical methods. In addition, special attention is paid to the wear of the cutting tool and its influence on the cutting process and machining results. The challenge in this work was to incorporate the size effects in micro-milling, namely the continuous change in material removal mechanism form chip removal towards micro-plowing in dependence on the cutting depth.

The second part of the presented method deals with analytical contact and friction modeling, taking into concern meaningful parameters derived from ISO 25178 are seen to be the key to more precise derivation of input values for inverse process simulation. This represents the basis for the predictive machining of functionalized forming dies with locally adapted tribological characteristics by process parameter manipulation utilizing the previously presented process model. The work flow of the presented methodology is depicted in Fig. 4.45.

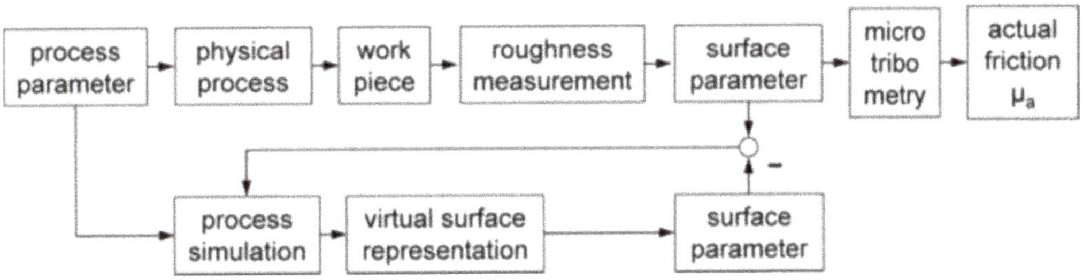

Fig. 4.45 Work flow of computational engineering for micro-milling

4.5.2 Variation, Dispersion, and Tolerance in Inverse Problems

Classical Tikhonov functionals (4.7) consist of two terms, a discrepancy and a regularization term. By minimizing both terms, weighted with a coefficient α on the regularization term, a suitable approximation of the true parameter is found; that is, the parameter that causes the given or observed data. Let V be the space in which the process model F maps onto and all data resides. The discrepancy term then describes the distance of given or observed noised data v^δ to the predicted data $F(u)$ for the parameter $u \in U$. The regularization term $\mathscr{R}_q(u)$ uses some a priori information and maps the parameter u onto the positive real numbers. A common choice for $\mathscr{R}_q$ is (4.8) based on an L^q distance, where u_0 is often set to 0. For prediction models, this is changed to $u_0 \neq 0$ and instead a density point of known data is used to avoid extrapolation in prediction.

$$\mathscr{R}_q(u) := \|u - u_0\|_q^q, \quad 1 \le q \le 2 \tag{4.8}$$

In this work, process models are not only restricted to value estimation of the outcome but can also provide a variance or dispersion, as was applied in Sect. 1.4.2. Furthermore, the given data can include a variance, such as through multiple measurements. Both cases are addressed by allowing the provided data to not be a single measurements but instead be a closed set of feasible data points. This set will be denoted as the tolerance area of the point v^δ. To compare two sets while maintaining good numerical properties, the distance measurement given in (4.9) is used. This distance is continuous and measures the overlapping area; that is, it is only equal to zero if S_1 is a subset of S_2. This is important to ensure that the determined parameter is within the tolerance area if this parameter exists.

$$d(S_1, S_2) = \int_{S_1} \min_{y \in S_2} \|x - y\| dx \tag{4.9}$$

In the following, $S_{F(u)}$ denotes the tolerance set of $F(u)$. The tolerance set of v^δ will be denoted by S_{v^δ}.

For quality management, the three sigma area is a common choice to define a process quality if an additive normal error is assumed [Whe92, Cas12]. To perform the parameter identification, the functional (4.10) is minimized. The minimizer of (4.10) is then the identified parameter.

$$T_{\alpha,\delta}^{p,q}(u) = d_p(S_{F(u)}, S_\delta) + \alpha \mathscr{R}_q(u) \tag{4.10}$$

The minimization of (4.10) is done numerically using a standard non-linear solver, such as gradient descent with adaptive step size, inner point algorithm or sequentially quadratic problem solver. Depending on the properties of F, the solver may need to be adapted to be able to minimize the functional [Boy04, Ber99]. In all cases, only a local minimum is identified, which may not be global minimum because all algorithms are of iterative nature and depend on the starting value. Different strategies, such as hill climbing and trust region, can be used to soften the dependency on the starting value but they have a higher numerical cost.

4.5.3 Computational Engineering

Computational engineering is applied to further comprehend and adjust the micro-milling process. Computational engineering deals with the development and application of computational models and simulations to solve complex physical problems arising in engineering analysis. An example of computational engineering for turning processes can be found in [Bra12].

In this present work, a mathematical model of the cutting process was designed and analyzed. This model is based on physical properties and observed measurements. It combines analytical and data based approaches to ensure that it is accurate and manageable at the same time. The obtained model allows us to simulate the cutting process and predict the outcome of the process for different process parameters and setups. In addition to prediction, it can also be used to find ideal process parameters for a predefined output via parameter identification using methods from inverse problems. This parameter identification can be used to find the cause for an observed data, such as the current wear on the cutting tool.

4.5.3.1 Process Model with Wear on Cutting Tool

The process model in this work is based on sweep volumes to map process parameters, including tool path, feed velocity v_f and cutting tool geometry, onto a finished surface. It is not a finite element or volume method and it differs from these approaches. Its main components are models of the work piece, tool and the tool path. The tool moves along its tool path and a sweep volume is generated. This sweep volume and a model of the workpiece are used to calculate the material removal process and change of the workpiece itself. While this method can be applied in the macro scale (see the Chapter *Predictive compensation measures for the prevention of shape deviations of mircomilled dental products* for one example), it is extended to reproduce size effects in micro-scale. These size effects are the occurrence of plowing when undershooting the minimum uncut chip thickness, as well as a considerably increased cutting tool deflection. More detail into the model generation [Pio11], validation, and simulation process is given in [Veh15b, Veh17a]. The process model focuses on the finishing step of micro-milling where the last layer of material is removed. This can be applied to all stages of micro-milling. Since size effects are most influential in the last finishing step, a macro model can be used for earlier stages to reduce computational effort and reduce computation times.

The cutting tools used in milling process are also affected by the process. The wear of the tools alters the characteristics of produced surfaces. For micro-milling the shape of the cutter highly influences the surface quality. Progressive tool wear ultimately leads to a decrease in the quality of the work result. Once product specifications (which are the geometrical accuracy and the roughnesses of work pieces) can no longer be obtained, the milling tool will be replaced to provide the desired work result.

In this section, a process model for micro-milling is outlined and the modeling of wear on the tool is explained. Further detail on how to find parameters with respect to wear are given, including how it can be integrated into the theory of inverse problems.

Process Model

The micro-milling process is modeled in three consecutive steps. The first step is the implementation of the process kinematics. The kinematic of a milling cutter is described by a composition of the rotational and translational motion of the tool. This composition describes how the tool is rotating around itself and moving along a given tool path. To account for dynamical changes occurring during the process, a dynamical force model is applied. This is accomplished with a system of ordinary differential equations. The model is presented in details in [Pio11]. Moreover, it is assumed, that the reader is familiar with kinematics of the micro-milling process. If otherwise, refer for example to [Alt06].

To obtain the surface, additional information is needed, such as the shape of the cutting tool, material and size of the workpiece and the material to be machined. Additionally, the material removal mechanism has to be included. For ball-end micro-scale machined surfaces, the minimum chip thickness has to be considered. This means that the chip formation is divided into three parts, see [Veh15b]. Below a certain minimum cutting depth, material is elastically deformed and thus no material is removed; that is, no chip is formed. Above a certain cutting depth, material is completely removed and the chip thickness is equal to the cutting depth. The third stage describes the transition between the former two stages. A visualization of the three different stages and its model are given in Fig. 4.46 according to [Veh15b]. Because the cutting depth is different at all points of the

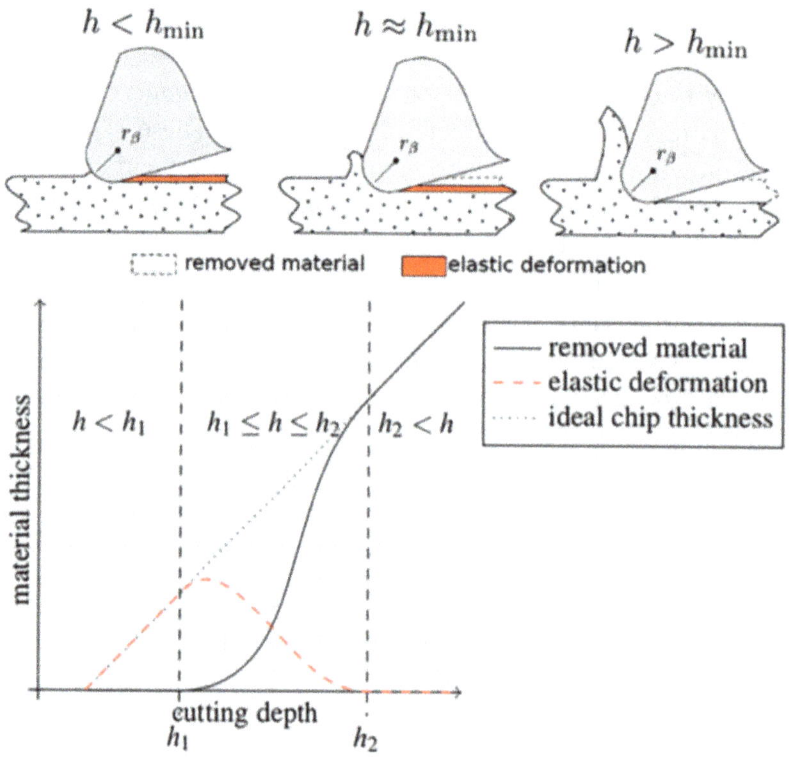

Fig. 4.46 Three stages of the micro-cutting process according to [Ara09], and visualization of the function for the implementation of the chip removal in dependence on the cutting depth into the process model

milling tool and only cutting depth smaller than the tool radius are used for the last milling step, all three stages are constantly present. This results in micro-structures on the produced surface, depending on the process parameters.

The forward model generates a detailed map of the generated surface. Thus, it can be used to compare characteristics of surfaces from different input parameters. To allow a comparison of different surfaces and their tribological characteristics a characterization of the surface is needed.

One way to compare surfaces is to divide deviations into six categories, as defined by DIN: 760:1982-06 from the Deutsche Institut für Normierung (DIN Standards). Deviations of the third and fourth categories describing the roughness of a surface are of special interest. This is further specified in DIN EN ISO 4287, where the so-called one-dimensional characteristics are defined. These include peak-to-valley value, median roughness, average roughness and more. These values are only statistical height parameters, whose calculation is based on every data point. Those parameters that do not include any information about the number of peaks, distribution of peaks and so on, are not sufficient for the application in this work.

For a more precise classification, the Abbott-Firestone-Curve or bearing area curve can be used; some alternative procedures are introduced later in this chapter. The Abbott-Firestone-Curve describes the ratio of the surface at a given height and allows to compare surfaces and their properties. From a mathematical perspective, it

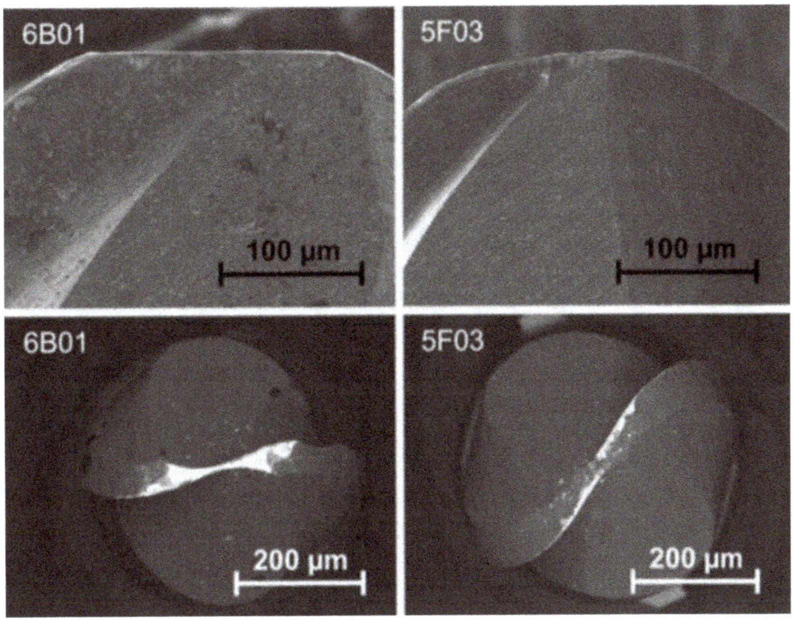

Fig. 4.47 SEM images of a cutting tool as an example of the wear on a ball-end tool

is a cumulative probability density function of the surface's profile height. For the rest of this chapter, it is assumed that all of the data are provided as bearing area curves.

During the manufacturing process, the cutting tool wears out. This results in a change of the cutting tool shape, as depicted in Fig. 4.47. The figure shows a ball-end cutting tool after use. Without wear, the front view would show a circle segment. In the top view (bottom two pictures) the white spots mark the wear. Due to the forces acting on the cutting tool during the manufacturing process, it starts to flatten out. A simple way to model this behavior is to flatten the tool shape over the time. A more accurate way is to consider the actual forces in each point and then model the material removal and deformation of the cutting tool. Two models for the wear over time are included. The first model uses a predefined wear depending on the time. The second model simulates the wear depending on the force applied on the cutting edges. The programming of the first model is less elaborate and it substantially reduces simulation time, whereas the second model is more accurate. To understand how wear of the cutting tool influences the process itself, use of the second model is advised. For ideal parameter identification and adjustment of process parameters during manufacturing, use of the first model is advised.

To identify the wear and exact shape of a cutting tool from a surface measurement, a parameter identification can be performed. This parameter-identification does not depend on the chosen model for wear on the cutting tool because no change in the wear is assumed. For measurements over a long time interval, separate parameter identifications for different points in time are done. The cutting tool in the model is described in polar coordinates and discretized over the radius to allow a

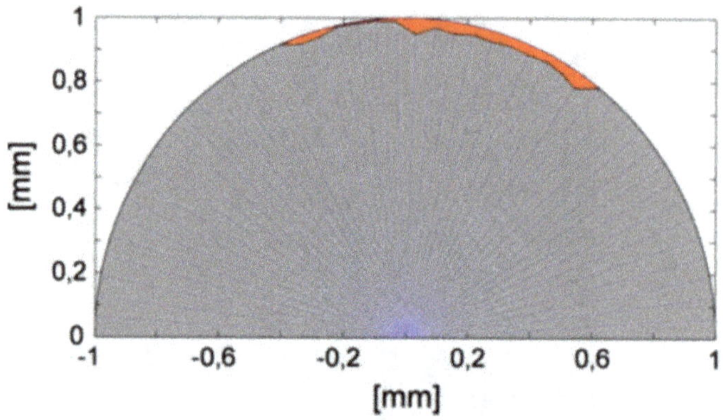

Fig. 4.48 Profile of a simulation of nonuniform wear on a ball-end tool

equidistant discretization on the cutting edge (see Fig. 4.48 for a nonuniform wear on a ball-end tool).

The mathematical operator is extended with the model of wear on the cutting tool and takes an additional input t. This additional input defines how long the cutting tool has been used prior and thus provides the wear on the tool at the beginning of the simulated process. The exact shape of the tool at the time t can also be mapped. In the following two sections, a simple wear model is applied incorporating the flattening of the cutting tool over time.

Minimize Deviation Over a Fixed Time Interval

For a fixed time-interval, $t \in [0, c]$, the overall deviation from the desired tolerance ε can be calculated via

$$\int_0^c \left\| F(p,t) - g^\delta \right\|_{L_2(\Omega),\varepsilon} dt,$$

where F is the process model, p process parameters, t the time the cutting tool is used; that is, the degree of wear on the tool, and g^δ the desired or measured surface as Abbott-Firestone-curve. This term measures how much the output will deviate over the whole time interval but does not hold any information at which point in time the output will not lie within the given tolerance. Therefore, a soft boundary is considered in the sense that a local minimum unequal to zero might have no point in time for the condition

$$\left\| F(p,t) - g^\delta \right\|_{L_2(\Omega),\varepsilon} = 0$$

The benefits of minimizing deviation over a fixed time interval is that it can be directly applied as a discrepancy term in the Tikhonov functional (4.7) and that the integral can be approximated via known rules, such as the Simpson rule.

Maximizing Machining Time Within Tolerance

To actually maximize the timespan in which a cutting tool can be used with the same parameter set before it needs to be replaced or new process parameters need to be determined, the ending time of the time-interval has to be flexible. The ending time c is then calculated through

$$c = \arg\max \left\{ c \middle| \int_0^c \|F(p,t) - g^\delta\|_{L_2(\Omega),\varepsilon} dt = 0 \right\}.$$

In the case that there is no $c \geq 0$ that fulfills the condition, $c = 0$ will be set. This provides the maximum time before the simulated output is outside the desired tolerance for the first time. As long as the parameter does have a discrepancy greater than zero in the initial case without wear (i.e. $t = 0$), the maximum stays zero. Thus, the argmax cannot be used to replace the discrepancy term in (4.7), as was the case for fixed time-interval, because parameter sets that minimize only the penalty term might be favored over terms that minimize discrepancy and penalty term.

This results in a min-max problem where the discrepancy and penalty have to be minimized while the time interval has to be maximized. While being more accurate for the application, this approach also yields higher computational effort due to finding the argmax for each functional evaluation.

4.5.3.2 Numerical Implementation

The necessary minimization and maximization are done numerically. Numerical methods are often used for parameter identification, especially for complex models. Various methods exist and can be used depending on the properties of the mathematical operator representing the model. An overview on existing methods can be found in [Boy14].

The operator F in this work is non-linear and thus iterative methods have to be applied. In each iteration, the Tikhonov functional and the maximum time-interval length has to be evaluated. In addition, the evaluation of the operator F is computationally expensive.

One solution for these problems is to apply a two term strategy to solve the min-max problem. First, the Tikhonov functional with incorporated tolerances is minimized to find a starting point for the min-max problem within the given tolerance. Using this starting point, the min-max problem is solved by maximizing the time interval and minimizing the penalty term to keep regularization. The former discrepancy term is now used as a non-linear boundary condition on the feasible set. This ensures that all requirements of the inverse problem are still fulfilled while optimizing the process parameter further. It also reduces the number of evaluation of the maximum time-interval length and thus reduces computational effort.

For each of the two steps, a different solver can be applied. To solve the initial inverse problem, the REGINN algorithm with a conjugate gradient method is used [Lech09]. This algorithm linearizes the operator and uses the conjugate gradient algorithm to solve the linear problem. A new iteration is then started with the found solution as the point where the operator is linearized.

If the algorithm is able to find a parameter set that produces a simulated surface with the given tolerance to the provided data, then the non-linear optimization for maximizing the end-time as described in Sect. 4.5.3.1 is performed. However, this step cannot be solved by using the REGINN algorithm because the full min-max problem does not fulfill all requirements and a gradient method is used instead. This method is slower but has fewer conditions on the mathematical operator, which allows the application.

Overall, the two step approach reduces the amount of performed operator evaluation by utilizing a fast converging algorithm to solve the non-linear inverse problems without wear on the tool, which then provides a good start estimation for the more complex min-max problem at hand.

4.5.4 Tribologically Active Textured Surfaces

Metal forming processes are generally subject to friction. Friction (e.g., in micro-deep drawing) predetermines process feasibility, work piece quality, or the achievable drawing ratio β_0 [Nie18]. In the micro-regime, friction becomes even more crucial to forming processes because of the size effects associated with a relative increase of adhesive forces [Vol10]. Friction in forming processes is a result to the effectiveness of lubrication, surface chemistry, velocity of relative motion between die and work piece, temperature, the dies' macro geometry, and fundamental friction mechanisms. These friction mechanisms are the adhesion, the plastic deformation [Bhu04], plowing (two- and three-body) [Suh87], and elastic hysteresis [Czi10]. The predominance of each of the mechanisms is directly determined by the contact state of the interfacing surfaces on microscopic level. The contact conditions are a function of the applied forces, the material properties, and the microscopic surface geometry. Friction control in forming processes usually involves lubrication, surface chemistry manipulation, and the adaption of the die's macro and micro-geometry. With regard to the development of dry forming processes, texturing of forming die's surfaces has become a major topic in research.

The following sections deal with the generation of textured surfaces using micro-milling, methodologies for function orientated surface characterization, and approaches for the selection of most suitable surface textures to achieve minimum friction in dry micro-deep drawing processes.

4.5.4.1 Micro-Milling to Generate Textured Surfaces

Micro-milling is subject to plowing because the uncut chip thickness becomes the same order of magnitude as the cutting edge radius r_β, resulting in a larger proportion work piece material to be elastically and plastically deformed rather than to be removed in the form of chips; see Fig. 4.46 [Ara09]. Even though plowing is associated with an increase of cutting tool wear and losses in dimensional accuracy, this particular effect can be utilized to generate a regular micro-textured surface, associated with friction reducing properties [Twa14]. This allows us to apply micro-milling for both the manufacture the forming die's geometry and the tribological active micro-structured surfaces carried out in a single process step.

The prerequisite for the textured surface generation through micro-milling is the application of micro-ball-endmills with diameters in the range of 0.5 to 2.0 mm and a cutting tool alignment normal to the machined surface. The actual design of the regular texture is determined by the properties of the machined material; that is, the material hardness, the cutting strategy (up- or down-milling), and the machining parameters width of cut a_e and feed per tooth f_z [Twa14].

4.5.4.2 Micro-Tribological Investigation

Reliable and significant tribological investigations are the premise for the selection of suitable tribological active surface textures and thus the successful development of, for example, dry micro-deep drawing processes applying micro-textured forming dies. The aims of this investigation are the assessment of short-term and long-term frictional behavior and wear. However, tribological assessment in the micro-regime is challenging due to size effects, such as the dominance of adhesion, and also the application of delicate samples and experimental setups. This led to the development and application of specialized equipment for micro-tribological investigations. With regard to design and complexity of these experiments, the DIN 50322 defines six categories of setups linked to sheet metal forming processes. These categories range from I, representative for the actual forming process in industrial environment, to VI for model like experiments with simple sample geometries and equipment. The accessibility to the surfaces under test and the simplicity of the setup is improved for higher category numbers, although the comparability and transferability of the investigation's results to the real forming process is limited. Precise micro-forming presses have been developed for category I tribological experiments [Sch08a, Beh15]. Equipped with additional linear encoders and piezo-electric force measurement systems, these machines allow for the determination of punch- and blank holder forces with a resolution of 0.01 N and the punch stroke detection with a resolution of 1 μm when micro-deep drawing. Work carried out in the Collaborative Research Center 747 comprised the first deep drawing experiments using rectangular micro-deep drawing dies with textured surfaces made by micro-milling, see Fig. 4.49. The aim of this work was to increase the achievable drawing ratio β_0 by providing favorable tribological conditions [Böh14]. However, a clear increase in forming die performance due to surface texturing was not observed. This was drawn back to inaccuracies in die machining and punch-die-alignment, and also manual positioning of the blanks over the drawing cavity [Böh16].

Furthermore, geometrically scaled down strip drawing tests were carried out in the Collaborative Research Center 747 (tribological experiment category V)

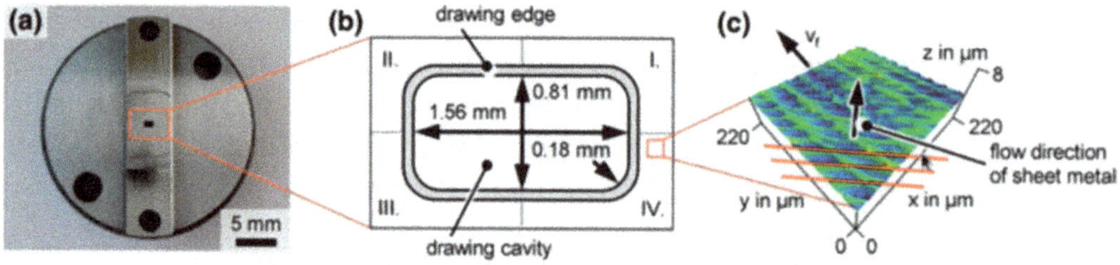

Fig. 4.49 Micro structured micro-deep drawing die **a**, design of drawing cavity **b**, and topography of micro-structured area **c** (in accordance with [Böh16])

[Bri10]. Various textured samples made from hardened tool steel that was machined by micro-milling, which were applied against strips of austenitic stainless steel. The texture design was determined by the tool diameter d and the line pitch a_e. A clear dependence of the confection of friction on the arithmetical mean height Sa of the textured samples was found. Compared to a polished reference sample, the coefficient of friction was reduced by about 20% when applying micro-textured samples with arithmetical mean heights of about Sa = 200 nm [Bri10].

Micro-tribometers, due to their simplicity in setup and the capability to reduce testing time compared to strip drawing experiments, are well suited for the tribological investigation of micro-textured surfaces, especially under dry conditions. A methodology for the reliable tribological investigation using a micro-tribometer in ball-on-plate configuration (tribological experiment category VI) was presented by the researchers of the Collaborative Research Center 747 [Böh18]. The spheres used as tribological counterparts were undergoing run-in procedures in preparation for tribological investigations. Here, equilibrium facet areas on the spheres are formed due to initial wear. The final size of the facet areas is determined by the sphere material and the applied normal force F_N. Once a steady state of the facet area is reached, reliable tribological investigations can be carried out without the impact of crucial changing wear and friction mechanism to the experimental result.

Tribological investigations using a micro-tribometer in ball-on-plate configuration with spheres from aluminum alloy Al99.9 exhibiting equilibrium facet areas were conducted under dry conditions. A polished reference sample and various textured samples made by micro-milling made form hardened tool steel were used in the experiments. The surface plots of four exemplary samples are shown in Fig. 4.50. The experimental setup and determined coefficients for each of the four surfaces are shown in Fig. 4.51.

The results of the micro-tribological investigations displayed in Fig. 4.51b, which clearly indicate that the surface roughness of the samples (i.e., the contact conditions at the sample-sphere-interface on microscopic level) has a distinct influence on the dry frictional properties. For the presented material combination, hardened tool steel and Aluminum Al99.9, the tribological active sample surface should ideally exhibit an arithmetical mean height Sa of about 150 nm to achieve minimum friction. A lower sample roughness leads to an increase in friction, which

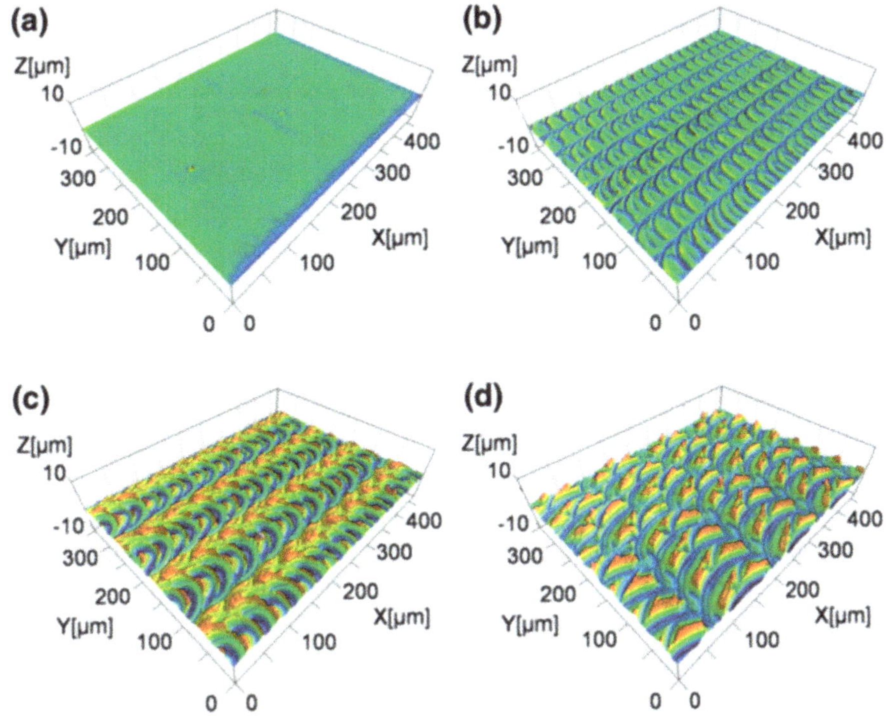

Fig. 4.50 Surface plots of samples applied in tribological testing: **a** polished reference sample R (Sa = 24 nm), **b** sample #1 (Sa = 148 nm), **c** sample #2 (Sa = 470 nm), and **d** sample #3 (Sa = 741 nm) (in accordance with [Böh16])

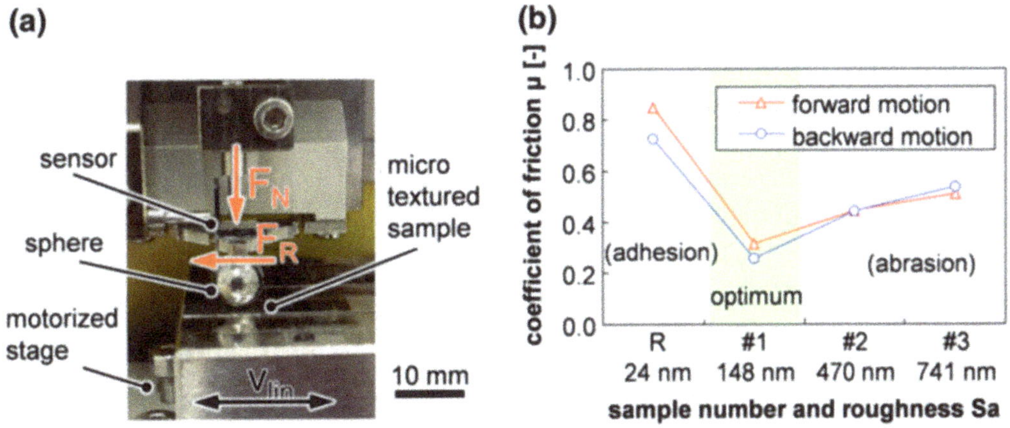

Fig. 4.51 **a** Ball-on-plate setup for frictional testing on a micro-tribometer and **b** average coefficients of friction for forward and backward motion as a function of the sample number and the associated arithmetical mean height Sa (in accordance with [Böh16, Böh18])

can be traced back to an increased impact of adhesion. Roughness greater than 150 nm is expected to promote interlocking of surface's asperities also associated with an increase of friction. However, a deeper understanding of the correlation of the surface topography and dry friction, and their mathematical description inevitably requires function orientated surface characterization ahead of common statistical roughness parameters, including suitable contact modeling.

4.5.4.3 Function Orientated Surface Characterization

The precise description of the interrelationship of the topography of a surface and their functional properties, such as the friction provoked in dry tribological contact, requires for sufficient roughness measurement and analysis techniques. The well-established, two-dimensional profilometry is not sufficient, especially when taking into concern tribological problems of micro-cold forming processes [Gei97, Ver01]. For example, profilometry does not allow to reliably distinguishing a dent from a scratch on a technical surface [Gei97]. Rather, areal measurement and characterization of surfaces is necessary. The first appropriate areal roughness measurement systems such as confocal microscopes or white light interferometers (WLI) were introduced in the ninetieths of the last century. Together with the ISO 25178 standard *Geometrical product specification (GPS)—surface texture: areal* for areal roughness analysis, released in 2012, there is today a powerful toolbox for sufficient and function orientated surface characterization.

The ISO 25178 standard distinguishes between two categories of parameters for surface characterization: areal field parameters and areal feature parameters [Lea13]. Areal field parameters comprise first of all of the statistical surface parameters (or height parameters) such as the arithmetical mean height Sa. Those parameters originate form two-dimensional profilometry and were transferred to areal surface characterization. All areal field parameters have in common, that their calculation is based on the use of every data point measured and filtered over the surface evaluation area. With regard to function orientated surface characterization the function related parameters (a sub-category of areal field parameters) are of greater relevance [Veh15b]. Function related parameters mainly comprise those parameters derived from the Abbott-Firestone curve or material ratio curve such as the material ratio corresponding to a section height Smr(c) or the Sk parameters determined by graphical construction. For surface characterization using areal feature parameters (these are e.g. the peak density Spd or the arithmetical mean of peak curvature Spc) only beforehand determined features on the surfaces are taken into account [Lea13]. These features are categorized by the areal features hills (H) and dales (D), line features such as ridge lines (R), and the punctual features peak (P) and valley (V). The determination of feature parameters from a filtered surface is carried out using watershed segmentation [Blu03]. To avoid over segmentation additional Wolf-pruning is applied [Wol84, Wol93]. Wolf-pruning removes all fractions of the surface of interest below a predetermined threshold height value. The threshold value is defined as percentage of the maximum height difference of the filtered surface Sz and is 5% by default. With regard to micro-contact and friction modeling the determination of the hills (H) and related parameters of surfaces are of greatest relevance, as they directly determine the contact state on microscopic level when interfacing with another surface and thus as

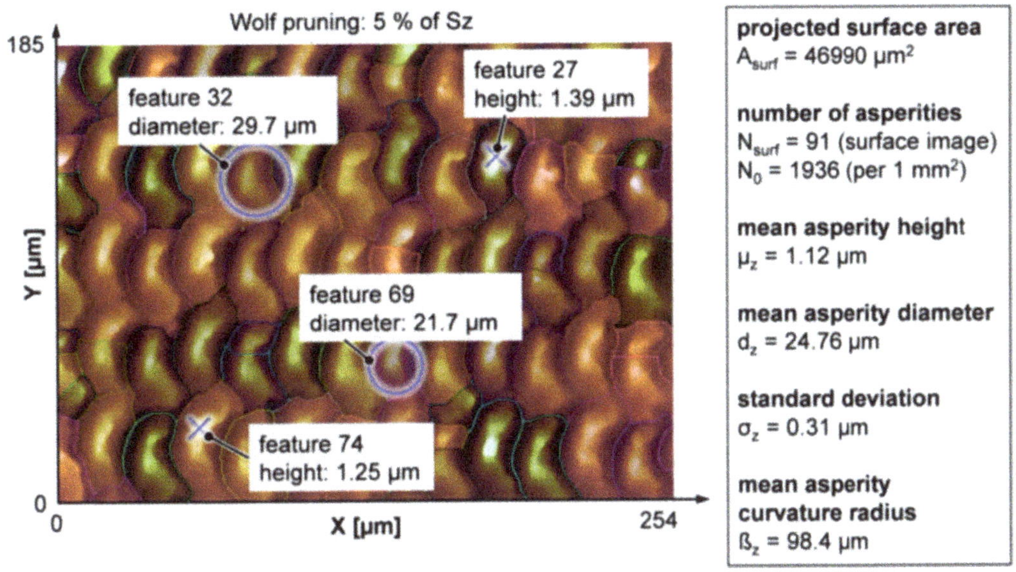

Fig. 4.52 Surface plot of an exemplary micro-textured surface generated by micro-milling and associated areal feature parameters

expected the dry frictional conditions. Figure 4.52 shows the exemplary segmentation of a textured surface generated by micro-milling to derive the feature parameters hill (H) and the associated surface characteristics.

4.5.4.4 Surface Micro-Contact Modeling

The rough nature of a surface provokes the formation of discrete, microscopic contact areas, when interfacing with another surface. Therefore, the fine finish of a technical surface has a considerable impact on its functional characteristics, such as friction and wear. The predetermination of these particular characteristic requires the precise description of the micro-contact state of two interfacing surfaces. This demand led to the development of various statistical micro-contact models throughout the past decades. Even though there are more sophisticated models, the contact model of Greenwood and Williamson remains the most famous and most cited model.

Greenwood and Williamson's model describes the solely elastic contact of a rough surface with an ideally flat surface [Gre66]. The micro-geometry of the rough surface is described by some idealized properties. It is assumed that the topography consists of a large number of spherical hills or asperities, which are all of the same curvature radius β_z (coinciding with the parameter Spc) and that their heights are distributed normally. Together with mean asperity height μ_z and standard deviation of the asperity height σ_z, the height distribution can be expressed by the specific probability density function Eq. (4.11).

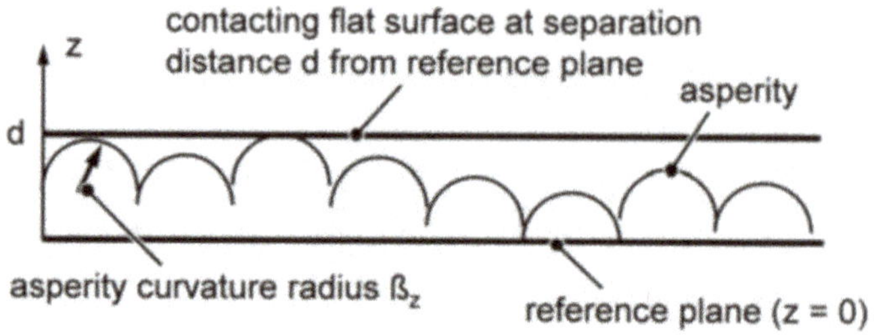

Fig. 4.53 Contact of a rough with an ideally flat surface

$$\Phi(z) = \frac{1}{\sqrt{2\pi\sigma_z^2}} e^{-\frac{(z-\mu_i)^2}{2\sigma_z^2}} \tag{4.11}$$

If the surfaces are pressed together until the ground level of the rough surface and the ideally flat surface are separated by a distance *d*, then every asperity of the rough surface will make contact with the ideally flat surface whose height was originally greater than *d*, see Fig. 4.53.

The probability of any asperity of the height *z* making contact with the flat surface can be calculated by Eq. (4.12).

$$P(z > d) = \int_d^\infty \Phi(z)dz \tag{4.12}$$

If the rough surface comprises of $\mathbf{N}_0$ (coinciding with the parameter *Spd*) asperities in total, then the number of contacts will be:

$$N = \int_d^\infty N_0 \boldsymbol{\Phi}(z)dz. \tag{4.13}$$

The real area of contact can be derived from Eq. (4.14).

$$A = \int_d^\infty N_0 \pi \beta_z (z-d) \boldsymbol{\Phi}(z)dz \tag{4.14}$$

Taking into concern Hertzian theory of non-adhesive elastic contact and the elastic material properties of both the surfaces in contact expressed by the effective elastic modulus E^*, the total load F_N is:

$$F_N = \int_d^\infty N_0 \frac{4}{3} E^* \sqrt{\beta_z} (z-d)^{\frac{3}{2}} \Phi(z)dz \tag{4.15}$$

The applicability of the Greenwood and Williamson model to the micro-contact modeling of actual technical surfaces is limited due to the over-simplified assumptions regarding the micro-geometry of the rough surface. In addition, the implementation of actual technical surfaces into the Greenwood an Williamson model based on well-established parameters from two-dimensional profilometry or from the more recently introduced areal field parameter is not feasible. Serendipitously, the particular surfaces of interest (that is, micro-textured surfaces generated by micro-milling) exhibit a regular arrangement of asperity-like surface features. The assessment of these surfaces using areal feature parameters allows their implementation into the Greenwood and Williamson model, and thus micro-contact modeling for moderate loads and predominantly elastic contact behavior.

Let us consider the four surfaces (#R, #1, #2, and #3) introduced in Sect. 4.5.3.2 for micro-tribological investigation. The parameters total number of asperities N_0, mean of curvature radius β_z, mean asperity height μ_z, and standard deviation of the asperity height σ_z were evaluated for all four surfaces using areal feature parameter analysis according to ISO 25178. The parameters were transferred to the Greenwood and Williamson model and, subsequently, the number of asperities in contact N and the real area of contact A were calculated for normal forces in a range from 1 N to 5 N assuming a projected interface area of $1\,\text{mm}^2$ and a material combination of hardened tool steel and aluminum Al99.5 ($\text{E}^* = 57{,}079\,\text{N/mm}^2$), see Fig. 4.54a and b. Additionally, the average force acting on an asperity F_i was calculated in dependence on the applied normal force and is shown in Fig. 4.54c).

A distinctive correlation of the friction coefficients of the surfaces measured in the micro-tribological experiments and the three evaluated parameters N, A and F_I can be derived from the diagrams in Fig. 4.54. An increase of the asperities in contact generally coincides with a decrease in friction, an increase in the real area of contact A generally coincides with an increase in friction. The same correlation was found for the average force acting on a single asperity F_I and the coefficient of friction μ. From the impressive consistency of the results shown, it can be concluded that the even though simple model of Greenwood and Williamson is

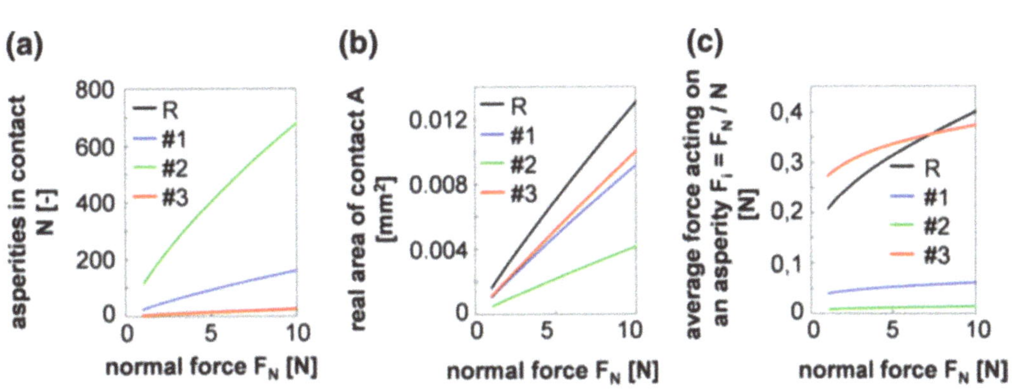

Fig. 4.54 Contact of a rough with an ideally flat surface according to [Gre66]

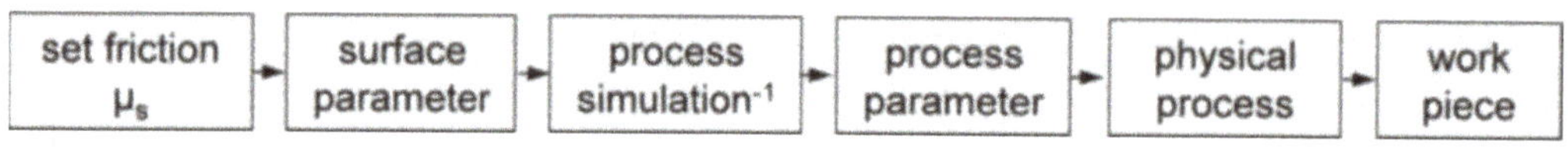

Fig. 4.55 Work flow of inverse modeling for optimized forming die manufacture

well-suited for the micro-contact modeling of textured surfaces generated by micro-milling in tribological contact. This approach builds a sound foundation for the development of scale dependent friction models, for example, based on the work of Bhushan and Nosonovsky helping to determine best suited textured surfaces for friction reduction in dry micro-sheet metal forming [Bhu04]. The application of the friction model of Bhushan and Nosonovsky with regard to micro-sheet metal forming was shown previously by and Shimizu et al. before [Shi15].

4.5.4.5 Inverse Modeling for Optimized Forming Die Manufacture

All of the introduced concepts and models are combined to find ideal process parameters for optimized forming dies. A desired friction μ_S is set and surface parameters derived from set friction. The process model is used as the forward operator in an inverse problem to find a suitable set of process parameters. These process parameters are then used in the real physical process to produce a work piece with the desired friction. The work flow is visualized in Fig. 4.55.

Overall this approach allows us to optimize manufactured forming dies by using process parameters that result in a desired friction.

4.6 Predictive Compensation Measures for the Prevention of Shape Deviations of Micromilled Dental Products

O. Riemer*, P. Maaß, F. Elsner-Dörge, P. Gralla, J. Vehmeyer, M. Willert, A. Meier and I. Zahn

Abstract Micro-milling is widely applied for the manufacture of dental prostheses because only this process guarantees geometrical accuracy and adequate surface topographies compared to novel methods such as selective laser melting. Dental prostheses require hard and tough materials like cobalt-chrome alloys or ceramics due to the demands of the application. When machining, static and dynamic interactions between the milling tool and the workpiece govern the cutting process and may worsen the work result, especially the geometrical accuracy. The deflection of the tool has a significant influence. However, deflections can be compensated through adapted manufacturing strategies by adjusting the process parameters or by optimizing the tool path, therefore increasing the quality of the final product. In the work presented, the deflection of the milling tool is compensated by readjusting tool positioning before the actual process to ensure the nominal tool position and eliminate form deviations of the workpiece. The interaction between tool and workpiece is investigated with help of micro- and macroscopic surface simulations, supported by experimental data and parameter identification. The optimization is performed through mathematical methods from the area of inverse problems, applying regularization with sparsity constraints.

Keywords Milling · Simulation · Optimization

4.6.1 Introduction

Micro-milling as part of micro-machining offers high efficiency and adaptability regarding machinable geometries and materials, while maintaining relatively high material removal rates [Veh15a]. Deploying these advantages, micro-milling is an established process for the manufacture of mold inserts and also dental prostheses (see Fig. 4.56), which require hard and tough materials, such as cobalt-chrome (CoCr) alloys, ceramics or titanium alloys.

Because mating surfaces of dental prostheses are often seated in narrow cavities with steep inclines, milling tools with large length-to-diameter ratios (up to l/d = 18) are necessary to machine these functional surfaces. However, these fine tools are prone to deflection as a direct outcome of the acting process forces. Shape deviations up to several tens of micrometers can be caused by the deflections and can eventually lead to a rejection of the prosthesis. Thus, applicants of micro-milling processes are obliged to enhance their technology if they wish to stay competitive.

Fig. 4.56 Dental bridge from CoCr alloy (Courtesy of Bego Medical GmbH)

A promising and universal method to meet these production challenges is to prepare an adjusted tool path via Computer Aided Manufacturing (CAM), which can help to reduce shape deviations. However, these preparations are difficult to reproduce because they always depend on the individual capabilities of the programmer and do not necessarily consider all factors responsible for shape deviations. To this day, apart from the workpiece geometry given by a CAD model, only the tool geometry and the machining strategy are considered within CAM systems, whereas material properties and the interaction of tool and workpiece leading to variable process forces and, therefore, hard to predict tool deflections are not yet considered in these systems.

Here, aiming for a better comprehension of the process itself and to investigate possible correlations between process forces, tool deflection and shape deviations, defined abstracted geometries are machined, which inhibit most the critical conditions in the manufacture of dental prostheses. The forces and the resulting shape are measured and compared for a variety of machining conditions. The obtained insights were then employed for a predictive optimization method using modeling and simulations elaborated in preliminary works [Pio15, Pio11].

The simulation based approach requires a calibration of the existing model. Subsequently, a validation and the application of the model for the process optimization are carried out. The implementation follows the diagram shown in Fig. 4.57.

First, a sufficiently precise forward model is required. Based on the work in the SFB 747-subproject *Flexible Manufacture of Tribologically Optimized Forming Tools* (Sect. 4.5), the data based model was expanded for the dental material CoCr and appropriate tools. The acquisition of the empirical data base was realized by means of a model geometry, which was developed together with the industrial partner BEGO Bremer Goldschlägerei Wilh. Herbst GmbH & Co. KG (BEGO). The calibration was carried out as an iteration, as can be seen in Fig. 4.57a; the actual optimization follows Fig. 4.57b. The calibrated model is applied for the calculation of the optimal machine control by minimizing

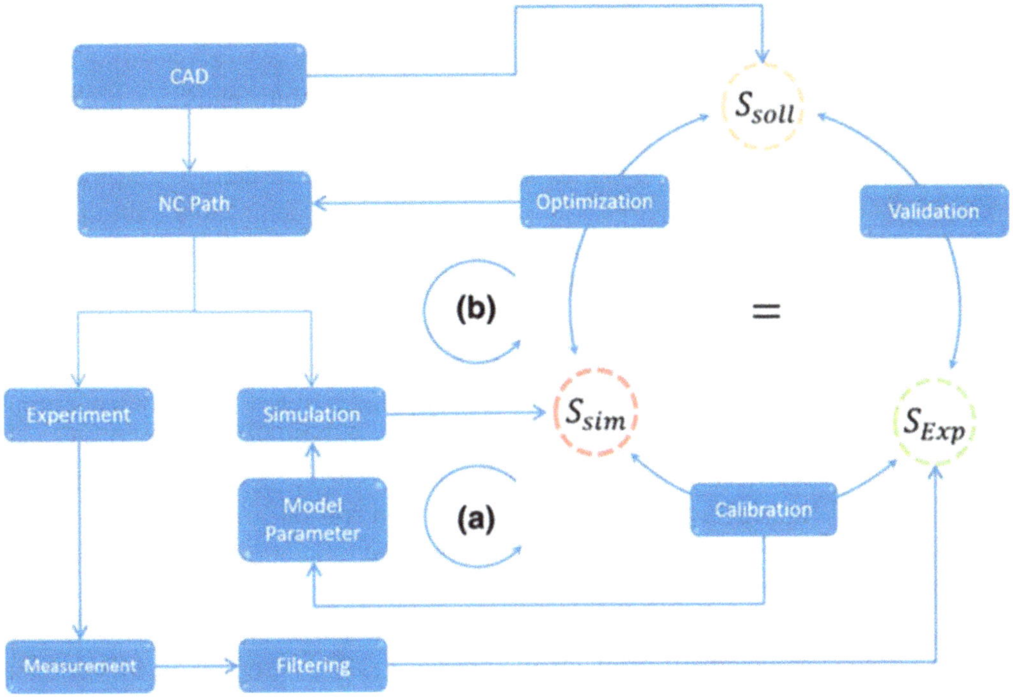

Fig. 4.57 Overview of the working steps and participation

$$\|A(u + d) - y_0\|. \tag{4.16}$$

A comparison between CAD geometry and actual measured geometry serves to evaluate the results and the model validation.

4.6.2 State of the Art and Aim

CAD/CAM-process chains for the manufacture of dental prostheses have been made possible through the expansion of cutting processes towards a miniaturization of tools and through an enhancement of machine tool precision. The main difference between micro-milling and conventional milling is the size of the applied tools and the associated structures that can be manufactured. Generally, tool or structure size range in dimensions smaller than a millimeter [Cam12].

Regarding machine tools, the manufacture of micro-components or workpieces inhibiting micro-features requires adapted machine behavior and process control. Because of the diameters of these small tools, high spindle speeds of several ten thousand up to a few hundred thousand rpm are essential to achieve adequate cutting velocities. Aside from high-frequency main spindles, means for higher position accuracy, higher stiffness and improved thermal stability of the machine system are decisive. Consequently, machine tools for micro-milling rely on technologies such as hydrostatic bearings, linear direct drives, precision scales, active cooling and controlled compensation of thermal expansion [Che10].

In contrast to conventional cutting, which employs a variety of cutting materials, micro-cutting and particularly micro-milling are almost solely performed with cemented carbide tools. For the machining of hard and tough materials, a physical vapor deposited (PVD) coating (e.g. TiAlN) is applied to reduce tool wear [Ara08]. Tool suppliers have to meet the demands for the provision of defined cutting edges because powder metallurgic tools are limited in their achievable cutting edge sharpness. In conjunction with the small chip thickness in micro-milling, effects can occur that may result in a plastic deformation (ploughing) in addition to mere cutting, in case the chip thickness falls below a critical minimum. Consequently, process forces can increase and often be the cause of unpredictable tool damage [Uhl14].

Tool deflection during micro-milling was examined by Dow et al. by machining slots with a linearly varying depth and circular grooves with a radius of 80 mm and a depth of 0.50 mm in hardened tool steel. In the first case, a non-linear relationship between the machining force and the tool's deflection was discovered. In the second case, the sweep angle in combination with the tool feed direction was determined as a significant factor for the tool's deflection. The eligibility of their developed model regarding compensation of the deflection was proven [Dow04].

In a study conducted by Kim et al., similar results were obtained. Here, a half-cylinder was machined by ball end milling with 10–16 mm diameter tools and, therefore, cannot be ranked among micro-milling processes [Kim03].

4.6.3 Applied Materials and Methods

Workpiece Materials

To parameterize the surface model established in Sect. 4.5 [Pio11, Veh12, Veh15a, Veh17b], the model inherent cold work steel was machined with a milling tool inhibiting a large length-to-diameter ratio of l/d = 12, cf. Tables 4.6 and 4.7. Subsequently, the adapted model was transferred to the machining of SLM-generated CoCr dental material.

Machine Tool, Force Measurement and Position Measurement

The machining experiments were for the most part carried out on a DMG Sauer Ultrasonic 20 linear micro-milling center, as depicted in Fig. 4.58. The tool path was programmed with the commercially available CAM software CimatronE. For a comparison of the achievable accuracy (cf. Fig. 4.64), experiments were also conducted on a Röders RXP500 using DentMILL as CAM software.

Forces were measured with a Kistler multicomponent dynamometer with a sampling rate of 10,000 Hz. A method for the synchronized recording of force and position was developed to enable a subsequent correlation of geometry, forces and the resulting shape deviations. Therefore, the force signals of the multicomponent dynamometer were recorded simultaneously with the analogue position signals of the x-, y- and z-axes and plotted synchronously to allow for an allocation of forces to distinguish the geometric features of the machined elements.

Table 4.6 Material data of the applied workpiece materials

Material	Cold-work steel	CoCr
Specification	1.2379 (X 155CrVMo-12-1)	Wirobond® MI+
Alloy elements	C (1.55)	Co (63.8)
	Si (0.4)	Cr (24.8)
	Mn (0.3)	W (5.3)
	Cr (11.8)	Mo (5.1)
	Mo (0.75)	Si (<1)
	V (0.82)	
Hardness	Max. 64 HRC	36.6 HRC
Density	7.7 g/cm^3	8.5 g/cm^3
Tensile strength	>2,000 N/mm^2 (hardened)	970 N/mm^2

Table 4.7 Milling tool specification for machining of hardened steel and CoCr

	Hardened steel	CoCr
Type	Cemented carbide ball end	Cemented carbide ball end
Tool radius r	0.25 mm	0.5 mm
Length l	6	12 mm
Number of flutes z	2	2
Coating	(Al, Ti, Si)N	TiAlN
Spindle speed n	36,000 min^{-1}	38,000 min^{-1}

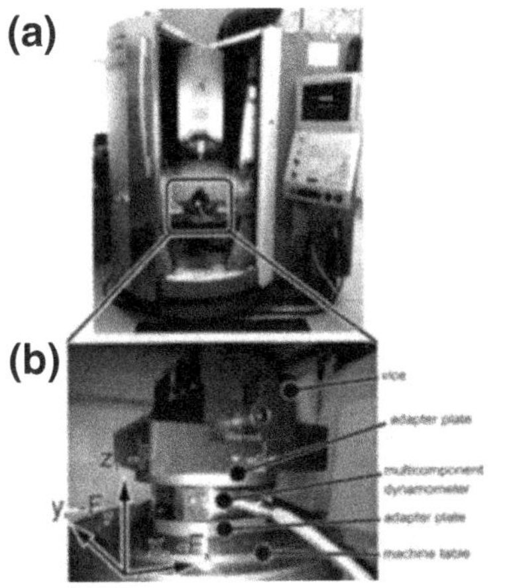

machine tool	DMG Sauer ULTRASONIC 20 linear
structure	5-axes-portal
max. rpm	60,000 min^{-1}
max. feed rate	40,000 mm/min
travel X / Y / Z	200 / 220 / 280 mm

Fig. 4.58 **a** Machine tool; **b** machining setup

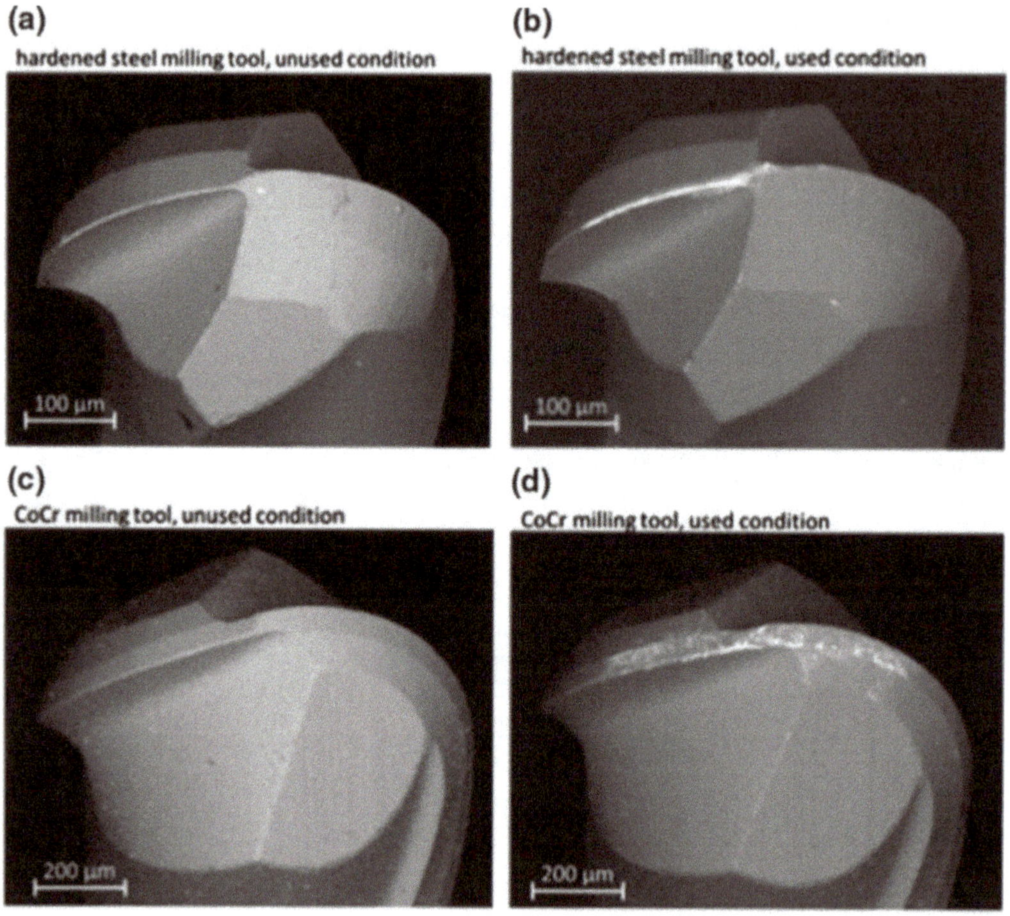

Fig. 4.59 SEM images of applied ball end mills in fresh and used condition

Applied Milling Tools

Cemented carbide two edged ball end mills with a TiAlN coating were applied for the experimental procedures, both for the hardened steel and CoCr workpiece materials. Because the different manufacturers could not provide corresponding tool geometries, a compromise was made by maintaining a matching length-to-diameter ratio of $l/d = 12$ for both materials, whereas the radius r and length l are larger by a factor of 2 for the CoCr milling tools, as given in Table 4.7 and shown in Fig. 4.59.

Tool wear was investigated by scanning electron microscopy after an approximate cutting distance of 3000 mm, as shown in Fig. 4.59. While the milling tools for hardened steel showed slight coating flaws in the delivered condition, which indicated by the brighter areas on the cutting edge, the tools for CoCr were in a faultless condition. Both tool types showed a similar wear characteristic after machining. On the cutting edge, the coating has worn off and the highest tool wear can be located in proximity to the tool center, as expected due to the contact conditions during machining. Additionally, the CoCr showed signs of wear on the rake face and flank but no critical thresholds were exceeded for any of the applied tools.

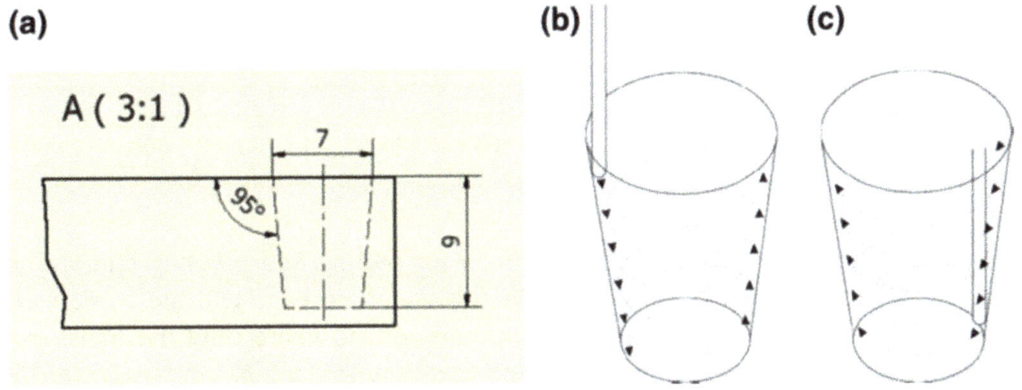

Fig. 4.60 **a** Geometry of cone; different machining strategies: **b** spiral pressing cut, **c** spiral pulling cut

Test Geometries, Machining Parameter/Strategy and Metrology

To make a comparison between the different CAM-programs (CimatronE and DentMILL) and machine tools (DMG Sauer and Röders) regarding the achievable accuracy, application oriented geometries in the form of concave cones dimensionally in the range of connective elements of dental implants were machined, cf. Fig. 4.60. Two different strategies were applied (spiral pressing cut and spiral pulling cut).

The workpiece geometry for the subsequent experimental investigations (parameterization etc.) was a linear arc ridge consisting of two plane faces and three segments of a circle with the same radius merging into each other, as shown in Fig. 4.61. These defined abstracted geometries represent the critical conditions in the manufacture of dental prostheses, while at the same time an easy metrological assessment is possible. These geometries were machined into the cobalt-chrome alloy manufactured by selective laser melting (SLM).

The arc ridges were machined according to different strategies. As described in [Rie16, Rie17b], the tool was moved alongside the ridge (longitudinal) and across

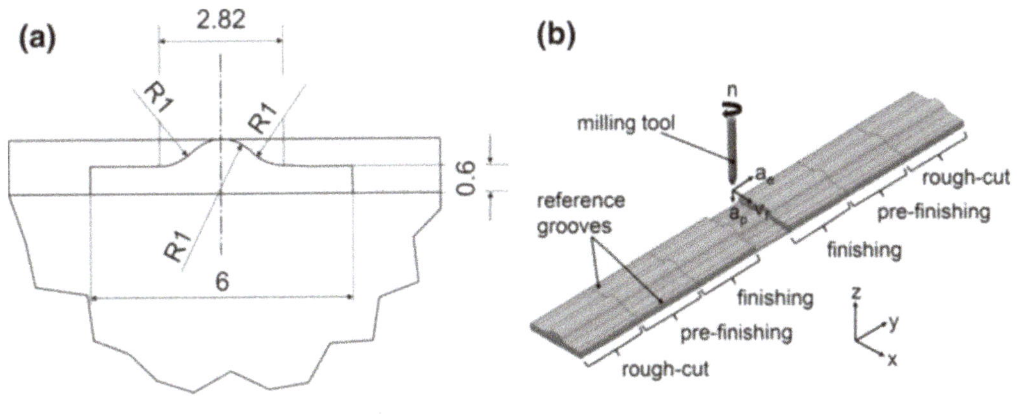

Fig. 4.61 **a** Geometry of the arc ridge for CoCr (geometry for steel was half the dimensions); **b** machining strategy

Table 4.8 Machining parameters

Material	Cold-work steel				CoCr			
a_e (mm)	0.010	0.015	0.020	0.025	0.04	0.08	0.12	–
v_f (mm/min)	500	500	500	500	1,000	1,300	1,600	1,900

(transverse) to reveal influence of deflections for pulling and pressing cut, and to analyze different contact conditions of the tools' cutting edge and the workpiece. Additionally, a series of experiments were carried out where only the transverse strategy with pulling cut was applied.

The finishing cut was the final cut after a rough cut and a pre-finishing cut. Rough cuts and pre-finishing cut were the same for each arc ridge and only the finishing cut was conducted with different parameters. According to Fig. 4.61, the symmetric arc ridge was divided into six segments. The outer segments were kept in the state after the rough cut, the segment next to it in the state after pre-finishing and only the inner segments were finished. This was done to guarantee a comparison of the shape before and after finishing.

For the test series with hardened steel as workpiece material, the spindle speed was kept constant at n = 36,000 min^{-1}. The same applied for the axial infeed and feed rate, which were kept constant at a_p = 4 μm, v_f = 500 mm/min respectively. Radial infeed a_e was varied according to Table 4.8.

For the test series with CoCr as workpiece material, the spindle speed was kept constant at n = 38,000 min^{-1}. The same applied for the axial infeed, which was kept constant at ap = 20 μm. Radial infeed ae and feed rate vf were varied according to Table 4.8. For all materials and parameter variations, up milling was chosen to incite significant force signals. The inclination angle of the tool was kept constant at $\beta = 0°$.

The tool path was programmed with CimatronE CAM software. In total, 24 arc ridges were micro-milled. The resulting shape was measured with a KLA Tencor profilometer P-15 and compared to the target geometry. The resulting shape after machining—that is, the measured surface profile (dotted orange line Fig. 4.62b)—was measured tactilely with the profilometer. Reference geometries in the form of two radial grooves, as depicted in Fig. 4.62, were machined in advance into the plane surfaces of the arc ridge to align the measured surface profile with the target geometry in x- and y-direction by removing the offset and thus guaranteeing an exact fit of target geometry and measured geometry.

The results for the steel experiments regarding force measurements show that surface roughness will be recapitulated briefly ahead because their relevance to the subsequent simulations and optimization is subordinate. Generally, the results show a qualitative resemblance to the results obtained with the CoCr material, which will be discussed in detail in the following chapters. The progression of the vertical force across the arc ridge geometry was almost symmetric, with the highest values registered on the plane surfaces and the peak. The force increased with radial infeed ae from F_z = 1.5 N up to F_z = 3 N. The same applies to the surface roughness,

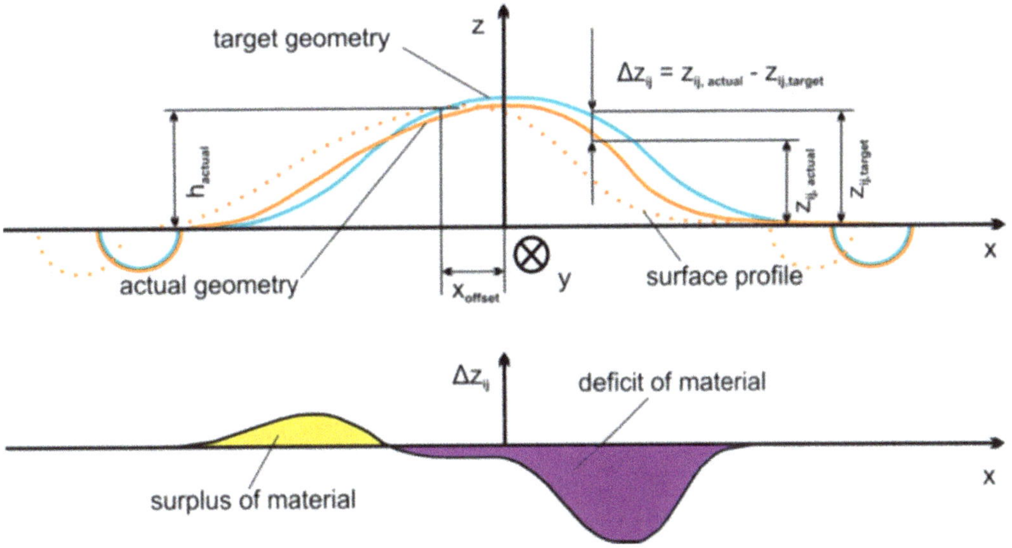

Fig. 4.62 Schematic principle of shape deviation evaluation

where a dependency between radial infeed and roughness value Ra can be confirmed. Ra increases from Ra = 225 nm up to Ra = 400 nm with increasing a_e according to Table 4.8. The obtained insights were applied for the parameterization of the model established in the SFB 747-subproject C2.

Simulation and Modeling

The simulation-based method consists of a geometric surface model and a kinematic process model. Therefore, geometric data processing for workpiece and tool volumes, and for concepts for offsetting path manipulators are applied [Veh17b]. The estimation of cutting forces is based on regression [Den14]. The surface model mainly provides the interface surface between the workpiece volume and the swept volume of the moving tool. The interface can be calculated for both microscopic and macroscopic scale. The differences are not primarily the sampling accuracy of the underlying computation grids but are more closely related to simplifications in modeling; that is, size effects like the minimum chip thickness, accuracy of cutting edge approximation and simplification in kinematics [Veh12]. From a mathematical point of view, both scales differ in the regularity and, therefore, in numerical treatment of the optimization.

The surface model is also applied to determine the geometry of the contact zone between tool and workpiece to estimate the load F. The interaction is modeled by a one mass oscillator, equation

$$m\ddot{\delta} + d\dot{\delta} + k\delta = F \tag{4.17}$$

with tool deflection δ and cutting force F. The physical constants m, d, k depend on the process dynamics and are treated as scalar model constants, identified with help of parameter identification methods.

Mathematical Optimization and Parameter Identification

The obtained mathematical model is used to find process parameters that will reduce the deviation of the surface. For parameter identification in calibration and to reduce deviation in the manufacturing process, mathematical optimization and inverse problem methods are used. The forward model is noted with A and takes the parameter $u \in U$ as input, which is mapped onto the measurement $v \in V$ by the model A. In parameter identification, F and v are known and a suitable u has to be found. This can be done by minimizing

$$\|A(u) - v\|, \tag{4.18}$$

which in the ideal case has a minimum of 0. If only the right-hand side of the equation

$$A(u) = v \tag{4.19}$$

is of interest, then the resulting problem is a optimization problem. If the parameter u itself is of interest, for example for indirect measurements or to find true model parameters, then the problem at hand is an inverse problem [Rie03, Lou01]. The boundaries between these two cases are not strict and some methods are used for both aspects of parameter identification. In the subproject C2 of the SFB 747 inverse methods were used to identify model parameters from experiments through indirect measurements. This still holds true for the model parameter selection in this subproject T4 because it is based on the same model. In addition, the field of inverse problems addresses the issue of noised data and ill-posed problems. Noised data means that instead of ideal data v or measurements, a small perturbed version v^δ is known. An ill-posed problem refers to characteristics of the model A, which will result in large deviations of the reconstructed u if v^δ has small perturbations. In combination, this leads to small error in measurements to large error in the parameter identification. To solve this issue, several different types of regularizations are used to handle the ill-posed nature of the model. In Fig. 4.63, the forward model and inverse problem are visualized.

The micro and macro models used in this work are both ill-posed and need regularization for parameter identification. Tikhonov's method with classic and

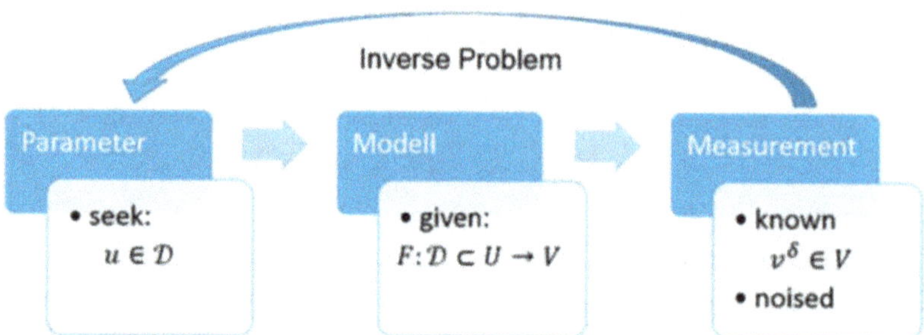

Fig. 4.63 Visualization of inverse problems

sparsity penalty terms are used to handle the ill-posed nature of the model at hand. For Tikhonov Regularization a functional

$$J_{\alpha,\delta}^{p,q}(u) = \frac{1}{p}\left\|A(u) - v^{\delta}\right\|_p^p + \alpha\frac{1}{q}\Re_q(u) \tag{4.20}$$

is minimized. This functional consists of two terms: a discrepancy term that minimizes the distance between simulated and given data and a penalty term on the parameter u that stabilizes the parameter identification. A classical choice for $\Re_q$ is $\Re_q(u) = \|u\|_q^q$ with $q = 2$ or sparsity regularization a certain structure of few active basis elements for the parameter in a defined basis of the space U is assumed. While a l_0-norm on the basis coefficient would best represent this idea, it is not very practical for numerical computation. Instead, a l_1-norm on the basis coefficients of is used as penalty term (for more information on sparsity see [Jin12]).

Since the model A is non-linear, iterative methods are used to find the desired minimizer. Depending on the parameter type different methods are used. For small dimension vector parameters, a sequential quadratic programming (SQP) and gradient descent solver are used [Boy14]. To compensate the deviation of surface, the cutting tool path needs to be adjusted. This leads to a high dimensional parameter and, therefore, too many cost intensive iterations for a SQP or gradient descent method. To reduce the dimension, a greedy approach is used by dividing the tool path into i intervals of equal length. Therefore, the parameter u is a composition of parameters u_k on the defined intervals. Instead to finding u at once each u_k is found in succession of the prior U_{k-1}. This leads to multiple minimization problems which are based on solutions found on the first intervals. This way the dimension of each minimization problem is reduced; however, the resulting composition of all u_k might not be a local minimum on the whole tool path.

Processing the row data is essential for comparison and application in optimization. Because simulation and measurement need to be aligned, a shift on the x, y, z -axis and rotation of the measurement is determined by comparing simulated, v_s, and measured, v_m, data by minimizing

$$\int_{\Omega} \left(v_s(s) - S(x,y,z;\ \mathrm{rot}_z(\varphi; v_m(s)))\right)^2 ds \tag{4.21}$$

over x, y, z, φ, where S is the shifting operator on the x, y, z-axis and $\mathrm{rot}_z(\varphi;\cdot)$ a rotation around the z-axis by φ in radian. This minimization is done once for a measurement. The noised data for parameter identification is then

$$v^{\delta} = S(x,y,z;\ \mathrm{rot}_z(\varphi; v_m)). \tag{4.22}$$

Through this adjustment, the shape of simulated and measured surface can be compared directly. While this is enough in practical applications, it hides symmetrical under and overcuts in the process. For the arc ridge, this means that under and overcut can be partly compensated by a single shift in y -axis. To understand an underlying cause of deviation in surface profile, this compensation is unwanted.

Therefore, two parallel cuts on the left and right of the arc ridge were made and used to find the shift and rotation of the measurement. Consequently, process deviations are being looked at rather than shape deviations. The actual shape deviations will be less or equal to the process deviations.

4.6.4 Results

Comparison of Machined Cones

The machined cones were measured with a coordinate measuring machine, a Ferranti Merlin MK 2 Twin Star. The diameter was measured at three different heights with each 36 measurement points and compared among each other and to the nominal value, cf. Fig. 4.64. The deviation from the ideal circular form was also determined and subjected to comparison.

While in no case the nominal value of the diameter could be attained, it is apparent that the spiral pressing cut provides better results than the spiral pulling cut, both regarding the diameter and regarding the shape deviation. Comparing the facilities at BEGO and LFM, both can approximately deliver the same results, which is crucial for a possible transition of optimization methods between the two partners.

Force/Position Measurements for CoCr

On the basis of priority, the results for the longitudinal strategy will not be discussed here but are described in detail in [Rie16, Rie17a]. While in x- and y-direction signals were for the most part overlaid by noise, the most explicit signals were obtained in z-direction for the vertical force F_z. Here, every milling line can be recognized (s. Fig. 4.65a) and detailed progression of F_z for a single line can be achieved by zooming in, see Fig. 4.65.

A simultaneous plot of the z-position together with F_z over time provides comprehension about the coherence of acting forces depending on the position of the tool, as depicted in Fig. 4.65b. Obviously, the force changed depending on the geometry of the arc ridge, and thus the contact conditions of the tool. To allow for a position specific evaluation of the force, the arc ridge was divided into five sections in the x-direction, compare Fig. 4.65b. Observing the process from the middle of the arc ridge in Sect. 3, the forces showed an almost symmetric progression. A comparison of the average vertical force F_z in all five sections is given in Fig. 4.65. High forces were recorded on the plane surfaces in arc ridge sects. 1 and 5 as well as on the convex arc in Sect. 3 but it decreased in the circle segments (concave arcs in Sects. 2 and 4). The forces for Sect. 4 are slightly higher than those for Sect. 2, which can be traced back to the downward movement of the tool in Sect. 4.

Regarding the influence of the machining parameters, the maximum vertical force reaches up to $F_{z,\,max}$ = 3 N for a radial infeed of a_e = 0.04 mm. For the higher radial infeed of a_e = 0.08 mm the maximum vertical force reaches up to $F_{z,\,max}$ = 9.5 N. Comparing the minimal and maximal values for the different feed rates, two circumstances are remarkable. In general, higher feed rates result in

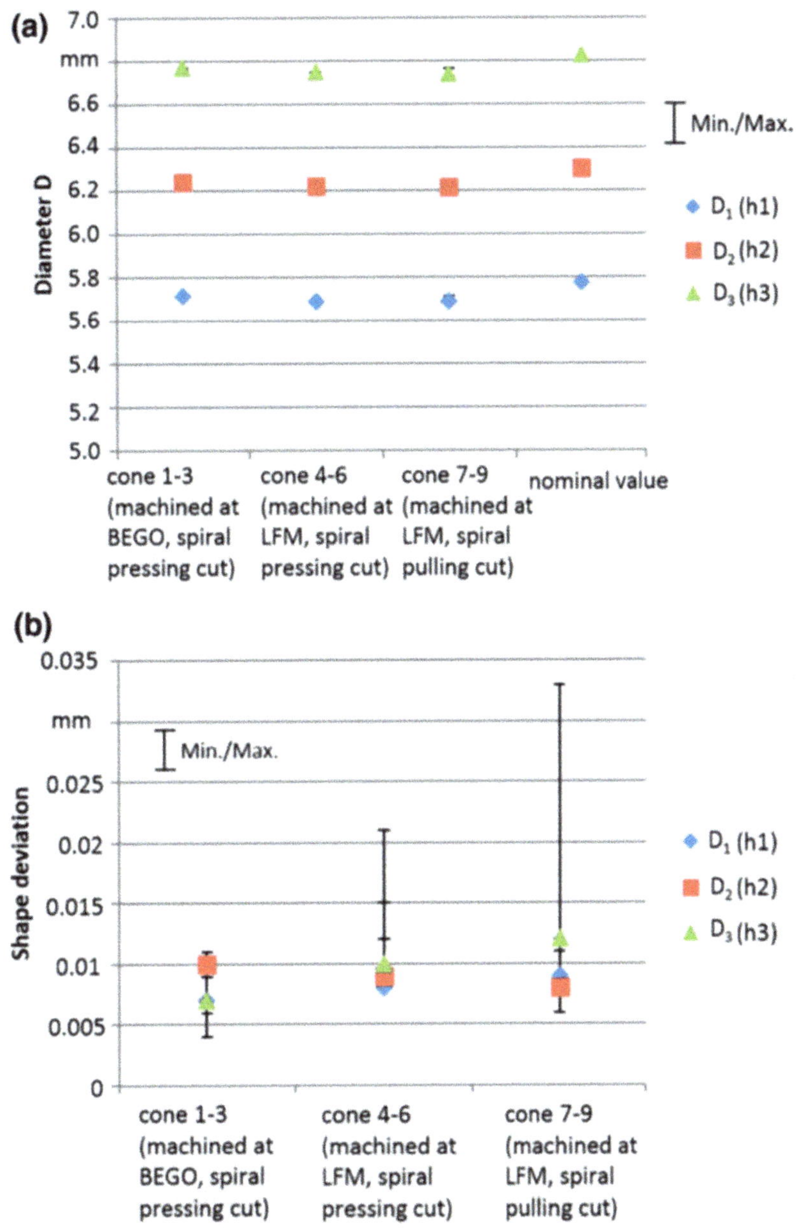

Fig. 4.64 **a** Measured diameter and comparison to target diameter; **b** measured deviation from an ideal circle

higher forces although the force values for feed rates $v_f = 1{,}300$ mm/min and $v_f = 1{,}600$ mm/min are very similar for both radial infeeds ($F_{z,\ min} = -0.5$ N; $F_{z,\ max} = 4.5$ N). For $v_f = 1{,}900$ mm/min. Meanwhile, the force for $a_e = 0.08$ mm is particularly high with a maximum of $F_{z,\ max} = 9.5$ N (Fig. 4.66).

The influence of the radial infeed a_e is more distinct. The difference between $a_e = 0.04$ mm and $a_e = 0.08$ mm is at least $\Delta F_z = 3$ N in-between the plane and convex arc section. In the concave arc sections, the forces are lowest and also differ only marginally for all parameter combinations.

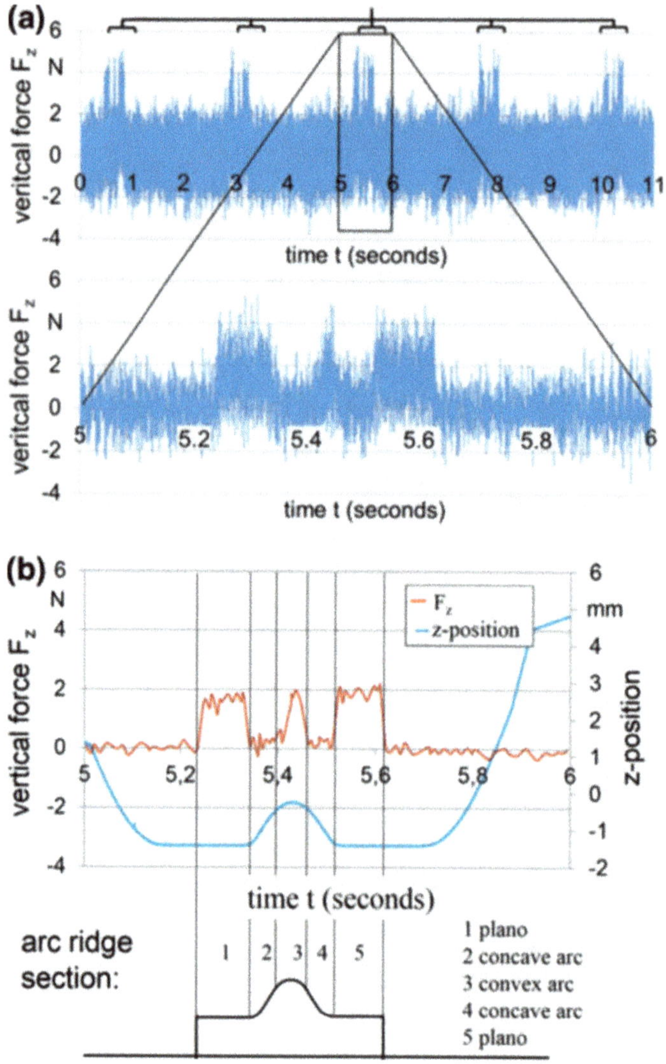

Fig. 4.65 **a** Force measurement plot of five milled lines and zoomed single milling line; **b** correlation of average vertical force F_z (blue) and z-position (red) and sections of actual arc ridge geometry shown below (not true to scale)

For all progression of forces, $F_{z,\ max}$ is registered in Sect. 3 of the arc ridge but not on the peak of the arc ridge. In all cases, the highest force has a displacement to the right (i.e. in the direction of Sect. 4), which is most likely caused by a bending of the milling tool.

Surface Roughness

The surface roughness was investigated by white light interferometry and the parameters Sa and Sq were derived (Gaussian filter ISO 11562; 80 μm cut-off) to evaluate the influence of the machining parameters on the surface quality. The average roughness ranges from Sa = 130 nm to Sa = 390 nm and, respectively, from Sq = 170 nm to Sq = 480 nm, as shown in Fig. 4.67. Except for the lowest feed rate v_f = 1,000 mm/min, the surface roughness increases with a higher radial infeed a_e . The influence of the feed rate itself does not become as apparent but a tendency for higher roughness when applying higher feed rates is visible. Correlating the surface roughness to the forces in the preceding chapter, it becomes obvious that higher forces indicate a higher surface roughness.

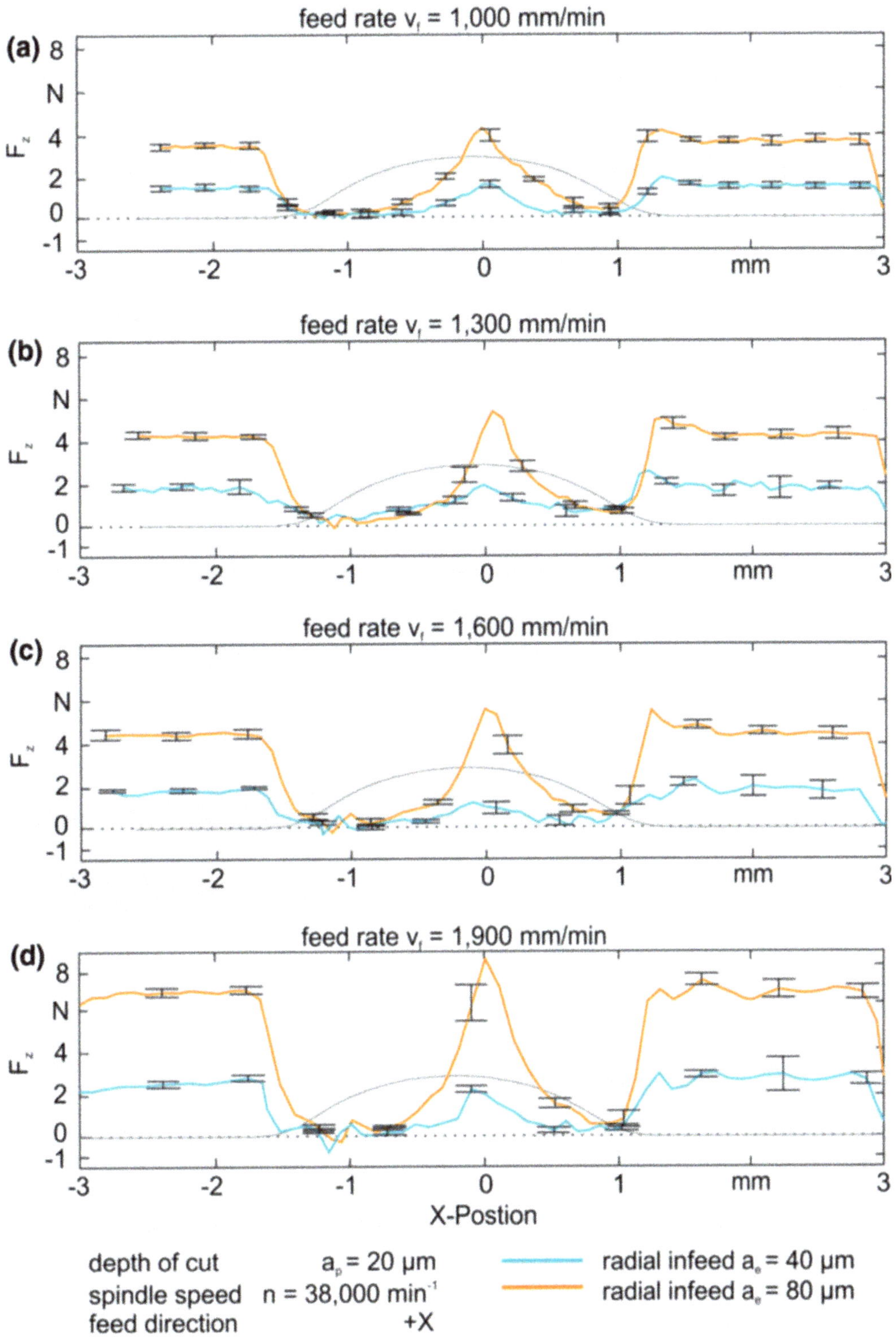

Fig. 4.66 Average vertical force F_z for different sets of v_f and a_e in the five sections of the arc ridge (transverse machining strategy)

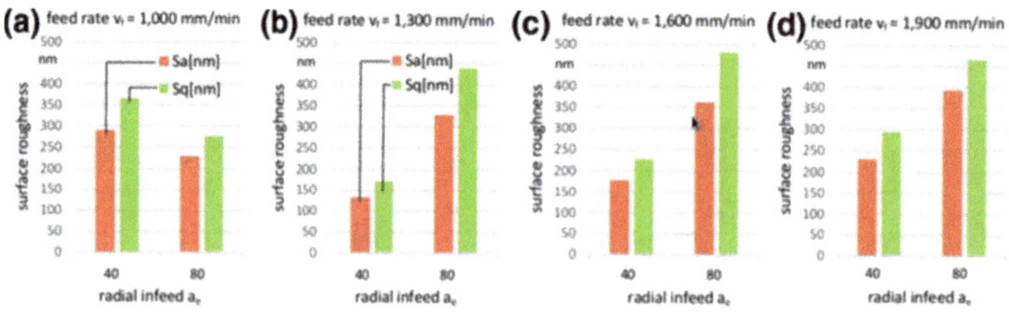

Fig. 4.67 Surface roughness Sa and Sq depending on the feed rate v_f and the radial infeed a_e (transverse machining strategy)

Fig. 4.68 Simulation setup for the transversal machining campaign

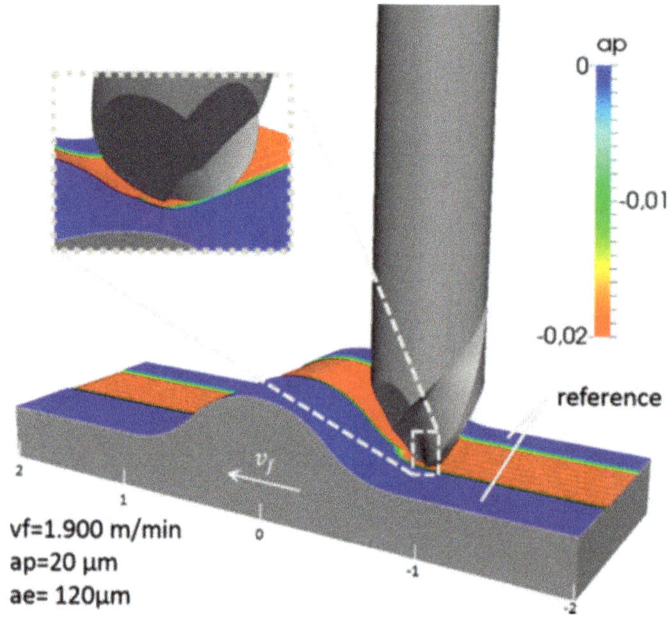

Simulation Results

In the first step, process dynamics were analyzed and, therefore, the transversal tool path was chosen. The simulation setup is illustrated in Fig. 4.68.

In Fig. 4.69a the simulated function of uncut chip thickness is shown and the removed volume along the ridge is visualized in Fig. 4.69b. Due to the cavity in the order of the tool diameter, the plotted function indicates peak forces at the transition from flat to sloped shape, which can be explained by the sudden engagement of the outer parts of the cutting edges. Furthermore, a slight difference between pulling and pressing cut can be stated from Fig. 4.69.

The simulated and the reference surface profiles are shown in Fig. 4.70a, the distance of both is plotted together with measurement results in Fig. 4.70b. The measurement is represented by a tolerance area, which is determined through three experiments. Measurements and simulation show a good agreement with neglected dynamics of the cutting tool; that is, $m = d = 0$ in the dynamic behavior of the

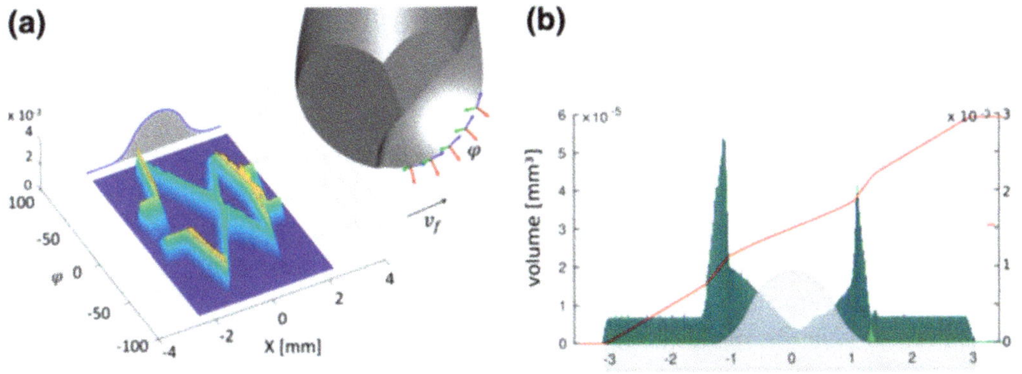

Fig. 4.69 **a** Simulated function of uncut chip thickness; **b** removed material volume along the ridge

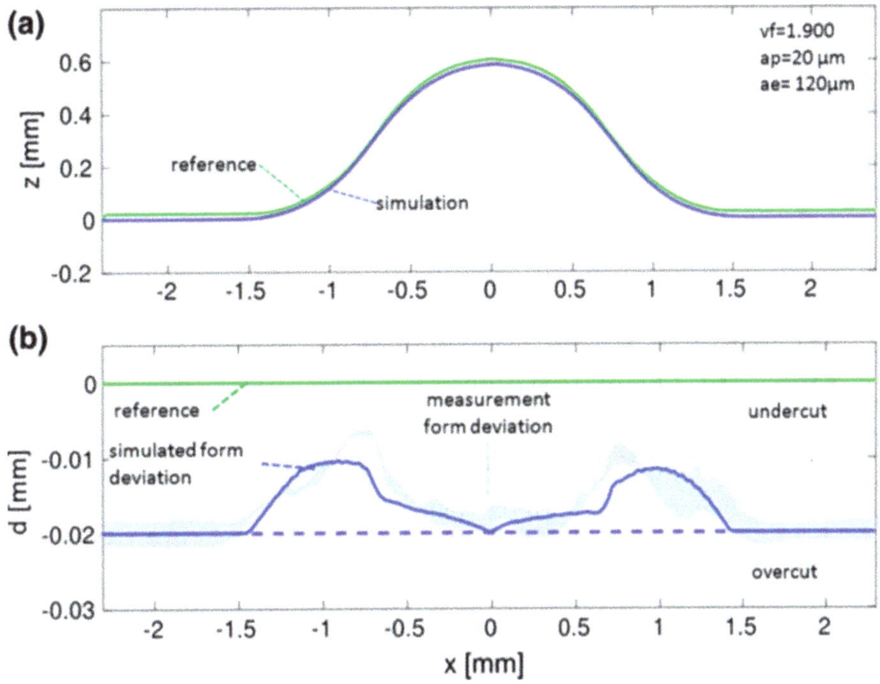

Fig. 4.70 **a** Simulated surface profile and reference; **b** form deviation (distance) of the reference and the final surface (simulation and measurement)

cutting tool. However, this does not mean that there is no deflection of the tool in the model. The dynamic behavior has a delay effect on the deflection with small effects on the surface and does not improve the conformity in the range of the chosen cutting parameters. The longitudinal milling strategy is generally insensitive to oscillations of the cutting tool. The static method is used for further simulations.

4.7 Thermo-Chemical-Mechanical Shaping of Diamond for Micro Forming Dies

Ekkard Brinksmeier, Oltmann Riemer* and Christian Robert

Abstract Diamond reveals outstanding properties such as high hardness, low wear against mechanical load, high chemical inertness against most aggressive chemical liquids and a very low friction coefficient against metals. Therefore, diamond appears as a suitable material for micro forming tools, especially when provided with micro-structured surfaces; thus, the results of forming processes will be improved compared to tools with smooth surfaces. For machining single crystal diamond, the thermo-chemical effect between diamond and selected carbon-affine metals like iron or nickel at high temperatures was exploited. Major findings of this investigation reveal that, besides graphitization, the diamond removal is governed by oxidation, while diffusion was not detected. This thermo-chemical-mechanical shaping process was successfully applied for micro-structuring of single crystal diamond.

Keywords Diamond • Thermo-chemical machining • Material removal • Micro structure

4.7.1 Principles of Diamond Machining by Using Thermo-Chemical Effect

Although diamond is mostly chemically inert, there are materials which react with diamond, causing excessive wear of the diamond. This effect is based on the reaction kinetics of diamond, which is metastable under normal conditions. Figure 4.71a shows the pressure–temperature phase and the transformation diagram for carbon [Che13]. The solid line (Berman–Simon line) which starts at 1.7 GPa and 0 K (point 1) represents the equilibrium phase boundary between stable diamond plus metastable graphite (region A) and stable graphite plus metastable diamond (region E). Region C shows the liquid phase of carbon. The dashed line above triple point 2 marks the temperature/pressure threshold of the very fast and complete solid–solid transformation of graphite to diamond [Bun96]. Above this dashed line, marked as region B, transformation yields cubic-type diamond. Region D represents a full graphite transformation. Furthermore, Fig. 4.71a shows the typical high-pressure–high-temperature regime (region f) for commercial synthesis of single crystal diamond by catalysis [Bun96] and the typical chemical vapor deposition (CVD) regime (region h) for the synthesis of diamond [Ass02]. Region g represents the fast temperature–pressure regime of diamond–graphite transformation.

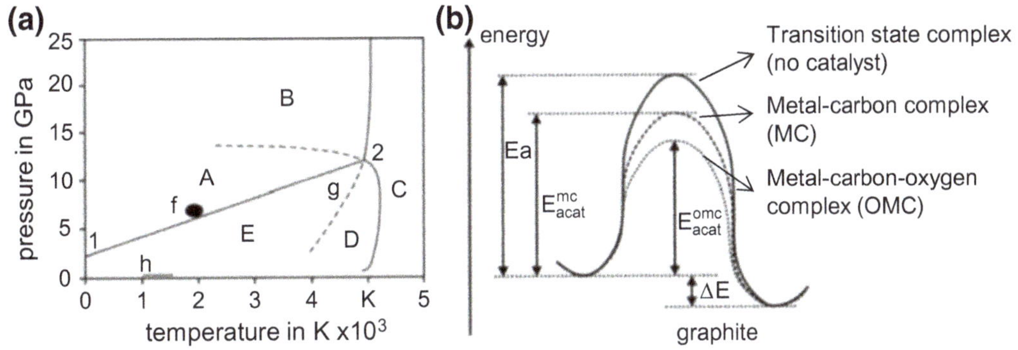

Fig. 4.71 Pressure–temperature phase and transformation diagram of carbon according to [Ass02, Bun96] (**a**); and energy reaction diagram of diamond to graphite [Pau96] (**b**)

At normal conditions (room temperature and normal pressure), diamond is thermodynamically metastable. Nevertheless, a transformation of diamond to graphite under normal conditions is not detectable, due to the required high activation energy (Fig. 4.71b; transition state complex). Even so, there are some catalytic metals like iron, nickel, molybdenum or vanadium that can decrease the required activation energy to graphitize diamond (Fig. 4.71b; metal–carbon complex). Paul et al. [Pau96] proposed that transition metals with unpaired d-shell electrons decrease the required activation energy. Furthermore, they suggested that the number of unpaired d-shell electrons is proportional to the transformation rate obtained during diamond machining experiments. Thus, for example, iron with four unpaired d-shell electrons should produce a higher transformation rate than cerium with one unpaired d-shell electron. Measurable quantities of diamond–graphite transformation start under normal pressure at around 700 °C. Once the diamond comes into contact with the transition metal at elevated temperature, the C-C bonds are stretched and afterwards destroyed. Unbound carbon atoms can subsequently diffuse into the transition metal, graphitize, react with ambient oxygen to form CO or CO_2 (Fig. 4.71b; metal–carbon–oxygen complex), or react with the metal to form carbides [Pau96].

This effect of increased diamond transformation in contact with transition metals has been used by different researchers to polish single [Tat16] and polycrystalline diamond [Wei01] by the technique of dynamic friction polishing. The principal setup of this technique is a fast rotating metal plate made of carbon-affine metals like iron, nickel, manganese or molybdenum, which is pressed against the diamond. To increase the contact temperature between the metal plate and the diamond, the contact force is increased or external heating is used to heat the metal plate or the diamond, respectively. A typical setup for dynamic friction polishing is shown in Fig. 4.72 to polish single crystal diamond [Suz03].

Chen et al. [Che09] utilized the thermo-chemical polishing technique to machine polycrystalline diamond composites with a diameter of 12.7 mm. The specimens contained about 75% diamond particles with a grain size of about 25 μm and an initial surface roughness of Ra = 1.7 μm. The remainder of the specimens consisted

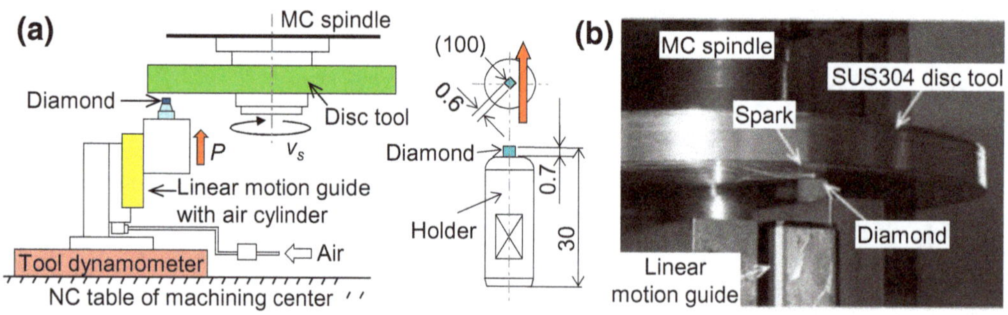

Fig. 4.72 Schematic illustration of dynamic friction polishing method (**a**) and polishing experiment on a machining center (**b**) [Suz03]

of SiC and Si and were pressed against a metal plate made from 1.4301 steel. The contact pressure between the diamond and metal plate was chosen between 2.2 MPa and 5.5 MPa, and the sliding speed of the rotating metal plate between 8 and 31 m/s. To characterize the technique of friction polishing without external heating for polycrystalline diamond composites, Chen et al. were able to develop a material removal map and determined three different working regimes, which are divided into zones with low material removal rates, safe removal and a cracking zone, depending on the contact pressure and sliding speed. By using experimental parameters (e.g. 25 m/s and 2.7 MPa), the diamond removal rate amounts to 0.084 µm/min and the achieved surface roughness of the diamond reaches down to Ra = 50 nm after 18 min polishing time.

To understand the influence of the ambient gas during dynamic friction polishing, Suzuki and co-workers [Suz03] investigated the diamond removal mechanism in various atmospheres. Non-heated polishing experiments were carried out on single crystal diamonds (0.6 mm × 0.6 mm × 5 mm) which were pressed with 100 MPa against a rotating metal plate (sliding speed v_s = 2500 m/min) made from 1.4301 steel. As the ambient gas, air, oxygen, argon and nitrogen were applied. Suzuki et al. could show that polishing in air and under argon atmosphere led to similar diamond removal rates of 900 µm/min and 920 µm/min respectively. Polishing under oxygen atmosphere enhanced the diamond removal rate to 1480 µm/min, whereas a nitrogen atmosphere reduced the diamond removal rate to 820 µm/min. These high diamond removal rates are attributable to the high contact pressure, which generates a high contact temperature. With respect to the experiments performed and the subsequently executed investigations, the authors concluded that there is no effect of oxidation under argon and nitrogen atmospheres. The basic mechanism in this experiment is a rapid diffusion of carbon into the stainless metal plate, while oxygen increases the diamond removal strongly.

To exploit the higher diamond wear affected by ambient oxygen, Wang et al. [Wan06] investigated thermo-chemical polishing assisted by a mixture of oxidizing agents, which are KNO_3, NaOH, KNO_3 + NaOH and KNO_3 + $LiNO_3$. The chemical vapor deposited diamond films exhibited a surface roughness of 8 and 17 µm before polishing. Aluminum was chosen as the material for the metal plate, which has no thermo-chemical effect on the diamond. Therefore, the diamond wear potential of the oxidizing agents can be identified separately. As the second plate material, iron was used to investigate the interaction between iron and the oxidizing agent. For all experiments the metal plate was preheated to 350 °C. Wang et al. could show that all oxidizing agents have a diamond removal effect after 3 h of polishing at a pressure of 0.1 MPa and a rotating speed of 81 rpm. The highest diamond wear was achieved with the mixture of $LiNO_3$ + KNO_3 with 1.2 mg/cm^2/h. When using the iron plate and the $LiNO_3$ + KNO_3 mixture, the diamond wear increased to 1.7 mg/cm^2/h with a final surface roughness of Ra = 0.4 µm. A similar approach to improve the diamond surface was pursued by Kubota et al. [Kub16]. They investigated the potential of various tool materials in an H_2O_2 solution to smooth single crystal diamond. It was shown that, by adding H_2O_2 solution, nickel generates the highest material removal rate of 33.35 nm/h at room temperature, followed by iron with 4.29 nm/h, copper with 1.33 nm/h and 1.4301 steel with 1.28 nm/h. The best surface roughness (Ra) for all tool materials was very similar, in the range between 0.157 nm (copper) and 0.165 nm (steel).

All the studies described confirm that diamond wear with diamond polishing of flat surfaces is possible. The achievable removal rates of diamond and the surface qualities depend on the applied experimental parameters. Due to the experimental setup of the dynamic friction polishing, a micro-structuring of single and polycrystalline diamond is not possible. To overcome this restriction, Imoto et al. [Imo17] proposed a non-moving technique of thermo-chemical patterning, which used graphitization and diffusion of carbon atoms into the tool material. Therefore, they pressed a micro-structured nickel mold (1 mm × 1 mm) for 15 min at approximately 30 MPa and 60 MPa, respectively, against a single crystal diamond at a temperature of 800 °C. To avoid oxidation of the nickel mold, argon gas was pumped into the experimental chamber. The machined diamond surface showed a rough, rugged and tessellated texture [Imo17]. This surface texture was caused by the nickel mold, which creates a rugged surface texture during heating at 800 °C. The transferred depth of the micro-structures into the diamond surface was determined to be 20 nm (± 5 nm) and 83 nm (± 12 nm) at a pressure of 30 MPa and 60 MPa, respectively.

The authors compared mischmetal (composition in Table 4.9) and pure iron (99.5%) in a contact test with single crystal diamond at elevated temperatures [Bri09]. The motivation for this investigation was the higher solid solubility of carbon in mischmetal than in iron, as described by [Jin93]. Thus, if diamond material removal is limited by the solubility of carbon in the metal, higher solubility should produce a higher removal rate.

Table 4.9 Composition of mischmetal alloy as provided by the manufacturer

Element	Ce	La	Fe	Nd	Pr	Mg	Zn
mass%	40.3	23.6	20.0	10.2	3.3	1.97	<0.5

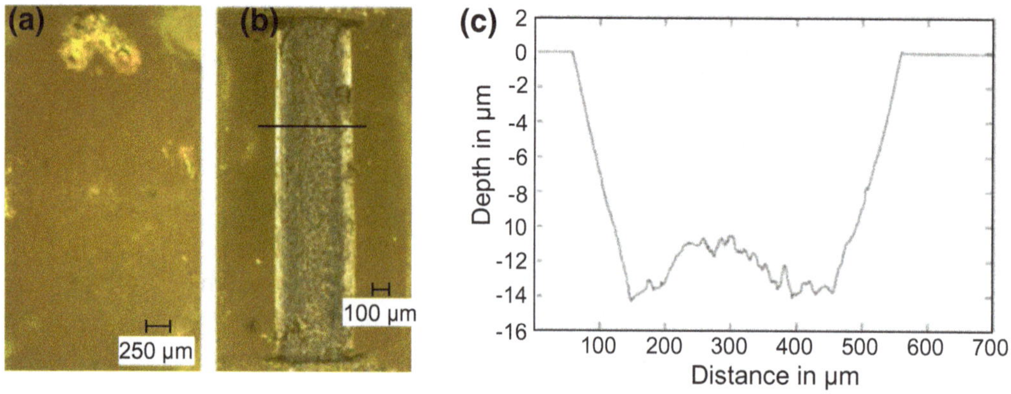

Fig. 4.73 Diamond surface after contact test with mischmetal at 750°C and 20 h (**a**) and iron at 850 °C and 5 h machining (**b**) and profile section of iron machined diamond (**c**) [Bri09]

It was found that mischmetal has no diamond removal effect for temperatures at 600 °C and holding times up to 24 h [Bri09]. Initial removal effects were determined at temperatures of 750 °C and holding times of 16 h or more. No removal of diamond material appeared over the complete contact area, and only certain locations on the diamond surface were etched. Contact tests with iron at 850 °C showed a diamond removal over the entire contact area with an increasing diamond removal as the holding time rose from 3 to 5 h. The authors showed that solid iron is a better catalyst material for defined shaping of single crystal diamond than mischmetal [Bri09] (Fig. 4.73).

4.7.2 Ultrasonic Assisted Friction Polishing

Previously performed investigations showed that machining of diamond with transition metals is possible. The methods used so far, i.e. dynamic friction polishing and non-moving thermo-chemical patterning, appear to be unsuitable processes due to the experimental setup or low diamond wear rates. To achieve higher diamond removal rates to shape complex diamond surfaces in a short time, the authors developed the technique of ultrasonic assisted friction polishing [Rob17a]. Therefore, a 20 kHz, 700 W ultrasonic generator was used to oscillate the metallic friction tool; a piezo ceramic converts the electrical voltage into a mechanical longitudinal vibration and the booster amplifies the amplitude. The friction tool was clamped in a non-amplifying sonotrode with an amplitude of about 8 µm in the z-direction. The machining setup is depicted in Fig. 4.74, which shows a schematic illustration (a) and the realized setup (b); for the machining, a single crystal diamond was fixed in a holder and pressed against the friction tool. The diamond holder was fixed on a dynamometer to control the force acting between the diamond and the friction tool. To prevent bending of the oscillating friction tool, a second diamond was fixed on the opposite side and pressed against the friction tool as a

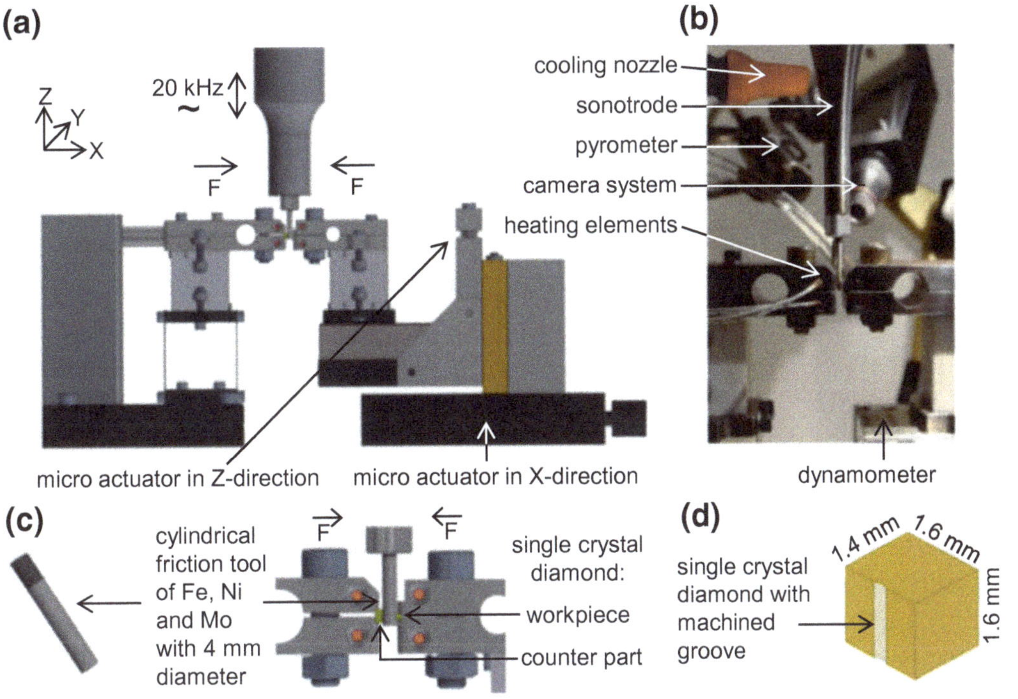

Fig. 4.74 Experimental setup of ultrasonic assisted friction polishing with pure metals: overall machining setup (**a**); realized setup (**b**); workpiece and tool engagement (**c**); machined single crystal diamond (**d**)

counterpart. In order to realize the required reaction temperature, heating elements were integrated in the holder close to the diamonds. The temperature of the machined diamond was determined by an optical pyrometer with a wavelength of 1.6 µm. For this wavelength, single crystal diamond is almost fully transmissive and thus the temperature in the contact zone between the diamond and metal is measurable. The experiments were performed on synthetic single crystal diamonds of type Ib with dimensions of $1.6 \times 1.6 \times 1.4\ mm^3$ in the preferred (100) direction. As described above, the metal friction tool is used as the catalyst and the unbinding carbon from the diamond lattice can diffuse into the metal tool with the possibility of forming carbides, reacting with ambient oxygen to form CO and CO_2 or to transform into a graphite layer.

4.7.2.1 Diamond Removal by Ultrasonic Assisted Friction Polishing Using Pure Metals

The objective of the first part of this work was to determine the capability of the three different pure metals, iron (min. 99% Fe), nickel (min. 99% Ni) and molybdenum (min. 99% Mo) as catalyst materials, to increase the diamond removal rate. With respect to the formerly discussed theory from Paul et al. [Pau96], molybdenum has five, iron four and nickel two unpaired d-shell electrons. Experimental investigations were performed with a friction tool with 4 mm diameter which was pressed against the single crystal diamond during machining.

After machining, grooves with different removal depths could be detected. A schematic illustration is shown in Fig. 4.74.

Initially, the authors studied the influence of the different removal mechanisms: oxidation, diffusion and carbide forming, graphitization and possible mechanical abrasive diamond removal by the ultrasonic assisted friction polishing. To separate the effective diamond removal mechanism, a concept of varying experimental setups was developed as depicted in Fig. 4.75. By this concept, a separation of the effective diamond removal mechanisms and, therefore, a determination of the diamond removal caused by the mechanisms becomes possible. The basis of this concept is the ultrasonic assisted friction polishing process, which is carried out under atmosphere and in a high vacuum chamber. During polishing in atmosphere, all the diamond removal mechanisms (oxidation, diffusion and carbide forming, graphitization as well as mechanical diamond removal) are acting at the same time. When polishing in high vacuum, oxidation of cracked carbon atoms is prevented. To avoid a mechanical abrasive diamond removal, pure contact tests with non-moving contact parameters at elevated temperatures were performed.

The friction polishing experiments in atmosphere were conducted in a contact temperature range between 20 and 450 °C and in high vacuum between 20 and 300 °C. This means that the measured contact temperature during machining was kept constant. Due to the ultrasonically assisted friction tool, the contact temperature may increase lightning-fast for infinitely small areas; these localized temperatures are not detectable. Therefore a median temperature in the contact zone was measured. For each start of machining, the diamond was heated up externally

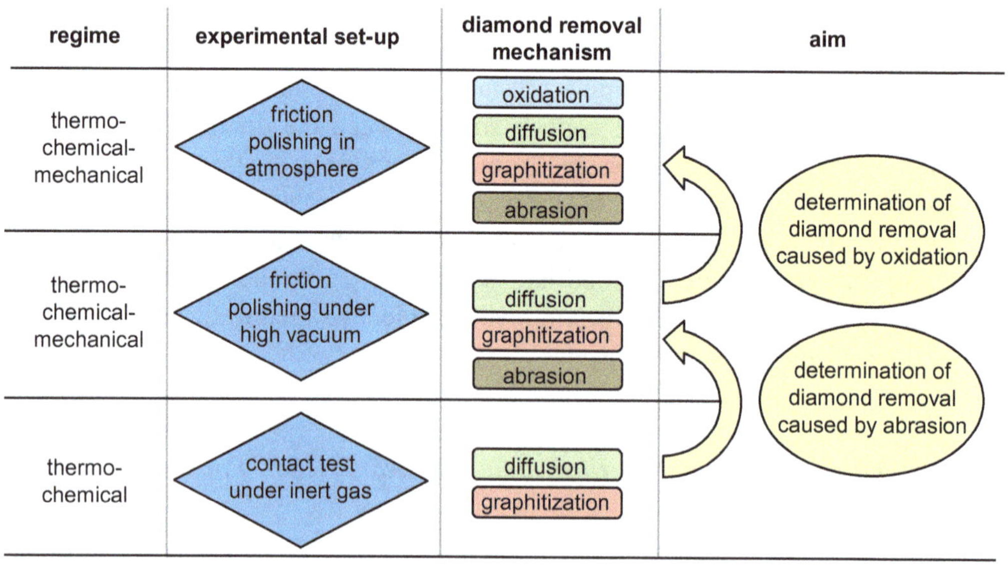

Fig. 4.75 Illustration of the experimental procedure for separating the diamond removal mechanism in machining of diamond

to the desired temperature and subsequently machined between 10 and 60 min with the moving friction tool. The contact load for all diamond removal tests with pure metals was kept at 5 N.

In addition, contact experiments were carried out in nitrogen atmosphere to avoid oxidation of carbon atoms in the heated chamber. During the contact tests, the diamond was fixed in a holder and a metal rod with 4 mm diameter was pressed against the diamond with a dead weight of 0.5 kg. Experiments were conducted in a temperature range between 600 and 850 °C with holding times between 0 and 60 min. Holding times of 0 min were used to determine the diamond removal which was achieved by solely heating and cooling the test sample. Diamond removal volumes without holding time were subtracted from each measurement with higher holding times and the same contact temperature.

After finishing all the machining experiments, the diamond was cleaned by etching with aqua regia at 100 °C for 60 min to dissolve residues from the diamond surface. Subsequently, a white light interferometer was used to measure the machined diamond surface; subsequently, the removed diamond volume at a length of 1 mm was evaluated by using SPIP software. The removed diamond volumes were plotted in 3-dimensional diagrams including a fit plane.

4.7.2.2 Experimental Results

Friction polishing experiments in ambient atmosphere with pure iron showed that, for all parameters, diamond removal was observed (Fig. 4.76). The removed diamond volume was measured between $6.18 \times 10^0\ \mu m^3$ for 20 °C, 10 min, and $7.04 \times 10^6\ \mu m^3$ for 450 °C, 50 min. These values lead to a median cavity depth of 0.25 μm and 19.76 μm, respectively. The results confirm the strong influence of the contact temperature. With rising temperature, the removed diamond volumes increased. Experiments without preheating showed the lowest diamond removal and a rough surface (condition A). With increasing temperature, the surface roughness decreased to a minimum (condition B) and rose with further increase of the contact temperatures (condition C). These different roughness values can be attributed to the varying regimes and varying diamond removal mechanisms, as well as the changing material properties of the friction tool condition for higher temperatures. Nevertheless, different mechanisms are acting in parallel. For experiments at lower contact temperature, the machined diamond shows a rutted surface with grooves in the direction of the vibration. In contrast to this, diamond surfaces machined at the contact temperature of 300 °C show a smooth surface with almost no surface fissuring. These results imply that at lower contact temperatures mechanically induced diamond wear takes place simultaneously with the thermo-chemical mechanism. The increasing surface roughness at elevated contact temperatures is due to the fact that the friction tool loses its mechanical strength. As a result, surface disruption occurred and the diamond surface mirrored the deteriorated surface of the tool. Further significant data points for iron as the tool material are summarized in Table 4.10 and will be discussed in the following.

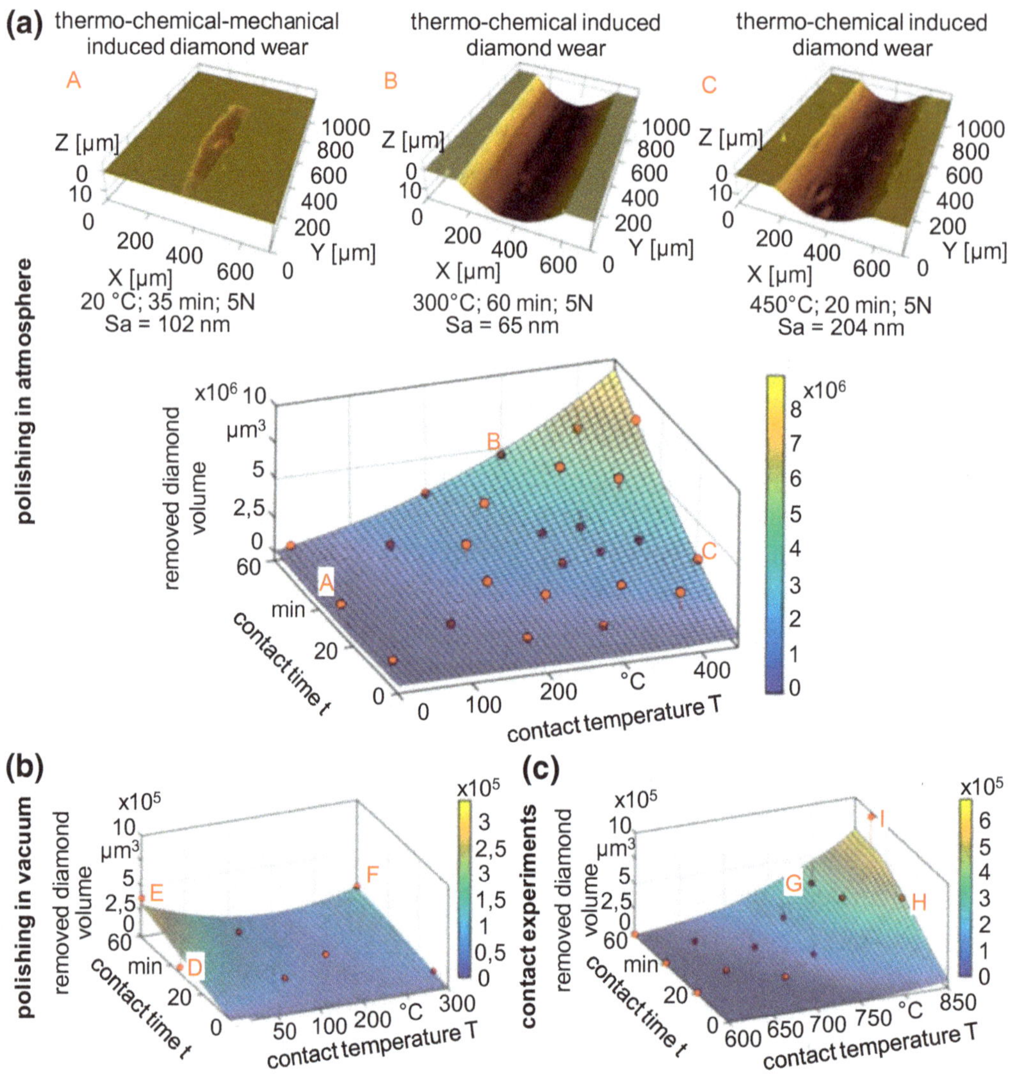

Fig. 4.76 Machined diamond surfaces after polishing in atmosphere and diamond removal diagram for polishing in atmosphere (**a**), polishing in high vacuum (**b**) as well as for contact test with pure iron (**c**)

Table 4.10 Significant data points with iron as tool material for different experimental setups and machining conditions

	Polishing in atmosphere			Polishing in high vacuum			Contact test		
Machining condition	A	B	C	D	E	F	G	H	I
Contact temperature [°C]	20	300	450	20	20	300	800	850	850
Contact time [min]	35	60	50	35	60	60	60	30	60
Surface roughness [nm]	102	65	204	82	196	68	1751	621	1004
Diamond removal x10^6 [µm³]	2.72	4.26	7.04	0.05	0.04	0.14	0.22	0.43	0.93

Table 4.11 Diamond removal rates for polishing in atmosphere with iron as tool material

Contact temperature [°C]	200	300	400
Contact time [min]	10/30/40/60	10/30/40/60	10/30/40/60
Diamond removal rate [$\mu m^3/min/\mu m^2$]	141/127/127/103	186/156/134/151	456/206/251/179

To analyze the effect of possible carbon diffusion into the tool material, the diamond removal rates were studied for experiments with contact temperatures of 200, 300 and 400 °C for machining times of 10, 30, 40 and 60 min (Table 4.11). To achieve comparable results for all contact temperatures, the diamond removal rates were related to 1 μm^2 of the machined diamond area. The experiments showed that the diamond removal rate after 10 min machining time is the highest for all contact temperatures compared to longer machining times.

With increasing machining time, the removal rate decreased slightly to similar diamond removal rates (Table 4.11). The slightly higher diamond removal rates for short machining times can be attributed to the cylindrical shape of the friction tool. At the beginning of the machining, the cylindrical friction tool is pressed against the diamond in line contact, which results in a higher contact load. After a short time, the tool is slightly flattened and the contact load decreases compared to the contact load with longer machining times. For experiments with a contact temperature of 400 °C, the diamond removal rates showed a much higher variability. Longer machining times showed no significant influence of the diamond removal rate, which is an indication that diffusion of carbon atoms into the tool material is negligible. For further proof, a metallurgical element analysis was carried out to identify the raised carbon content in the subsurface of the friction tool. For the analysis, a friction tool which was used at 300 °C and 60 min was chosen. The tool was cut and 10,000 measurements in an area of 0.5 mm × 0.1 mm from the contact zone into the tool material were taken as the depth profile. The investigation found no carbon atoms in the tool material, which underlines that diffusion is insignificant.

To investigate the influence of oxygen, polishing experiments were carried out in a high vacuum condition with 1×10^{-6} Pa and contact temperatures between 20 and 300 °C. Higher contact temperatures led to breakouts on the diamond surface, starting at the edge. In contrast to the experiments in atmosphere, the highest diamond removal was identified for machining conditions D and E without preheating. The removed diamond volume was two times higher than the diamond removal in atmosphere. Preheating of the diamond to contact temperatures reduced the diamond removal. With a further rise to 300 °C, the diamond removal increased again and the formation of graphite on the diamond surface was detected by Raman analysis, which is shown in Fig. 4.77. For lower contact temperature, the formation of graphite could not be detected.

In addition to polishing experiments, contact tests in a nitrogen atmosphere were conducted to analyze the possibility of diamond removal by graphitization and diffusion. The experiments were conducted at contact temperatures between 600 and 850 °C. Experiments with lower contact temperatures showed no (600 °C) or a

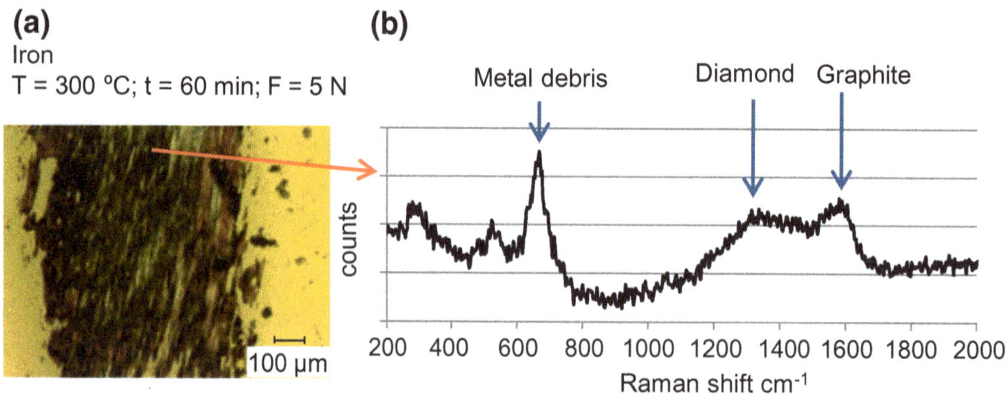

Fig. 4.77 Polished diamond surface in high vacuum (**a**) and Raman analysis (**b**)

very low (650°C) diamond removal. With increasing contact temperature up to 800 °C, the diamond removal increased steadily. Contact tests at a temperature of 800 ° C (condition G) showed a lower diamond removal compared to vacuum experiments with a highest contact temperature of 300 °C (condition F). When increasing the contact temperature up to 850 °C, the diamond removal increased significantly. These results reveal that the formation of CO and CO_2 is the significant mechanism for ultrasonic assisted polishing of single crystal diamond in contact with pure iron. Experiments in atmosphere show up to 30 times higher diamond removal rates compared to experiments in high vacuum. With increased contact temperatures, the formation of graphite has a more relevant role under the elimination of oxygen. Lower contact temperatures in ambient atmosphere and in high vacuum lead to a strong abrasive diamond removal (Table 4.12).

In further experiments, nickel was used as the tool material for friction polishing at ambient atmosphere (Fig. 4.78). The removed diamond volume was measured between 3.63×10^4 µm^3 for 20 °C, 10 min and 3.25×10^6 µm^3 for 300 °C, 60 min. These parameters led to a median cavity depth of 0.75 µm and 8.25 µm, respectively.

In contrast to experiments without preheating and with iron as the tool material, nickel showed an almost nine times higher removed diamond volume (condition A). With rise of the preheating temperature and time, the removed diamond volume increases to the maximum at condition B and decreases slightly with a further rise of

Table 4.12 Significant data points with nickel as tool material for different experimental setups and machining conditions

	Polishing in atmosphere			Polishing in high vacuum			Contact test		
Machining condition	A	B	C	D	E	F	G	H	I
Contact temperature [°C]	20	300	450	20	200	300	750	850	850
Contact time [min]	35	60	50	35	60	60	50	30	60
Surface roughness [nm]	505	109	2344	127	106	/	153	240	239
Diamond removal x10^6 [µm³]	0.23	3.25	2.98	0.18	0.10	0.01	0.01	0.06	0.07

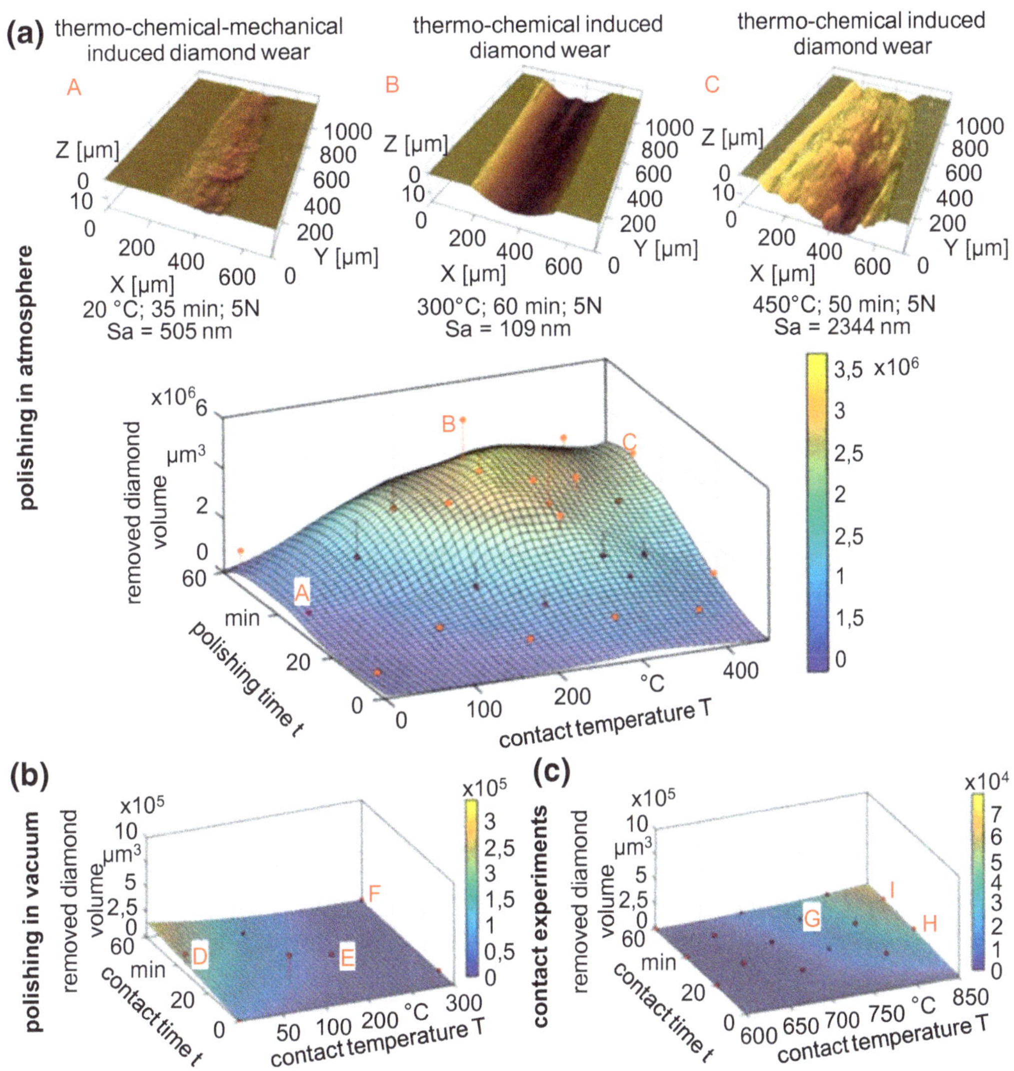

Fig. 4.78 Diamond removal for polishing in atmosphere (**a**), polishing in high vacuum (**b**) and for contact test (**c**) with pure nickel as well as machined diamond surfaces after polishing in atmosphere

preheating (condition C). The reduction of the ability to remove diamond is probably caused by the formation of nickel oxides. Experiments without preheating showed a rough surface and declined strongly with increasing temperature to a minimum at condition B. A strong rutted diamond surface was identified for condition C. Compared to the experiments with iron, nickel led to a higher diamond surface roughness after machining. For all the investigated contact temperatures, the diamond removal rates identified show a similar curved shape. For shorter machining times (10 min), the diamond removal rates showed high values. With increase of the machining time to 60 min the diamond removal rates decrease significantly for all the contact temperatures used. This means that nickel quickly loses the capability to remove diamond (Table 4.13).

Table 4.13 Diamond removal rates at polishing in atmosphere with nickel as tool material

Contact temperature [°C]	200	300	400
Contact time [min]	10/30/40/60	10/30/40/60	10/30/40/60
Diamond removal rate [$\mu m^3/min/\mu m^2$]	227/106/124/72	285/244/201/127	212/113/147/120

In addition to friction polishing experiments at ambient atmosphere, the ability of nickel to remove diamond in the absence of oxygen in a high vacuum was investigated in a temperature range between 20 and 300 °C. Similar to experiments with iron, the removed diamond volume decreased substantially with increasing preheating temperature (Fig. 4.78b). The contact test for experiments with a contact temperature of 600 °C revealed a negligible removed diamond volume (Fig. 4.78c). With increased contact temperature, the removed volume increased slightly to 0.0125×10^6 μm^3 for condition G, which is eight times lower than the equivalent experiments with iron (0.0926×10^6 μm^3). A rise of contact temperature up to 850 °C (conditions H and I) increased the removed diamond volume further. Overall, the diamond removal for nickel is about one order of magnitude lower than for iron (Table 4.13).

Friction polishing experiments at ambient atmosphere with molybdenum revealed the lowest capability to machine diamond (Fig. 4.79 and Table 4.15). Experiments without preheating exhibited no diamond removal at all. With increasing preheating temperature, the removed diamond volume increased slightly to a maximum at condition C, which is more than 20 times lower than the results obtained from the use of iron. The maximum diamond removal rates were identified for the shortest machining time of 10 min. With higher preheating temperatures, the diamond removal rates decreased for all the contact temperatures used. Furthermore, the removed diamond volumes exhibited a strong dependence on the contact temperature and machining time and showed for all parameters strong grooves aligned in the direction of vibration. The surface roughness was measured with Sa = 104 nm (condition A) and increased with rising preheating temperature to 551 nm (condition C). Higher contact temperatures did not lead to an increase of the surface roughness (Table 4.14).

Table 4.14 Significant data points with molybdenum as tool material for different experimental setups and machining conditions

	Polishing in atmosphere			Polishing in high vacuum		
Machining condition	A	B	C	D	E	F
Contact temperature [°C]	125	200	450	20	200	300
Contact time [min]	50	30	50	35	30	60
Surface roughness [nm]	104	305	551	190	202	351
Diamond removal $\times 10^6$ [μm^3]	0.02	0.04	0.31	0.19	0.18	0.33

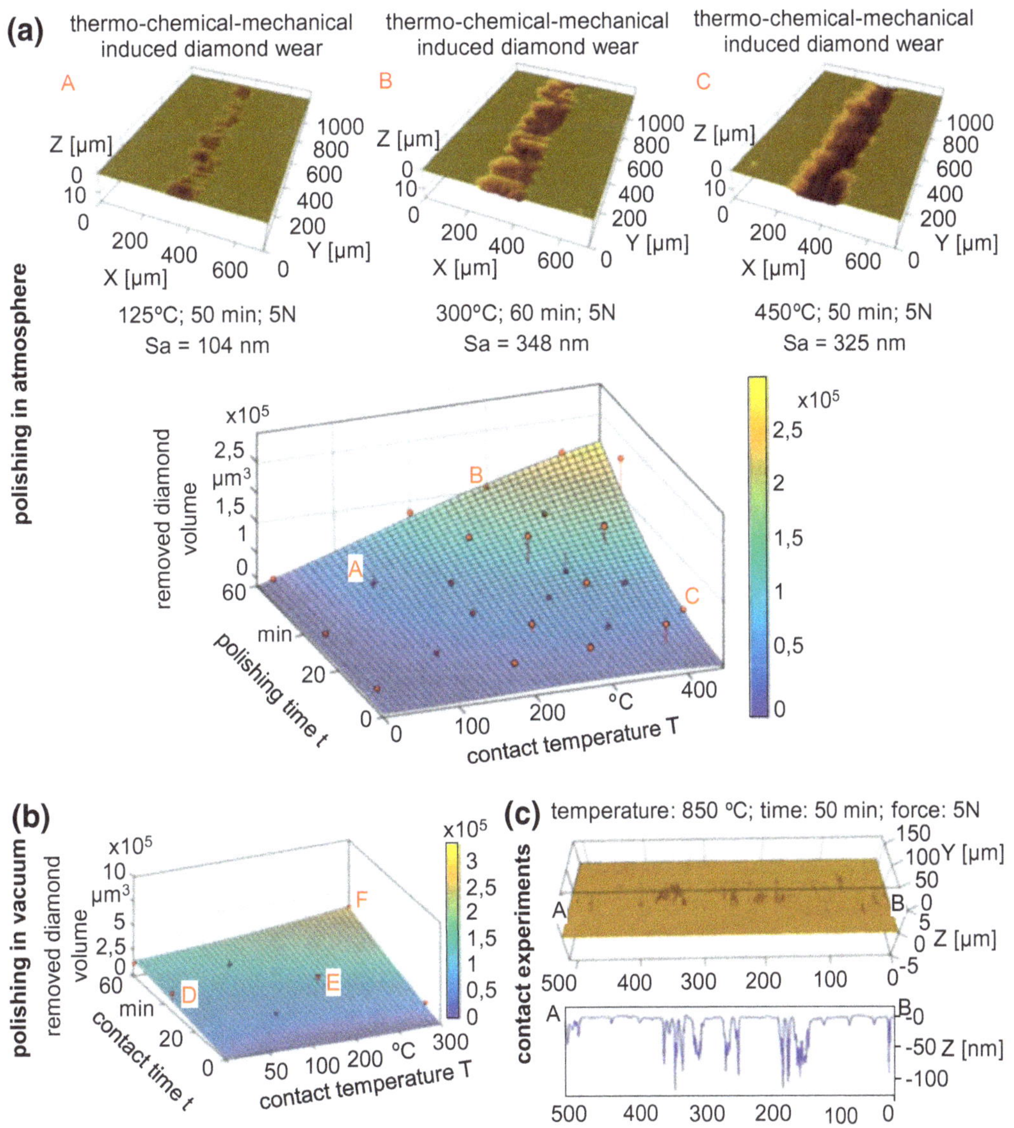

Fig. 4.79 Diamond removal for polishing in atmosphere (**a**) and polishing in high vacuum (**b**) with pure molybdenum. Diamond surface and profile after molybdenum contact test (**c**) with holding temperature of 850 °C and holding time of 60 min

Table 4.15 Diamond removal rates for polishing in atmosphere with molybdenum as tool material

Contact temperature [°C]	200	300	400
Contact time [min]	10/30/40/60	10/30/40/60	10/30/40/60
Diamond removal rate [µm³/min/µm²]	26/22/20/14	38/16/40/19	50/15/27/27

In contrast to the experiments at ambient atmosphere (Table 4.15), molybdenum showed a high removed diamond volume in high vacuum. The contact test with molybdenum showed no effect on the diamond surface. Only for the highest contact temperature of 850 °C and the longest holding time of 50 min a nearly negligible amount of removed diamond volume was detected with a maximum cavity depth of 112 nm.

These results exhibit that oxygen and diamond graphitization do not play a dominant role in removing diamond in a lower temperature range. Only in a higher temperature range oxygen does play a significant role. At lower temperature zones mechanically induced diamond wear is the key factor.

4.7.3 Micro-Structuring of Single Crystal Diamond Using Ultrasonic Assisted Friction Polishing

The second part of this report addresses the capability of the ultrasonic assisted friction polishing process to micro-structure single crystal diamond for linear [Rob17b] and circular [Rob17c] polishing. Therefore, 1.4310 steel was used as a favorable tool material due to its high hardness, higher elevated temperature strength and better machinability against pure metals. The composition of the stainless steel 1.4310 is shown in Table 4.16.

Ultrasonic assisted friction polishing tests were carried out to investigate the capability of steel 1.4310 to remove diamond in the same procedure as described above for pure metals. Experiments were conducted with contact temperatures in a range of 200 to 400 °C with a machining time between 20 and 120 min and a contact load of 5 N. Additionally, diamond removal experiments were carried out with contact loads of 2 N, 5 N, 8 N, 12.5 N and 25 N to investigate the influence of the contact load. These experiments were performed at contact temperatures of 125, 200, 300 and 400 °C and a machining time of 40 min.

4.7.3.1 Experimental Results

Diamond removal tests in contact with steel 1.4310 showed similar results as for the experiments on pure iron. Figure 4.80 shows that the diamond removal is strongly dependent on the temperature and time. The maximum volume of removed diamond was identified for condition C. In addition to a predictable removed diamond

Table 4.16 Composition of stainless steel 1.4310 (data provided by supplier)

Element	C	Si	Mn	P	S	Cr	Mo	N	Ni
Min mass%	0.05	–	–	–	–	16.00	–	–	6.00
Max mass%	0.15	2.00	2.00	0.045	0.015	19.00	0.80	0.110	9.50

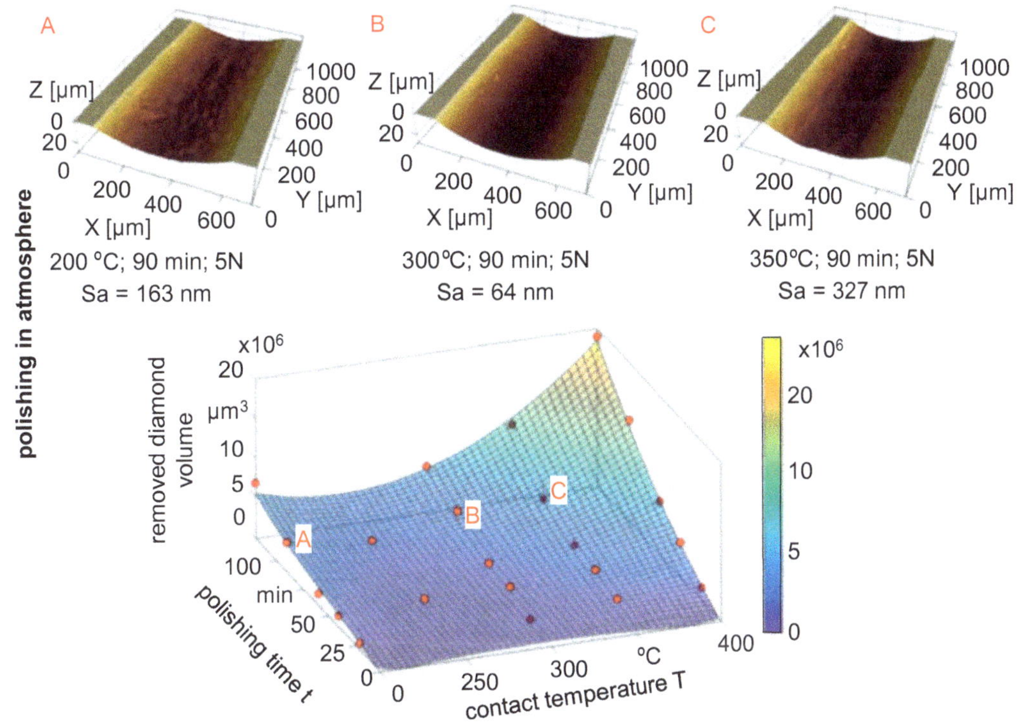

Fig. 4.80 Diamond removal for polishing in atmosphere with steel 1.4310

volume, knowledge of the minimum diamond surface roughness was important to use efficient parameters to micro-structure single crystal diamond. The minimum surface roughness was detected in the temperature range between 250 and 300 °C with the lowest surface roughness at condition B.

In addition to the contact temperature and machining time, the contact load plays an important role in micro-structuring single crystal diamond. To identify the influence of the contact load on the removed diamond volume and surface roughness, experiments with five different contact loads (2 N, 5 N, 8 N, 12.5 N, 25 N) were conducted.

Figure 4.81 shows that the removed diamond volume is strongly dependent on the contact load. In the temperature range between 125 and 300 °C, the removed diamond volume increased slowly with the increasing contact temperature, apart from the experiments with 25 N. With an increase of the preheating temperature to 400 °C, the removed diamond volume increased to around 4.2×10^6 μm^3 for all the contact loads used (except for 25 N). Experiments with a contact load of 2 N and 5 N showed the lowest surface roughness with Sa = 45 nm (200 °C) and 52 nm (300 °C), respectively. At the high pressure of 25 N, the friction tool lost its strength and a very rough diamond surface with Sa = 603 nm was generated.

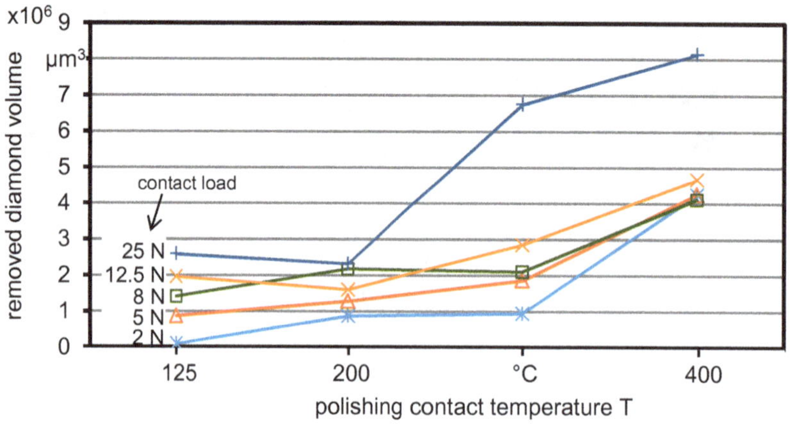

Fig. 4.81 Diamond removal depending on contact load and contact temperature by polishing with steel 1.4310 as tool material

4.7.3.2 Setup for Micro-Structuring Single Crystal Diamond

To micro-structure single crystal diamond, the shape of the friction tool was changed. The transformed tool featured two flat and parallel sides, where one side exhibited a small pin of 1 × 1 × 1 mm^3. The parallel exterior side of the pin, which was perpendicular to the centerline of the friction tool, was micro-structured by micro milling. The experimental setup is shown in Fig. 4.82.

Two different kinds of micro-structure, convex and concave, were chosen as shown in Fig. 4.83 (left column a–d). Figure 4.83a represents a linear convex micro-structure of 600 µm × 200 µm × 120 µm (length, width, height) which is aligned in the direction of the vibration. The friction tool with a concave micro-structure possesses the same parameters and is shown in Fig. 4.83b. To characterize a "smear effect" based on the tool motion during the ultrasonic assisted friction polishing, an L-shaped structure was chosen (Fig. 4.83c and d) with the same dimensions. With respect to the polishing and contact load experiments, a preheating temperature of 250 °C, a machining time of 120 min and a contact load of 4 N for the concave and 2 N for the convex micro-structured friction tools was chosen.

The experimental results exhibit for all four micro-structures the possibility of shaping single crystal diamond by ultrasonic assisted friction polishing (Fig. 4.83

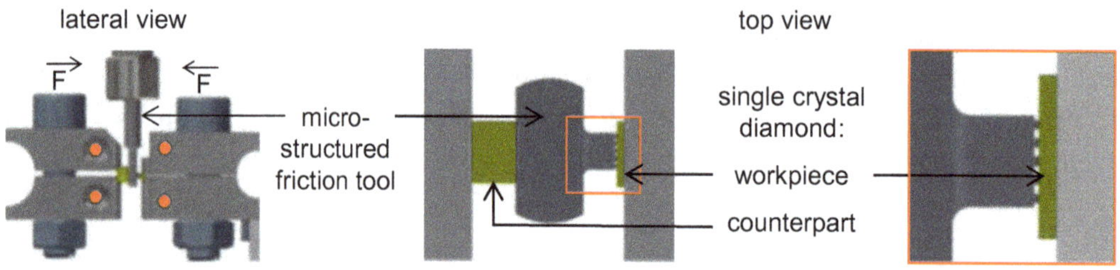

Fig. 4.82 Experimental setup of ultrasonic assisted friction polishing for micro-structuring single crystal diamond

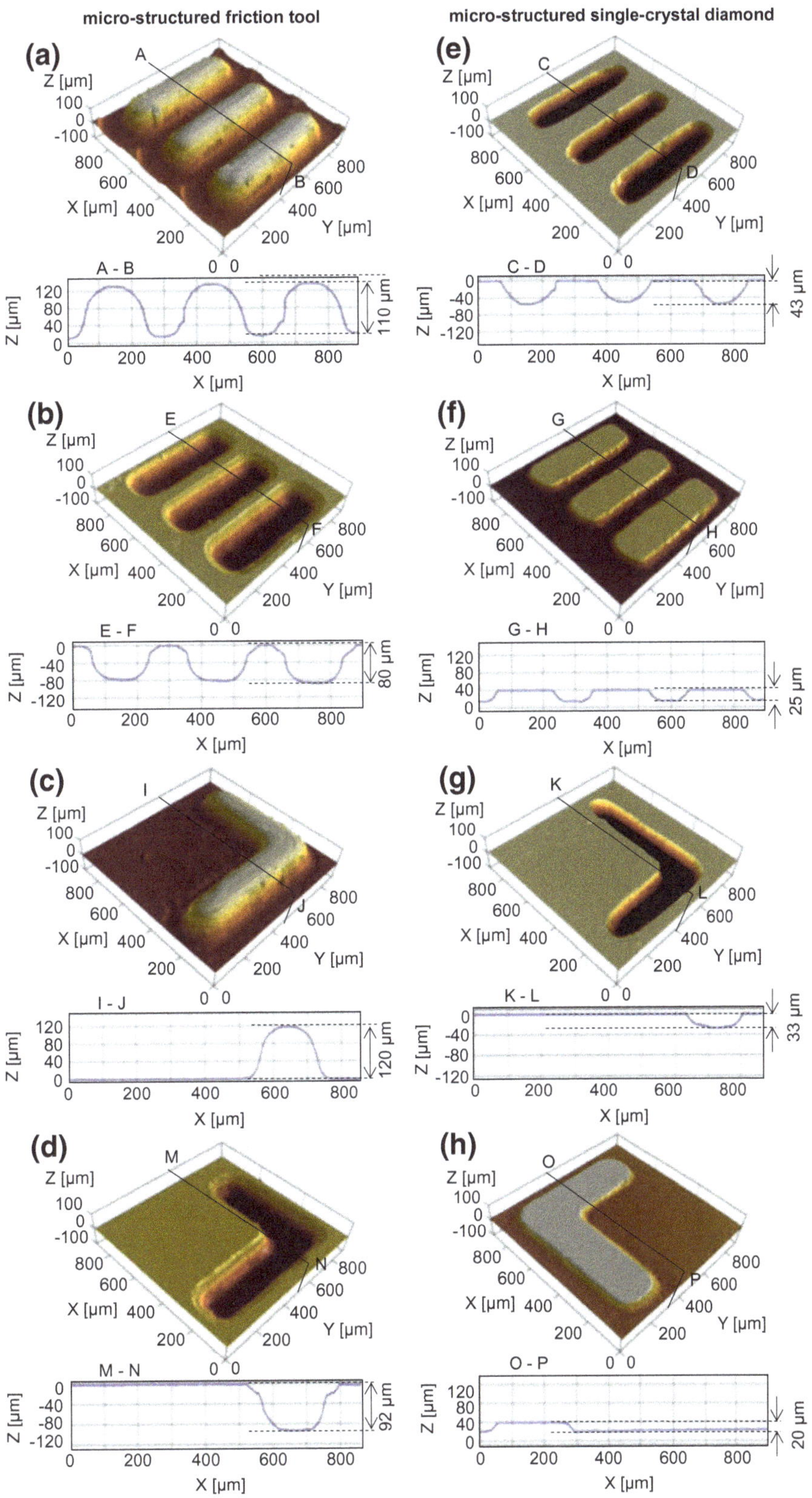

Fig. 4.83 Micro-structured friction tool **a–d** and micro-structured single crystal diamond **e–h**

right column e–h). It can be seen that the shape of the transferred micro-structure on the top of the diamond has sharp edges and has the same dimensions as the friction tool. This confirms that the micro-structure is transferred accurately into the diamond and the micro-structured friction tool can keep the geometry during processing. Convex micro-structured diamond surfaces show a deeper micro-structure (43 µm + 33 µm) than concave micro-structured diamond surfaces (25 µm + 20 µm). This result is based on a better oxygen feed to the convex micro-structured friction tool than to the concave micro-structured friction tool with a plane surface which has a tight contact with the diamond surface.

References

[Alt06] Altintas, Y.: Manufacturing Automation: Metal Cutting Mechanics, Machine Tool Vibrations, and CNC Design. Cambridge University Press (2006)

[Ara08] Aramcharoena, A., Mativengaa, P.T., Yangb, S., Cookeb, K.E., Teerb, D.G.: Evaluation and selection of hard coatings for micro milling of hardened tool steel. Int. J. Mach. Tools Manuf. **48**(14), 1578–1584 (2008)

[Ara09] Aramcharoen, A., Mativenga, P.T.: Size effect and tool geometry in micromilling of tool steel. Precis. Eng. **33**(4), 402–407 (2009)

[Ass02] Asmussen, J., Reinhard, D.K. (eds.): Diamond Films Handbook. Marcel Dekker, New York Basel (2002). ISBN: 978-0824795771

[Bae11] Bäuerle, D.: Laser Processing and Chemistry. Springer, Heidelberg, p. 482 (2011)

[Beh15] Behrens, G., Ruhe M., Tetzel, H., Vollertsen F.: Effect of tool geometry variations on the punch force in micro-deep drawing of rectangular components. Product. Eng. Res. Dev. **9**, 195–201 (2015)

[Ber99] Bertsekas, D.P.: Nonlinear Programming, 2nd edn. Athena Scientific (1999)

[Bhu04] Bhushan, B., Nosonovsky, M.: Comprehensive model for scale effects in friction due to adhesion and two- and three-body deformation (plowing). Acta Mater. **52**, 2461–2474 (2004)

[Blu03] Blunt, L., Jiang, X. (eds.): Advanced techniques for assessment surface topography: development of a basis for 3D surface texture standards "surfstand". Kogan page science (2003)

[Bob08] Bobe, U.: Die Reinigbarkeit technischer Oberflächen im immergierten System. Univ. Diss., München, Deutschland: Technische Universität München (2008)

[Böh14] Böhmermann, F., Preuß, W., Riemer, O.: Manufacture and functional testing of micro forming tools with well-defined tribological properties. In: Proceedings of the 29th ASPE Annual Meeting, vol. 29, pp. 486–491 (2014)

[Böh16] Böhmermann, F., Riemer, O.: Tribological Performance of Textured Micro Forming Dies, Dry Metal Forming OAJ FMT 2, pp. 67–71 (2016)

[Böh18] Böhmermann, F., Riemer, O.: Methodology for reliable tribological investigations applying a micro tribometer in ball-on-plate configuration. In: Proceedings of the 5th International Conference on New Forming Technology (ICNFT), MATEC Web of Conferences, vol. 190, pp. 15011-1–15011-6 (2018)

[Boy14] Boyd, S., Vandenberghe, L.: Convex Optimization. Hrsgs. Cambridge Press (2014)

[Bra12] Brandt, C.: Regularization of Inverse Problems for Turning Processes. Dissertationsschrift, Universität Bremen, Logos Verlag (2013)

[Bri09] Brinksmeier, E., Klopfstein, M.J.: Thermochemical material removal of diamond by solid iron and mischmetal. Product. Eng. **3**(3), 225–229 (2009)

[Bri10] Brinksmeier, E., Riemer O., Twardy S.: Tribological behavior of micro-structured surfaces for micro-forming tools. Int. J. Mach. Tools Manuf. **50**(4), 425–430 (2010)

[Bun96] Bundy, A.P., Basset, W.A., Weathers, M.S., Hemley, R.J., Mao, H.K., Goncharov, A.F.: The pressure-temperature phase and transformation diagram for carbon; updated through 1994. Carbon **34**(2), 141–153 (1996)

[But03] Butt, H.-J., Graf, K., Kappel, M.: Physics and Chemistry of Interfaces. Wiley-VCH, Weinheim (2003)

[Cam12] Camara, M.A., Rubio, J.C., Abrao, A.M., Davim, J.P.: State of the art on micromilling of materials: a review. J. Mater. Sci. Technol. **28**, 673–685 (2012)

[Cas12] Casella, G., Berger, R.L.: Statistical Inference. China Machine Press (2012)

[Che09] Chen, Y., Zhang, L.-C.: Polishing of polycrystalline diamond by the technique of dynamic friction, part 4: Establishing the polishing map. Int. J. Mach. Tools Manuf. **49**(3–4), 309–314 (2009)

[Che10] Cheng, K., Huo, D., Wardle, F.: Design of a five-axis ultra-precision micromilling machine-UltraMill. Part 1: holistic design approach, design considerations and specifications. Int. J. Adv. Manuf. Technol. **47**(9), 867–877 (2010)

[Che13] Chen, Y., Zhang, L.-C.: Polishing of Diamond Materials. Springer Verlag London (2013). ISBN: 978-1-84996-407-4

[Czi10] Czichos, H., Habig, K.H.: Tribologie-Handbuch-Tribometrie, Tribomaterialien, Tribotechnik. Vieweg + Teubner Verlag (2010)

[Dat87] Datta, M., Romankiw, L.T., Vigliotti, D.R., von Gutfeld, R.J.: Laser etching of metals in neutral salt solutions. Appl. Phys. Lett. **24** (1987)

[Den14] Denkena, B., Vehmeyer, J., Niederwestberg, D., Maaß, P.: Identification of the specific cutting force for geometrically defined cutting edges and varying cutting conditions. Int. J. Mach. Tools Manuf. **82**, 42–49 (2014)

[Dow04] Dow, T.A., Miller, E.L., Garrard, K.: Tool force and deflection compensation for small milling tools. Precis. Eng. **28**, 31–45 (2004)

[Eck17a] Eckert, S., Köhnsen, A., Vollertsen, F.: Surface finish using laser-thermochemical machining. In: Proceedings of the Laser in Manufacturing Conference LIM 2017, pp. 1–8 (2017)

[Eck17b] Eckert, S.; Messaoudi, H.; Mehrafsun, S.; Vollertsen, F.: Laser-thermochemical induced micro-structures on titanium. Journal of Materials Science and Surface Engineering 5(7), (2017) 685–691, doi:10.jmsse/2348-8956/5-7.3

[Eck18] Eckert, S., Vollertsen, F.: Mechanisms and processing limits of surface finish using laser-thermochemical polishing. CIRP Ann. Manuf. Technol. **67**(1), 201–204 (2018). https://doi.org/10.1016/j.cirp.2018.04.098

[Flo13] Flosky, H.; Vollertsen, F.: Wear behavior of a micro blanking and deep drawing tool combination with different drawing ratio. In: IDDRG 2013 Conference, Zurich (2013)

[Flo14a] Flosky, H., Veenaas, S., Feuerhahn, F., Hartmann, M., Vollertsen, F.: Flaking during a micro blanking and deep drawing process. In: Proceedings of the 9th International Conference on Micro Manufacturing (ICOMM2014), Bremen (2014)

[Flo14b] Flosky, F.; Vollertsen, F.: Wear behavior in a combined micro blanking and deep drawing process. CIRP Ann. Manuf. Technol. **63**, 281–284 (2014)

[Flo14c] Flosky, H., Krüger, M., Vollertsen F.: Temperature measurement with thermocouples during micro deep drawing process. In: Proceedings of the 4th International Conference on Nanomanufacturing (nanoMan2014) (2014)

[Flo15] Flosky, H., Krüger, M., Vollertsen, F.: Verschleißmessung in der Mikroumformung mittels Silikonabformverfahren an einem kombinierten Schneid-Ziehring. In: Tagungsband 7. Kolloquium Mikroproduktion. Aachen (2015)

[Flo16] Flosky, H., Feuerhahn, F., Böhmermann, F., Riemer, O., Vollertsen, F.: Performance of a micro deep drawing die manufactured by selective laser melting. In: Proceedings of the 5th International Conference on Nanomanufacturing (nanoMan2016), Macau, China (2016)

[Gar18a] N.N.: Gartner says smartphone sales surpassed one billion units in 2014. https://www.gartner.com/newsroom/id/2996817. Accessed 18 Apr 2018

[Gar18b] N.N.: Gartner says worldwide sales of smartphones recorded first ever decline during the fourth quarter of 2017. https://www.gartner.com/newsroom/id/3859963. Accessed 18 Apr 2018

[Gei01] Geiger, M., Kleiner, M., Eckstein, R., Tiesler, N., Engel, U.: Microforming. CIRP Ann. Manuf. Technol. **50**, 445–462 (2001)

[Gei97] Geiger, M., Engel, U., Pfestorf, M.: New developments for the qualification of technical surfaces in forming processes. Ann. CIRP **46**(1), 171–174 (1997)

[Geo03] McGeough, J.A., Pajak, P.T., De Silva, A.K.M., Harrison, D.K.: Recent research & developments in electrochemical machining. Int. J. Electr. Mach. **8**, 1–14 (2003)

[Goc08] Goch, G., Lübke, K.: Tschebyscheff approximation for the calculation of maximum inscribed/minimum circumscribed geometry elements and form deviations. CIRP Ann. Manuf. Technol. **57**(1), 517–520 (2008)

[Gre66] Greenwood, J.A., Williamson, J.B.P.: Contact of nominally flat surfaces. Proc. R. Soc. A **295**(1442) 300–319 (1966)

[Gup16] Gupta, K., Neelesh, K.J., Laubscher, R.F.: Electrochemical hybrid machining processes. In: Hybrid Machining Processes. Springer Briefs in Applied Sciences and Technology, pp. 9–32. Springer, Cham (2016). https://doi.org/10.1007/978-3-319-25922-2_2

[Gut86] Gutfeld von, R.J.: Laser-enhanced plating and etching for microelectronic applications. IEEE Circuits Devices Mag. **2**(1), 57–60 (1986)

[Han16] Han, W., Kunieda, M.: Influence of machining conditions on machining characteristics of micro-rods by micro-ECM with electrostatic induction feeding method. In: Proceedings of the 18th CIRP Conference on Electro Physical and Chemical Machining (ISEM XVIII), Procedia CIRP 42, pp. 819–824 (2016)

[Hau15] Hauser, O., Mehrafsun, S., Vollertsen, F.: Einfluss der Abtragsbahnfolge auf die resultierende Geometrie bei der laserchemischen Bearbeitung von Stellite 21. In: Hopmann, Ch., Brecher, Ch., Dietzel, A., Drummer, D., Hanemann, T., Manske, E., Schomburg, W.K., Schulze, V., Ullrich, H., Vollertsen, F., Wulfsberg, J.-P. (eds.) Tagungsband 7. Kolloquium Mikroproduktion, IKV Aachen, pp. 8–14 (2015). ISBN: 978-3-00-050-755-7

[Hau15] Hauser, O., Mehrafsun, S., Vollertsen, F.: Electrolytes for sustainable laser-chemical machining of titanium, Stellite 21 and tool steel X110CrMoV8-2. Appl. Mech. Mater. **794**, 262–269 (2015)

[Hau15a] Hauser, O., Mehrafsun, S., Vollertsen, F.: Electrolytes for sustainable laser-chemical machining of titanium, Stellite 21 and tool steel X110CrMoV8-2. In: Wulfsberg, J.P., Röhlig, B., Montag, T. (eds.) Progress in Production Engineering, WGP Kongress (2015). TransTech Publications Pfaffikon/CH, pp. 262–269 (2015)

[Hu10] Hu, Z., Huferath-von Luepke, S., von Kopylow, C., Vollertsen, F.: Characteristics of wear behavior of micro deep drawing tools. In: International Conference on Advances in Materials and Processing Technologies (AMPT2010), pp. 335–340 (2011)

[Hu11a] Hu, Z.; Vollertsen, F.: Auslegung von Mikrotiefziehwerkzeugen zur Bestimmung der Werkzeuglebensdauer. In: 5. Kolloquium Mikroproduktion, Karlsruhe (2011)

[Hu11b] Hu, Z., Wielage, H., Vollertsen, F.: Economic micro forming using DLC- and TiN-coated tools. J. Plast. (JTP) 51–57 (2011)

[Hu12] Hu, Z., Schubnov, A., Vollertsen, F.: Tribological behavior of DLC-films and their application in micro deep drawing. J. Mater. Process.Technol. **212**, 685–688 (2012)

[Imo17] Imoto, Y., Yan, J.: Thermochemical micro imprinting of single-crystal diamond surface using a nickel mold under high-pressure conditions. Appl. Surf. Sci. **404**, 318–325 (2017)

[Jin12] Jin, B., Maass, P.: Sparsity regularization for parameter identification problems. Inverse Probl. **28**(12), 23001 (2012)

[Jin93] Jin, S., Chen, L., Graebner, J., McCormack, M., Reiss, M.: Thermal conductivity in molten-metal-etched diamond films. Appl. Phys. Lett. **63**(5), 622–624 (1993)

[Jus07] Justinger, H., Hirt, G.: Scaling effects in the miniaturization of deep drawing process. In: International Conference on New Forming Technology (2nd ICNFT07), pp. 167–176 (2007)

[Kim03] Kim, G.M., Kim, B.H., Chu, C.N.: Estimation of cutter deflection and form error in ball-end milling processes. Int. J. Mach. Tools Manuf. **43**, 917–924 (2003)

[Klo13] Klocke, F., Zeis, M., Klink, A., Veselovac, D.: Technological and economical comparison of roughing strategies via milling, EDM and ECM for titanium- and nickel-based blisks. In: Proceedings of the 1st CIRP Global Web Conference on Interdisciplinary Research in Production Engineering, Procedia CIRP 2, pp. 98–101 (2013)

[Kra08] Kray, D., Fell, A., Hopman, S., Mayer, K., Willeke, G.P., Glunz, S.W.: Laser Chemical Processing (LCP)–a versatile tool for microstructuring applications. Phys. A **93**, 99–103 (2008)

[Kub16] Kubota, A., Shin, N., Touge, M.: Improvement of material removal rate of single-crystal diamond by polishing using H_2O_2 solution. Diam. Relat. Mater. **70**, 39–45 (2016)

[Kwo12] Kwon, H., Baek, W.-K., Kim, M.-S., Shin, W.-S., Yoh, J.J.: Temperature-dependent absorptance of painted aluminum, stainless steel 304, and titanium for 1.07 μm and 10.6 μm laser beams. Opt. Lasers Eng. **50**, 114–121 (2012)

[Lea13] Leach, R. (ed.): Characterisation of Areal Surface Texture. Springer (2013)

[Lech09] Lechleiter, A., Rieder, A.: Towards a General Convergence Theory for Inexact Newton Regularization. Springer (2009)

[Lou01] Louis, A.: Inverse und schlecht gestellte Probleme. Hrsgs. Teubner, Stuttgart (2001)

[Lue12] Lübke, K., Sun, Z., Goch, G.: Three-dimensional holistic approximation of measured points combined with an automatic separation algorithm. CIRP Ann. Manuf. Technol. **61**(1), 499–502 (2012)

[Man08] Manabe, K., Shimizu, H., Koyama, H., Yang, M., Ito, K.: Validation of FE simulation based on surface roughness model in micro-deep drawing. J. Mater. Process. Technol. **204**, 89–93 (2008)

[Meh12] Mehrafsun, S., Stephen, A., Vollertsen, F.: Comparison of laser-thermal and laser-chemical machining. In: Fang, F., Kuriyagawa, T. (eds.) 3rd International Conference on Nanomanufacturing (nanoMan2012,) July 25–27. Wako, Saitama, Japan (2012)

[Meh13] Mehrafsun, S., Vollertsen, F.: Disturbance of material removal in laser-chemical machining by emerging gas. CIRP Ann. **62**(1), 195–198 (2013)

[Meh16a] Mehrafsun, S., Harst, S., Hauser, O., Eckert, S., Klink, A., Klocke, F., Vollertsen, F.: Energy–based analysis of material dissolution behavior for laser-chemical and electrochemical machining. In: 3rd CIRP Conference on Surface Integrity (CIRP CSI). Procedia CIRP 45, pp. 347–350 (2016)

[Meh16b] Mehrafsun, S., Messaoudi, H., Vollertsen, F.: Influence of material and surface roughness on gas bubble formation and adhesion in laser-chemical machining. In: Proceedings of the 5th International Conference on Nanomanufacturing (nanoMan 2016), pp. 1–10 (2016)

[Meh18] Mehrafsun, S., Messaoudi, H.: Dynamic process behavior in laser chemical micro machining of metals. J. Manuf. Mater. Process. **2**, 54 (2018). https://doi.org/10.3390/jmmp2030054

[Mes14] Messaoudi, H., Mehrafsun, S., Vollertsen, F.: Influence of the etchant on material removal geometry in laser chemical machining. In: Proceedings of the 4th International Conference on Nanomanufacturing (nanoMan 2014), pp. 1–5 (2014)

[Mes17a] Messaoudi, H., Eckert, S., Vollertsen, F.: Thermal analysis of laser chemical machining: part I: static irradiation. Mater. Sci. Appl. **8**, 685–707 (2017). https://doi.org/10.4236/msa.2017.810049

[Mes17b] Messaoudi, H., Hauser, O., Matson, A., Mehrafsun, S., Vollertsen, F.: Fertigungsqualität laserchemisch hergestellter Mikroumformwerkzeuge, In: Vollertsen, F., Hopmann, C., Schulze, V., Wulfsberg, J. (eds.) Fachbeiträge 8. Kolloquium Mikroproduktion. BIAS Verlag, pp. 171–178 (2017). ISBN: 978-3-933762-56-6 (online)

[Mes18] Messaoudi, H., Böhmermann, F., Mikulewitsch, M., von Freyberg, A., Fischer, A., Vollertsen, F.: Chances and limitations in the application of laser chemical machining for the manufacture of micro forming dies. In: Proceedings of the 5th International Conference on New Forming Technology (ICNFT2018), MATEC Web of Conferences 190, pp. 15010 (2018)

[Mes18a] Messaoudi, H., Brand, D., Vollertsen, F.: Removal characteristics of high-speed steel in laser chemical machining. In: 14th International Symposium on Electrochemical Machining Technology INSECT2018, pp. 47–53 (2018)

[Mik17] Mikulewitsch, M., von Freyberg, A., Fischer, A.: Adaptive Qualitätsregelung für die laserchemische Fertigung von Mikroumformwerkzeugen, In: Vollertsen, F., Hopmann, C., Schulze, V., Wulfsberg, J. (eds.) Fachbeiträge 8. Kolloquium Mikroproduktion. BIAS Verlag, pp. 21–26 (2017). ISBN: 978-3-933762-56-6 (online)

[Nak09] Nakashima, S., Sugioka, K., Midorikawa, K.: Fabrication of microchannels in single-crystal GaN by wet chemical-assisted femtosecond-laser ablation. Appl. Surf. Sci. **255**, 9770–9774 (2009)

[Nie18] Nielsen, C.V., Bay, N.: Review of friction modeling in metal forming processes. J. Mater. Process. Technol. **255**, 234–241 (2018)

[Now94] Nowak, R., Metev, S., Sepold, G.: Laser chemical etching of metals in liquids. Mater. Manuf. Process. **9**(4), 429–435 (1994)

[Now96] Nowak, R., Metev, S.: Thermochemical laser etching of stainless steel and titanium in liquids. Appl. Phys. A **63**, 133–138 (1996)

[Paj06] Pajak, P.T., De Silva, A.K.M., Harrison, D.K., McGeough, J.A.: Precision and efficiency of laser assisted jet electrochemical machining. Precis. Eng. **30**, 288–298 (2006)

[Pau96] Paul, E., Evans C.J.: Chemical aspects of tool wear in single point diamond turning. Precis. Eng. **18**, 4–19 (1996)

[Pio11] Piotrowska-Kurczewski, I., Vehmeyer, J.: Simulation model for micro-milling operations and surface generation. Adv. Mater. Res. **223**, 849–858 (2011)

[Pio15] Piotrowska, I., Vehmeyer, J., Gralla, P., Böhmermann, F., Elsner-Dörge, F., Riemer, O., Maaß, P.: Reduzierung der Formabweichung beim Mikrofräsen, Tagungsband 7. Kolloquium Mikroproduktion in Aachen vom 16. bis 17. November 2015, IKV Aachen, pp. 23–30 (2015)

[Qia14] Qiang, H., Chen, J., Han, B., Shen, Z-H., Lu, J., Ni, X-W.: Study of underwater laser propulsion using different target materials. Opt. Express **22**(14), 17532–17545 (2014). https://doi.org/10.1364/oe.22.017532

[Rie03] Rieder, A.: Keine Probleme mit Inversen Problemen–Eine Einführung in ihre stabile Lösung. Hrsgs. Vieweg & Sohn Verlag/ GWV Fachverlag GmbH, Wiesbaden (2003)

[Rie16] Riemer, O., Elsner-Dörge, F.: Forces and form deviations in micro-milling of CoCr alloys for dental prostheses. In: The 5th International Conference on Nanomanufacturing (nanoMan), pp. 15–17 August 2016, Macau, China (2016)

[Rie17a] Riemer, O., Elsner-Dörge, F., Willert, M., Meier, A.: Forces, form deviations and surface roughness in micro-milling of CoCr alloys for dental prostheses. Int. J. Nanomanufacturing (not yet published)

[Rie17b] Riemer, O., Elsner-Dörge, F.: Charakteristische Kraftverläufe und resultierende Geometrieabweichungen beim Mikrofräsen von Strukturen in einer Kobalt-Chrom-Legierung, Tagungsband 8. Kolloquium Mikroproduktion (not yet published)

[Rob17] Robert, C., Messaoudi, H., Riemer, O., Brinksmeier, E., Vollertsen, F.: Modifikation der Oberflächenfeingestalt monokristalliner Diamanten. In: Vollertsen, F., Hopmann, C., Schulze, V., Wulfsberg, J. (eds.) Fachbeiträge 8. Kolloquium Mikroproduktion. BIAS Verlag, pp. 187–194 (2017). ISBN: 978-3-933762-56-6

[Rob17a] Robert, C., Riemer, O., Brinksmeier, E.: Mikrostrukturierung von monokristallinem Diamant mittels Reibungspolieren. wt Werkstattstechnik online 6, pp. 467–471 (2017)

[Rob17b] Robert, C., Riemer, O., Brinksmeier, E.: Micro-structuring of single crystal diamond by ultrasonic assisted friction polishing. In: Proceedings of the 17th International Conference of the European Society for Precision Engineering and Nanotechnology, pp. 187–188 (2017). ISBN: 978-0-9957751-0-7

[Rob17c] Robert, C., Riemer, O., Brinksmeier, E.: Bestimmung der Diamantabtragsmechanismen beim ultraschall-unterstützten Reibungspolieren, 8. Kolloquium Mikroproduktion. BIAS Verlag (2017). ISBN: 978-3-933762-56-6

[Sch08a] Schulze Niehoff, H.: Entwicklung einer hochdynamischen, zweifachwirkenden Mikroumformpresse. BIAS-Verlag, Bremen, 2008, Stahltechnik Band 33 (Dissertation)

[Sch08b] Schulze Niehoff, H.: Entwicklung einer hochdynamischen, zweifachwirkenden Mikroumformpresse. Dissertation, BIAS Verlag (2008)

[Sev18] Seven, J., Heinrich, L., Flosky, H.: Inline tool wear measurement in lateral micro upsetting. In: Vollertsen, F., Dean, T.A., Qin, Y., Yuan, S.J. (eds.) 5th International Conference on New Forming Technology (ICNFT2018), Bremen, MATEC Web Conference, vol. 190, pp. 15009 (2018)

[Shi15] Shimizu, T., Yang, M., Manabe, K.: Classification of mesoscopic tribological properties under dry sliding friction for microforming operation. Wear **330–331**, 49–58 (2015)

[Sil11a] De Silva, A.K.M., Pajak, P.T., McGeough, J.A., Harrison, D.K.: Thermal effects in laser assisted jet electrochemical machining. CIRP Ann. **60**(1), 243–246 (2011)

[Sil11b] De Silva, A.K.M., Pajak, P.T., McGeough, J.A., Harrison, D.K.: Thermal effects in laser assisted jet electrochemical machining. CIRP Ann. Manuf. Technol. **60**(1), 243–246 (2011)

[Ste09a] Stephen, A., Walther, R., Vollertsen, F.: Removal rate model for laser chemical micro etching. In: Ostendorf, A., Graf, T., Petring, D., Otto, A. (eds.) Proceedings of the 5th International Lasers in Manufacturing Conference (LIM 2009). München, pp. 615–619 (2009)

[Ste09b] Stephen, A., Walther, R., Vollertsen, F.: Removal rate model for laser chemical micro etching. In: Proceedings of the 5th International WLT Conference on Lasers in Manufacturing, pp. 615–619 (2009)

[Ste10] Stephen, A., Vollertsen, F.: Mechanisms and processing limits in laser thermochemical machining. CIRP Ann. **59**(1), 251–254 (2010)

[Ste11] Stephen, A.: Vollertsen, F., Bergmann, R.B. (eds.) Elektrochemisches Laser-Jet-Verfahren zur Mikrostrukturierung von Metallen. Strahltechnik 46. BIAS Verlag Bremen (2011)

[Suh87] Suh, N.P., Saka, N.: Surface engineering. CIRP Ann. **36**(1) 403–408 (1987)

[Suz03] Suzuki, K., Iwai, M., Uematsu, T., Yasunaga, N.: Material removal mechanism in dynamic friction polishing of diamond. Key Eng. Mater. **238–239**, 235–240 (2003)

[Tat16] Tatsumi, N., Harano, K., Ito, T., Sumiya, H.: Polishing mechanism and surface damage analysis of type IIa single crystal diamond processed by mechanical and chemical polishing methods. Diam. Relat. Mater. **63**, 80–85 (2016)

[Twa14] Twardy, S.: Funktionsgerechte Fertigung von Mikroumformwerkzeugen durch Mikrofräsen. Dissertation Universtiy of Bremen (2014)

[Uen08] Ueno, F., Kato, C., Motooka, T., Ichikawa, S., Yamamato, M.: Corrosion phenomenon of stainless steel in boiling nitric acid solution using large-scale mock-up of reduced pressurized evaporator. J. Nucl. Sci. Technol. **45**(10), 1091–1097 (2008)

[Uhl14] Uhlmann, E., Oberschmidt, D., Kuche, Y., Löwenstein, A.: Cutting edge preparation of micro milling tools. In: 6th CIRP International Conference on High Performance Cutting, HPC2014, Procedia CIRP, vol. 14, pp. 349–354 (2014)

[Veh12] Vehmeyer, J., Piotrowska-Kurczewski, I., Twardy, S.: A surface generation model for micro cutting processes with geometrically defined cutting edges. In: Proceedings of the 37th International MATADOR Conference, pp. 149–152 (2012)

[Veh15a] Vehmeyer, J., Piotrowska, I., Böhmermann, F., Riemer, O., Maaß, P.: Least-squares based parameter identification for a function-related surface optimisation in micro ball-end milling. In: Proceedings of the 15th CIRP Conference on Modelling of Machining Operations (15th CMMO), Procedia CIRP, vol. 31, pp. 276–281 (2015)

[Veh15b] Vehmeyer, J., Piotrowska, I., Böhmermann, F., Riemer, O., Maaß, P.: Least-squares based parameter identification for a function-related surface optimisation in micro-ball-end milling. In: Proceedings of the 15th CIRP Conference on Modelling of Machining Operations (15th CMMO), Procedia CIRP, vol. 31, pp. 276–281 (2015)

[Veh17a] Vehmeyer, J.: Geometrische Modellierung und funktionsbezogene Optimierung der inhärenten Textur von Mikrofräsprozessen, Dissertationsschrift, Universität Bremen, Logos Verlag Berlin (2017)

[Veh17b] Vehmeyer, J.: Geometrische Modellierung und funktionsbezogene Optimierung der inhärenten Textur von Mikrofräsprozessen, Logos Verlag Berlin (2017), Universität Bremen, Dissertation (2016)

[Ver01] Vermeulen, M., Scheers, J.: Micro-hydrodynamic effects in EBT textured steel sheet. Int. J. Mach. Tools Manuf. **41**, 1941–1951 (2001)

[Vol08] Vollertsen, F.: Categories of size effects. Prod. Eng. Res. Dev. **2**, 377–383 (2018)

[Vol10] Vollertsen, F., Biermann, D., Hansen, H.N., Jawahir, I.S., Kuzman, K.: Size effects in manufacturing of metallic components. CIRP Ann. Manuf. Technol. **58**(2), 566–587 (2009)

[Vol15] Vollertsen, F., Flosky, H., Seefeld, T.: Dry metal forming–a green approach. In: Tekkaya, A.E., Homberg, W., Brosius, A. (eds.) 60 Excellent Inventions in Metal Forming, pp. 119–123 (2015)

[Vol18] Vollertsen, F.: Schwerpunktprogramm SPP 1676. BIAS GmbH. http://www.trockenumformen.de/. Accessed 21 June 2018

[Wan06] Wang, C.Y., Chen, C., Song, Y.X.: Mechanochemical polishing of single crystal diamond with mixture of oxidizing agents. Key Eng. Mater. **315–316**, 852–855 (2006)

[Wei01] Weima, J.A., Fahrner, W.R., Job, R.: Experimental investigation of the parameter dependency of the removal rate of thermochemically polished CVD diamond films. J. Solid State Electrochem. **5**, 112–118 (2001)

[Wen12] Wendler-Kalsch, E., Gräfen, H.: Korrosionsschadenkunde. Nachdr. der 1. Aufl. 1998. Springer Vieweg, Berlin (2012)

[Whe92] Wheeler, D.J., Chambers, D.S.: Understanding Statistical Process Control. SPC Press (1992)

[Whi11] Whitelaw, J.H.: Convective heat transfer, thermopedia: guide to thermodynamics. Heat Mass Transf. Fluids Eng. (2011). https://doi.org/10.1615/AtoZ.c.convective_heat_transfer

[Wol84] Wolf, G.W.: A mathematical model of cartographic generalization. Geo-processing 2 (1984) 271–286

[Wol93] Wolf, G.W.: Data structures for the topological characterization of topographic surfaces. In: Proceedings of the 6th Colloque Européen de Géographie Théorique et Quantitative, vol. 6, pp. 24–34 (1993)

[Yav94] Yavas, O., Oltra, R., Kerrec, O.: Enhancement of pulsed laser removal of metal oxides by electrochemical control. Appl. Phys. A **63**, 321–326 (1994)

[Zha12] Zhang, P., Renken, V., Goch, G.: Cross-plane quality control concept of the bearing ring production. Materialwissenschaft und Werkstofftechnik **43**(1–2), 37–41 (2012)

[Zha15] Zhang, P., Goch, G.: A quality controlled laser-chemical process for micro metal machining. Product. Eng. **9**(5), 577–583 (2015)

[Zha17] Zhang, P., von Freyberg, A., Fischer, A.: Closed-loop quality control system for laser chemical machining in metal micro-production. Int. J. Adv. Manuf. Technol. **93**, 3693–3703 (2017)

Chapter 5
Quality Control and Characterization

Peter Maaß, Iwona Piotrowska-Kurczewski, Mostafa Agour, Axel von Freyberg, Benjamin Staar, Claas Falldorf, Andreas Fischer, Michael Lütjen, Michael Freitag, Gert Goch, Ralf B. Bergmann, Aleksandar Simic, Merlin Mikulewitsch, Bernd Köhler, Brigitte Clausen and Hans-Werner Zoch

P. Maaß · I. Piotrowska-Kurczewski (✉)
Zentrum für Technomathematik, Bremen, Germany
e-mail: iwona@math.uni-bremen.de

M. Agour (✉) · C. Falldorf · R. B. Bergmann · A. Simic (✉)
BIAS—Bremer Institut für angewandte Strahltechnik GmbH, Bremen, Germany
e-mail: agour@bias.de

A. Simic
e-mail: simic@bias.de

M. Agour
Faculty of Science, Department of Physics, Aswan University, Aswan, Egypt

A. von Freyberg · A. Fischer · M. Mikulewitsch (✉)
BIMAQ—Bremen Institute for Metrology, Automation and Quality Science, University of Bremen, Bremen, Germany
e-mail: m.mikulewitsch@bimaq.de

B. Staar · M. Lütjen · M. Freitag
BIBA—Bremer Institut für Produktion und Logistik GmbH, Bremen, Germany

G. Goch
The University of North Carolina at Charlotte, Charlotte, USA

R. B. Bergmann
Physics and Electrical Engineering and MAPEX Center for Materials and Processes, University of Bremen, Bremen, Germany

B. Köhler (✉) · B. Clausen · H.-W. Zoch
Leibniz-Institut für Werkstofforientierte Technologien, Bremen, Germany
e-mail: koehler@iwt-bremen.de

F. Vollertsen et al. (eds.), *Cold Micro Metal Forming*, Lecture Notes in Production Engineering, https://doi.org/10.1007/978-3-030-11280-6_5

5.1 Introduction to Quality Control and Characterization

Peter Maaß and Iwona Piotrowska-Kurczewski*

The increasing demand for miniaturization is having an immense impact on manufacturing technologies and is leading to the development of many novel and innovative production processes for high-precision micro parts. Industrially feasible micro production requires fast quality assurance control mechanisms, which need to take into account the micro-specific properties.

In this chapter several innovative technologies for quality control of production processes for high-precision micro parts will be presented. The proposed innovative solutions also address the need to ensure the quality of final products, which should be manufactured with approved (specified) materials and within certain performance requirements. Therefore, careful quality control is required during each production step.

Micro production processes and related techniques for quality control and characterization need to take into account the size effects that occur at the micro level, which lead to different scaling of forces and additional manufacturing errors that do not occur in macro processes. As a consequence, the conventional control methods for macro processes may fail, or become much more complicated or unreliable when applied to micro processes. In order to guarantee efficient quality assurance, possible error classes must be redefined and evaluated.

Moreover, in order to achieve an upscaling in the number of produced parts, the semi-finished products must be examined for mechanical properties. Micro production is characterized by a low aspect ratio of components to microstructure, which can dominate the mechanical behavior by means of the surface defects. It turns out that fatigue strength is an important component property.

This chapter presents some technical and methodological solutions for quality control and characterization measurement, with respect to size effects and up-scaling in micro production.

Digital holographic measuring systems (DHM) have the potential for complete (100%) and fast quality testing. Such a system is described in Sect. 5.2 together with fast algorithms for geometric evaluation and surface defect detection. The proposed method is demonstrated by inspecting cold formed micro cups.

This system is suitable for the automated examination and classification of micro deep drawn parts in an industrial environment. Together with automatic error detection, which uses the knowledge of the shape of the object, it allows the reliable inspection of components with a measuring speed of about one part per second. Further reduction of time is possible by additional optimization of the error detection algorithm. Section 5.3 shows the inspection of the interior of micro parts in the industrial environment.

To guarantee consistent production parameters for semi-finished products, which are subsequently used for the production of micro components, a characterization of their physical properties is of particular importance. Due to the often oligocrystalline character of these semi-finished products and components, it is necessary to use a suitable testing technique for static and dynamic investigations, as the mechanical properties are not transferable from the macroscopic point of view.

In addition, the micro semi-finished products and components often show inhomogeneities induced by the manufacturing process. On the one hand, these are directly reflected in the microstructure and on the other hand they have an effect on quantities such as hardness or residual stresses, which play a decisive role in the application. Mechanical testing, conventional metallography, SEM, EBSD, ultra-microhardness testing and X-ray residual stress analysis were used as measuring and analysis techniques suitable for the sub-millimeter range. In Sect. 5.4 the possibilities and limitations of two of these methods are illustrated using the example of mechanical testing and electron backscatter diffraction (EBSD).

Micro forming tools with high strength and hardness can be produced by means of laser chemical machining (LCM). The in situ geometry measurement of micro-structures during this process is based on the fluorescence of a liquid medium and principles of confocal microscopy. Section 5.5 focuses on that measurement technique, based on confocal fluorescence microscopy. It turns out that the model-based approach is suitable to detect the geometry parameter step height with an uncertainty of 8.8 μm for a step submerged in a fluid layer with a thickness of 2.3 mm.

5.2 Quality Inspection and Logistic Quality Assurance of Micro Technical Manufacturing Processes

Mostafa Agour*, Axel von Freyberg, Benjamin Staar, Claas Falldorf, Andreas Fischer, Michael Lütjen, Michael Freitag, Gert Goch and Ralf B. Bergmann

Abstract Quality inspection is an essential tool for quality assurance during production. In the microscopic domain, where the manufactured objects have a size of less than 1 mm in at least two dimensions, very often mass production takes place with high demands regarding the failure rate, as micro components generally form the basis for larger assemblies. Especially when it comes to safety-relevant parts, e.g. in the automobile or medical industry, a 100% quality inspection is mandatory. Here, we present a robust and precise metrology method comprised of a holographic contouring system with fast algorithms for geometric evaluation and surface defect detection that paves the way for inspecting cold formed micro parts in less than a second. Using a telecentric lens instead of a standard microscope objective, we compensate scaling effects and wave field curvature, which distort the reconstruction in digital holographic microscopy. To enhance the limited depth of focus of the microscope objective, depth information from different object layers is stitched together to yield 3D data of its complete geometry. The 3D data map is converted into a point cloud and processed by geometry and surface inspection. Thereby, the resulting point cloud data are automatically decomposed into geometric primitives in order to analyze geometric deviations. Additionally, the surface itself is checked for scratches and other defects by the use of convolutional neural networks. The developed machine learning algorithm makes it possible not only to distinguish between good and failed parts but also to show the defect area pixel-wise. The methods are demonstrated by quality inspection of cold formed micro cups. Defects larger than 2 μm laterally and 5 μm axially can be detected.

Keywords Digital holographic microscopy · Geometry · Surface defect

5.2.1 Introduction

Quality inspection is an essential tool for quality assurance during production. Generally, the precise geometrical and surface inspection of a test object plays a decisive role for developing and/or optimizing the manufacturing process. In the microscopic domain, where the manufactured objects have a size of less than 1 mm in at least two dimensions, very often mass production takes place with high demands regarding the failure rate, as micro components generally form the basis for larger assemblies. Especially when it comes to safety-relevant parts, e.g. in the automobile or medical industry, often a 100% quality inspection is mandatory [Kop13].

Achieving 100% quality inspection is especially challenging in the micro-domain, e.g. when measurement uncertainty has to be in the range of a few microns, as there is usually a trade-off between precision and speed [Ago17]. Processes like micro cold forming, however, allow for production rates of multiple parts per second [Flo14]. Presently, no solutions exist that are fast, precise and suitable for in-process measurements at the same time [Ago17b]. Actually, most industrial quality inspection processes of micro parts rely on manual sampling using confocal microscopy. This means also that no automatic data processing regarding the geometry and surface inspection is applied because such measurements are affected by strong image noise.

Due to these industrial requirements, methods such as tactile, confocal microscopy and phase retrieval [Fal12a] by means of multiple illumination directions [Ago10] are not suited for the inspection task since they are much too slow. In contrast to these methods, interferometric methods are established for a full field measurement. Examples of interferometric techniques include digital holography (DH) [Sch94], white-light interferometry (WLI) [Wya98] and computational shear interferometry (CoSI) [Fal15]. These methods are based on the determination of the optical path difference (OPD) of light scattered by or transmitted through a test object. Since available cameras can only measure intensities of light, the OPD must be encoded consequently. Based on the encoding technique utilized and the geometrical complexity of the test object, measurement uncertainties down to a fraction of the illumination's wavelength can be achieved [Mar05].

Here, we present a robust, fast and precise metrology method comprised of a holographic contouring system, as well as fast algorithms for geometric evaluation and surface defect detection, that paves the way for the inspection of metallic micro cups in less than a second. The holographic system is composed of four two-wavelength digital holographic microscopy setups. It uses four directions of illumination in order to enable the simultaneous observation of the whole object surface and for speckle noise reduction as well as two-wavelength contouring to collect form data. Spatial multiplexing, coherent gating, is used to capture four holograms by each camera in a single shot [Ago17]. Moreover, it utilizes a telecentric microscope objective instead of a standard microscope objective to compensate scaling effects and wave field curvature, which distort reconstruction in digital holographic microscopy. In order to enhance the limited depth of focus of the microscope objective, depth information from different object layers is stitched together to yield 3D data of its geometry utilizing the auto-focus approach presented in [Ago18].

The 3D data map is converted into a point cloud. The resulting point cloud data are automatically decomposed into simple geometric elements with a holistic approximation approach in order to analyze geometric deviations. In addition, surface defects are detected with a convolutional neural network. The measurement and data evaluation approaches are demonstrated by quality inspection of cold formed micro cups. As a result, defects larger than 2 μm in the lateral and 5 μm in the axial (depth) dimension can be detected.

The present section is organized in the following three subsections: Sect. 5.2.2 describes the measurement approach for dimensional inspection that allows the geometric features of deep-drawn micro cups to be characterized. Section 5.2.3 describes the dimensional inspection that allows for an automated dimensional analysis of prismatic surface data, where the surface can be a combination of different simple geometric base bodies (cylinder, plane, torus, etc.). Section 5.2.4 describes the detection of surface defects using convolutional neural networks (CNNs), which achieve high accuracy with only a few training samples.

5.2.2 Optical 3D Surface Recording of Micro Parts Using DHM

5.2.2.1 Holographic Contouring

Common engineering objects are much larger than the optical wavelength used for inspection, and the surfaces exhibit peaks and steps larger than one quarter of the wavelength of light. Such surfaces are therefore optically rough, and the measured phase values vary within the interval $[-\pi, \pi]$. The evaluation process becomes ambiguous and the measurement is not unique. The solution to this dilemma is the use of a synthetic wavelength much larger than the optical (natural) wavelength. In this approach, two phase distributions are retrieved from two measurements associated with two different wavelengths λ_1 and λ_2. Calculating the phase difference $\Delta\varphi$ between the reconstructed phase distributions corresponding to the two measurements, the 3D height map z_p of the test object can be directly calculated by utilizing [Fal12]

$$\Delta\varphi = -\frac{2\pi}{\Lambda} z_p(1 + \cos\alpha). \tag{5.1}$$

Here, α is the angle between the observation and illumination direction and Λ is the synthetic wavelength

$$\Lambda = \frac{\lambda_1\lambda_2}{|\lambda_2 - \lambda_1|}. \tag{5.2}$$

The resulting phase difference for a single measurement using Λ is given by Eq. 5.1. A measurement that covers the sample surface results in a map that contains fringes and is referred to as the phase contouring map. Adapting the difference between the two wavelengths λ_1 and λ_2 is required to enable the investigation of objects with step heights of several millimeters.

5.2.2.2 *Digital Holographic Microscopy*

The digital holographic microscope schematically sketched in Fig. 5.1 is used to capture the required two holograms with two different wavelengths. The setup contains a long-distance microscope objective (LDM) with a magnification of 10×, a numerical aperture (NA) of 0.21 and a working distance (WD) of 51 mm. Optical fibers are used to illuminate the test object and to provide the required reference wave. A beam splitter (BS) is used to combine the object wave, light diffracted from the surface under investigation, and the reference wave, resulting in a hologram which is captured across the camera plane. It is noted that there exists an angle α between the observation and illumination direction (see Eq. 5.1). Numerically, the phase distribution which corresponds to each measurement is reconstructed utilizing the spatial phase shifting method, where the carrier frequency is controlled by shifting the position of the reference wave with respect to the optical axis [Ago17]. Thus, the setup is used to register the phase information in a plane which is close to the surface of the object under test.

In the following, the 3D surface measurements of the micro cup based on digital holographic microscopy will be discussed; this is the backbone of geometric inspection.

Figure 5.2 shows an image of an experimental setup consisting of four DHMs based on the sketch of the individual DHM shown in Fig. 5.1. Four laser diodes are

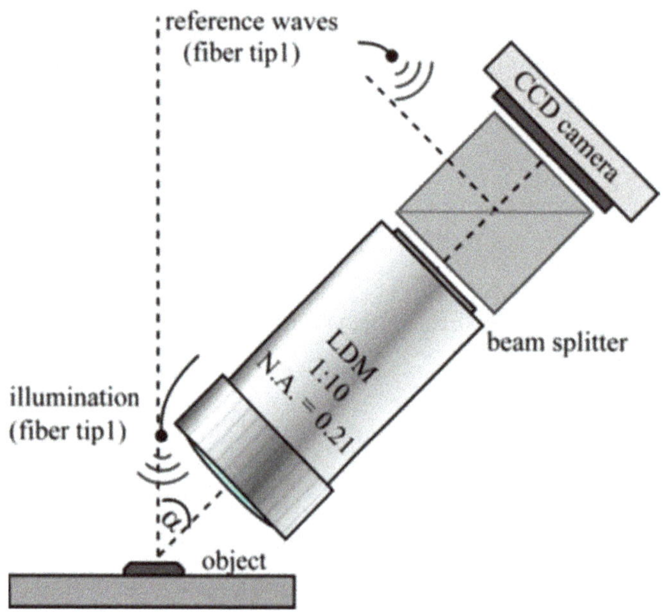

Fig. 5.1 Digital holographic microscope setup (DHM): To image the surface of the test object onto the utilized camera sensor, a long-distance microscope objective (LDM) with a 10× magnification, a numerical aperture N.A. = 0.21 and a working distance of 51 mm is employed. Optical fibers serve to illuminate the object under test and provide a spherical reference wave. For simplicity, only one illumination and one reference is shown. A beam splitter combines the object and the reference waves, producing a hologram across the camera plane

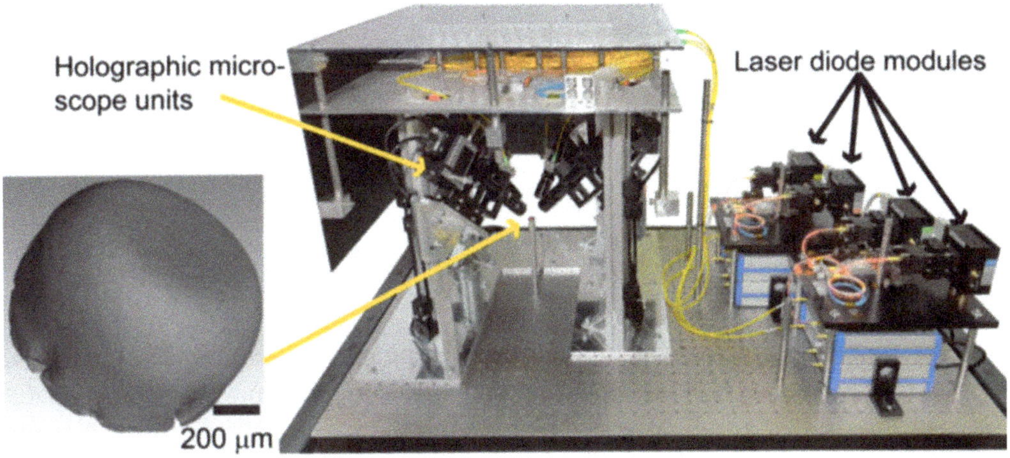

Fig. 5.2 Digital holographic microscopy (DHM) setup based on the scheme of the individual DHM shown in Fig. 5.1. The setup consists of four DHM units distributed around the test object shown in the coherence tomography image in the inset with a diameter of approximately 1 mm and a depth of 0.5 mm. Each unit delivers a measurement of a part of the test object. These four parts are then used to reconstruct the whole 3D shape of the test object. The long-distance microscope (LDM) is an object side telecentric objective with a numerical aperture of 0.21, a magnification of 10× and a working distance of 51 mm. The camera sensor has 2750 × 2200 pixels with a pixel pitch of 4.54 μm

used, two with $\lambda = 638.13$ nm and the other two with $\lambda = 644.08$ nm. According to Eq. 5.2, a synthetic wavelength of 69.07 μm results for the evaluation. Utilizing a fiber switch and a 1 × 2 fiber splitter, object and reference waves are formed. A digital hologram captured using the DHM is utilized for recovering the phase and the real amplitude of a monochromatic wave field across the object plane. The hologram generated across the output plane of the DHM is given by

$$H(X) = U_O(X) + U_R(X), \tag{5.3}$$

where U_O and U_R represent the complex amplitudes of light diffracted from the test object and a reference wave across the recording plane. Here, $X = (x_i, x_j)$ denotes the spatial coordinate vector at the recording plane. Equation 5.3 can be generalized as

$$H_{\lambda n}(X) = \sum_{n=1}^{N} (U_{O_{\lambda n}}(X) + U_{R_{\lambda n}}(X)), \tag{5.4}$$

representing the N holograms captured separately via the coherence-gating principle [Ago17]. The intensity of the hologram recorded by the DHM's camera is

$$I(X) = |H_{\lambda n}(X)|^2 = A(X) + \sum_{n=1}^{N}\left[\left(U^*_{O_{\lambda n}} \cdot U_{R_{\lambda n}}\right)(X) + \left(U_{O_{\lambda n}} \cdot U^*_{R_{\lambda n}}\right)(X)\right]. \quad (5.5)$$

Here, $|\ldots|^2$ denotes the intensity of the wave field

$$A(X) = \sum_{n=1}^{N}\left(|U_{O_{\lambda n}}|^2 + |U_{R_{\lambda n}}|^2\right) \quad (5.6)$$

that represents the dc term which is the incoherent superposition of the object and reference waves and * refers to the complex conjugation. Using the spatial phase-shifting method, the complex amplitude which corresponds to each wavelength can be determined. Based on the shifted reference point sources with regard to the optical access, the diffraction terms of Eq. 5.5 are separated. To show this, it is necessary to transform the intensity from the spatial domain to the frequency domain by implementing the fast Fourier transform (FFT). Performing the FFT ($\Im$) on Eq. 5.5 results in

$$\begin{aligned}\hat{I}(\upsilon) = \hat{A}(\upsilon) + \sum_{n=1}^{N}\Big[&\Im\left\{\left(U^*_{O_{\lambda n}} \cdot U_{R_{\lambda n}}\right)\right\}(\upsilon) \otimes \delta\left(\upsilon + \upsilon_{0,\lambda_n}\right) \\ &+ \Im\left\{\left(U_{O_{\lambda n}} \cdot U^*_{R_{\lambda n}}\right)\right\}(\upsilon) \otimes \delta\left(\upsilon + \upsilon_{0,\lambda_n}\right)\Big],\end{aligned} \quad (5.7)$$

where $\upsilon = (\upsilon_i, \upsilon_j)$ is a vector in the frequency domain, $\delta(\ldots)$ refers to the Dirac delta functions and indicates the corresponding shifted $\upsilon_{0,\lambda n}$ across the Fourier domain, which shows the effect of the shifted reference point sources, and $\otimes$ is the convolution symbol.

Figure 5.3a shows an example of a hologram captured by illuminating a micro cup from four different directions using the DHM setup. And Fig. 5.3b shows the spectrum of the single hologram, which contains object information for the four directions of illumination. It is noted here that there is no cross-talk between the four different holograms, which are recorded in a single shot camera image. Each object's information related to a certain illumination direction is shifted according to Eq. 5.7 to an exact position $\upsilon_{0,\lambda n}$.

It is noteworthy that the test object is illuminated from four different directions and four holograms are recorded on a single shot using four reference waves by applying the digital holography multiplexing principle [Ago17]. These holograms are used to reduce speckle noise in two-wavelength contouring. Accordingly, each holographic unit from the four units will capture two successive multiplexed holograms. The two successive multiplexed holograms are captured, one for each wavelength. In the following, the results that were obtained using the four observation directions will be presented and discussed. The time required for the capturing process and for the switching between the two wavelengths is 120 ms. Using

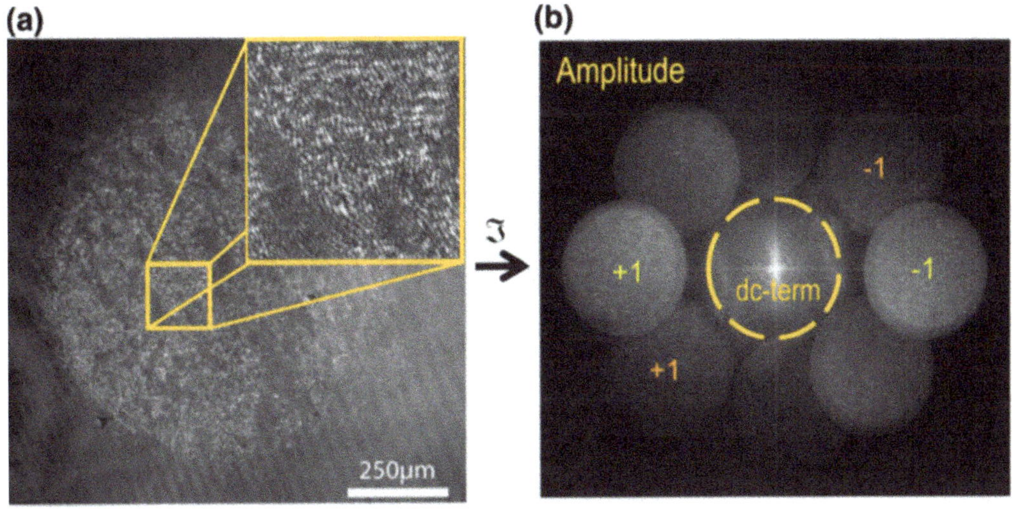

Fig. 5.3 **a** shows a single hologram, which contains object information for 4 directions of illumination and **b** shows the corresponding spectrum with four ± first order and the central dc components

the spatial carrier frequency method [Ago15], one can numerically reconstruct the phase distributions $\phi_{\lambda 1}$ and $\phi_{\lambda 2,}$ which correspond to the two measurements.

Figure 5.4 shows the recovered complex amplitude. The phase difference $\Delta = \phi_{\lambda 2} - \phi_{\lambda 1}$ between the two reconstructed phase distributions across the capturing plane that represents the countering map across that plane is shown in Fig. 5.4b. As can be seen in Fig. 5.4b, fringes are only sharp across the area of the object in focus, which can be clearly recognized from the real amplitude shown in Fig. 5.4a. This is as expected, since the microscope objective has a limited depth of focus. In contrast to microscopy, digital holography offers the extension of the objective depth of focus, and since holograms give access to the complex amplitude, digital refocusing across the whole object by means of numerical propagation is possible. Thus, in order to completely reconstruct a sharp contouring propagation, autofocus algorithms are used. For fast evaluation, an automated process was proposed and implemented within a graphics processing unit (GPU). The autofocus algorithm is implemented by scanning within small windows throughout all the propagated planes to define where the object is in focus by estimating the standard deviations of Δ, which are relatively high within these windows, where the object is out of focus because of the speckle decorrelation. The result of this process is shown in Fig. 5.5 for only one observation direction. In both contouring maps, Figs. 5.4b and 5.5b, a surface defect (dent) is clearly shown.

The contouring phase map shown in Fig. 5.5b is then unwrapped. Then the values are substituted into Eq. 5.1 to determine the 3D height map. The result is shown in Fig. 5.6. Eleven seconds are required for data transfer and hologram analysis to obtain the 3D point cloud which is used as input for the geometry and surface defect analysis. Defects with lateral extensions from 2 µm and a minimal

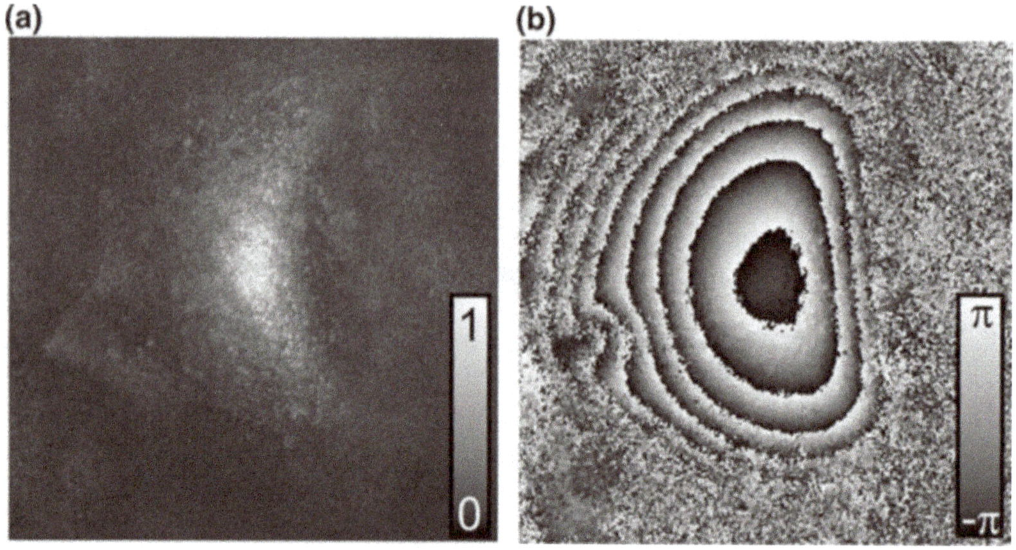

Fig. 5.4 **a** Image of the real amplitude of the reconstructed complex amplitude across the capturing plane of the recorded hologram for $\lambda = 638.13$ nm. **b** Image of the phase difference $\Delta = \phi_{\lambda 2} - \phi_{\lambda 1}$ between the two reconstructed phase distributions across the capturing plane. The image size is 2200×2200 pixels with a pixel pitch of 4.54 μm

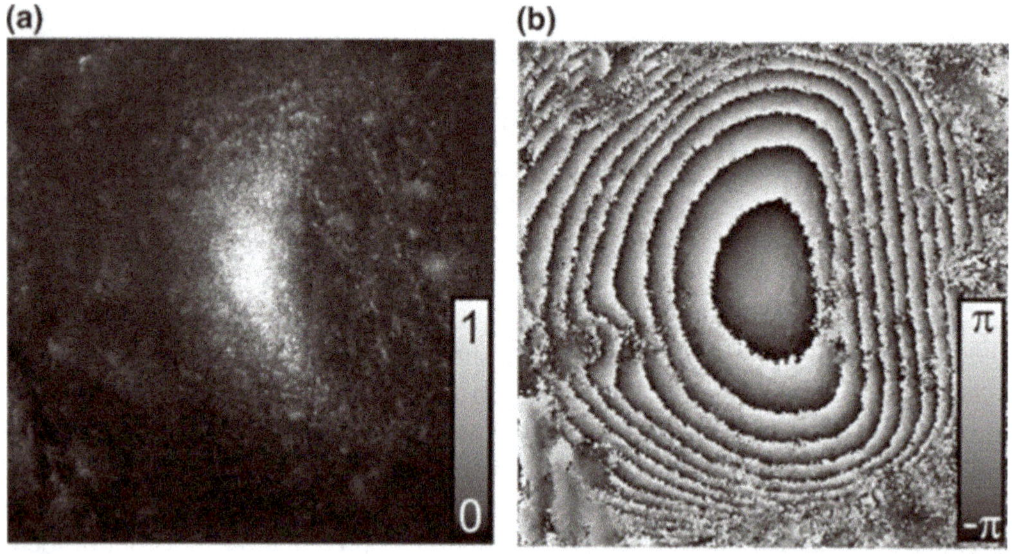

Fig. 5.5 **a** Image of the real amplitude, which represents a sharp image of the micro cup under test with respect to the observation direction. **b** Image of the phase difference distribution, which represents a sharp contouring phase map across the whole object

depth of at least 5 μm can be detected [Ago17]. Such a height map is converted to a 3D point cloud which is used as the input for both the dimensional and the surface inspection process, which will be discussed in the next section.

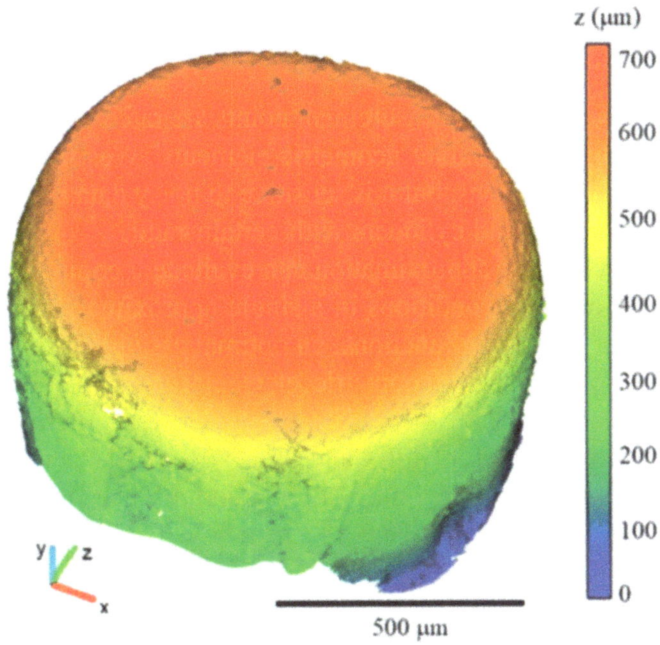

Fig. 5.6 The 3D height map calculated after unwrapping the countering phase maps obtained from the four holographic systems

5.2.3 Dimensional Inspection

Dimensional inspection implies the evaluation of surface data with respect to dimensional, form and position deviations of certain geometric features. These deviations are compared to the specified tolerances in order to decide whether the workpiece meets the quality requirements or not. The following subsections give a brief survey of the state of the art in evaluating point clouds and present the holistic approximation as the method of choice for the dimensional inspection of optically acquired surfaces of micro parts.

5.2.3.1 State of the Art

Optical measurement data contain a high number of surface coordinates of one or multiple observation directions and represent either a free-form surface or a combination of several geometric elements. The evaluation of free-form surfaces, on the one hand, consists in aligning the measurement data to the nominal CAD data [Sav07] and calculating and visualizing the deviations of each measurement point. For this kind of quality inspection, several commercial solutions exist. On the other hand, in order to analyze the optical measurement data regarding geometric parameters like dimensional or shape deviations, the measured points have to be segmented. This means assigning the individual measurement points to the approximating geometric elements. However, a manual segmentation is time-consuming and not suitable for automated analysis within a mass production. Only by an automated segmentation of the measurement data, the individual measuring points can be assigned in a reproducible and optimal way to the corresponding geometric elements. Two approaches are known for such an automated segmentation:

1. Neighboring measurement points are rated based on their curvature and assigned to corresponding geometric elements [Wes06]. This method can provide accurate solutions, but it is sensitive to noisy data and not able to distinguish between spheres and cylinders with certain radii.
2. A holistic approximation can evaluate a composed set of data under the present boundary conditions in a single approximation task [Goc91]. By the definition of separating functions, an optimal assignment of the measurement points to the corresponding geometric elements (segmentation) can be carried out simultaneously. The method is presented for different applications, e.g. for a 2D combination of lines and circles [Lüb10], or for micro punches as a 3D combination of a cylinder, a torus and a plane [Lüb12].

It was proved that the holistic approximation with automated segmentation (second approach) is only slightly sensitive regarding the initial values of the approximation and at the same time converges reliably within wide ranges [Lüb10]. Furthermore, this method was successfully tested for the evaluation of micro-measurements [Zha11], and it allows the automatic detection of outliers by a combination with statistical methods [Gru69]. Thus, the second approach is particularly suited for noisy optical measurement data. However, the algorithms have not yet been implemented for the evaluation of optical data acquired with DHM.

5.2.3.2 Method

The holistic approximation will be described for the evaluation of micro cups, whose surface is acquired by DHM. The part's geometry is a combination of a cylinder with radius r_c, a (quarter) torus with wall radius r_w and a plane (see Fig. 5.7). These radii form a vector of shape parameters $\vec{a}_g$, while the position of

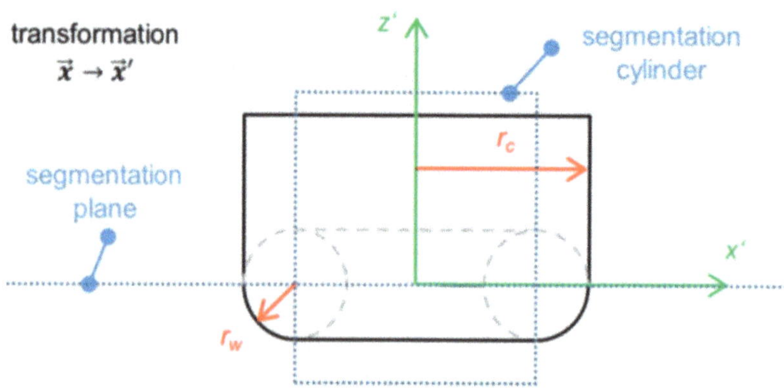

Fig. 5.7 Cross-section of micro cup model composed from geometric primitives (cylinder, torus, and plane) in the workpiece coordinate system (WCS) with segmentation elements according to the geometric model

the elements is included in a transformation vector $\vec{a}_p$. The detailed principle of the holistic approximation is described in [Goc91] for 2D combinations and in [Lüb12] for a 3D application. The approximation is performed by minimizing the L_2 norm

$$\min_{\vec{a}_p, \vec{a}_g} \left[\sum_{i=1}^{n_{cyl}} \left(d_{i,cyl}\left(\vec{a}_p, \vec{a}_g\right)\right)^2 + \sum_{i=1}^{n_{tor}} \left(d_{i,tor}\left(\vec{a}_p, \vec{a}_g\right)\right)^2 + \sum_{i=1}^{n_{pla}} \left(d_{i,pla}\left(\vec{a}_p, \vec{a}_g\right)\right)^2 \right]^{1/2}, \tag{5.8}$$

where n_{cyl}, n_{tor} and n_{pla} are the numbers of points assigned to the cylinder, the torus and the plane, respectively. A single point has the index i, and its orthogonal distance to the assigned geometric element is d_i. During the approximation, not only the degrees of freedom (transformation $\vec{a}_p$ and shape parameters $\vec{a}_g$) are optimized, but also the assignment of the measurement points to the geometric elements. This implies that the numbers of elements in Eq. 5.7 vary during the iterative calculation. The geometric assignment itself is based on a geometric model, which is presented as a cross-section in Fig. 5.7. It consists of a cylinder with radius r_c, whose axis represents the z-axis, a quarter of a torus in the x–y-plane with wall radius r_w and ring radius $r_r = r_c - r_w$ as well as a plane parallel to the x–y-plane at $z = -r_w$. This model contains certain geometric constraints, e.g. coaxiality of the cylinder and the torus axis, which are again perpendicular to the plane, as well as tangential transitions between all elements. These constraints result from the workpiece design and reduce the degrees of freedom to five transformation parameters $\vec{a}_p = [\Delta x, \Delta y, \Delta z, \varphi_x, \varphi_y]$ and two shape parameters $\vec{a}_g = [r_c, r_w]$. As the geometry is axially symmetric, the rotation φ_z around the z-axis remains disregarded.

Out of the geometric model, the decision rules shown in Fig. 5.8 are derived and implemented in the algorithm. All transformed points with a positive z-coordinate belong to the cylinder. The remaining points are distinguished by their polar radius, points with a radius $r_i < (r_c - r_w)$ are assigned to the plane, and the residual points are assigned to the torus.

5.2.3.3 *Verification and Measurement Results*

For the verification of the algorithms, the geometry of the measuring object was simulated as a combination of a cylinder, a torus and a plane. The cylinder radius was defined to $r_{c,0} = 412$ μm and the wall radius of the torus was $r_{w,0} = 126$ μm. These dimensions were chosen according to the application. The cylinder was formed by 300,000 equidistant points, the torus by approximately 100,000 points and the plane by 379,000 points, each element with a uniformly distributed noise in the normal direction of the nominal surface with different intervals $[-a_e/2, a_e/2]$ in seven steps between $a_e = 0.0 \ldots 5.0$ μm. Each case was simulated $n = 100$ times

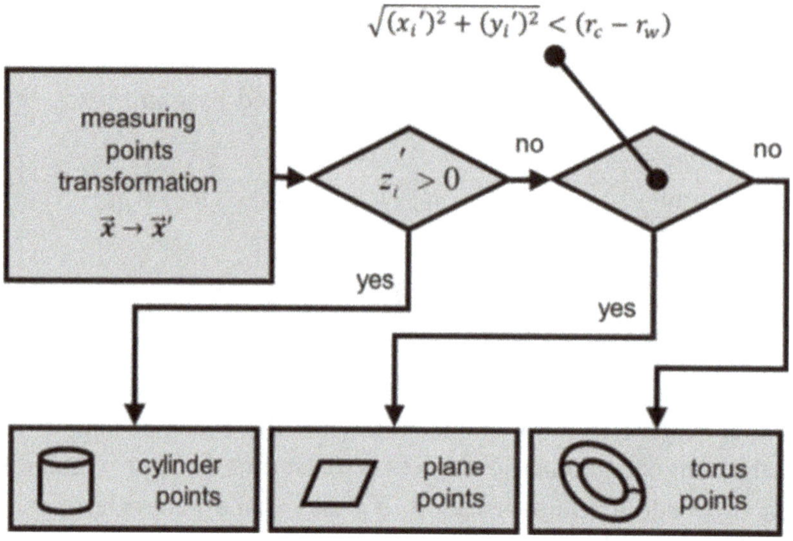

Fig. 5.8 Decision rules for assigning the measured points to the geometric primitives. All points with a positive z-coordinate are assigned to the cylinder; the remaining points are distinguished by their polar radius. Points with a radius $r_i < (r_c - r_w)$ belong to the plane, the residual points are assigned to the torus

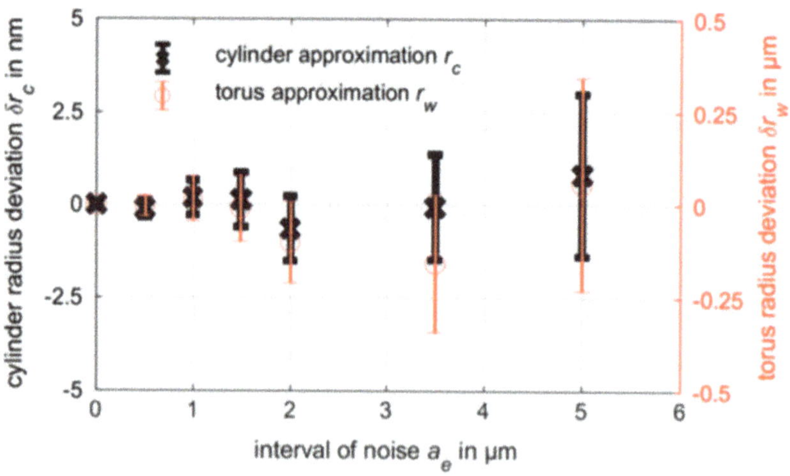

Fig. 5.9 Approximation results for simulated data (ca. 800,000 points) with different intervals of noise a_e: mean values of the radius deviations for cylinder $\boldsymbol{\delta r_c = r_c - r_{c,0}}$ and torus $\boldsymbol{\delta r_w = r_w - r_{w,0}}$ with standard deviations

and automatically evaluated by the holistic approximation. The results of the holistic approximation of the simulated data are presented in Fig. 5.9.

To analyze systematic deviations, a one sample t-test is performed with the hypothesis that the approximated radii r_c and r_w are equal to the set values $r_{c,0}$ and $r_{w,0}$ in the simulation. For this purpose, the radius deviations $\delta r_c = \mathrm{r_c} - r_{c,0}$ and $\delta r_w = r_w - r_{w,0}$ are introduced and the coverage factors

$$t_c = \frac{\overline{\delta r_c} \cdot \sqrt{n}}{\sigma_c} \quad \text{and} \quad t_w = \frac{\overline{\delta r_w} \cdot \sqrt{n}}{\sigma_w} \tag{5.9}$$

are calculated for the mean radius deviations $\overline{\delta r_c} = \overline{r_c} - r_{c,0}$ and $\overline{\delta r_w} = \overline{r_w} - r_{w,0}$, respectively, based on the mean approximated radii $\overline{r_c}$, $\overline{r_w}$ of the cylinder and the torus as well as their standard deviations σ_c, σ_w.

The maximum value for the cylinder radius is $t_{c,max} = 0.74$, whereas it is $t_{w,max} = 0.92$ for the torus radius. According to the t-distribution for a probability of 95% ($\alpha = 0.05$) and a degree of freedom of $f = n-1 = 99$, the critical value is $t_{crit} = 1.984$. As both calculated coverage factors are below this critical value, it can be stated that the verification results do not disagree with the hypothesis with a probability of error of 5%. Thus, it can be assumed that no systematic influence within the holistic approximation leads to significant deviations of both the approximated radii.

The random deviations can be characterized by the standard deviations of the calculated radii. In absolute numbers, the standard deviation of the cylinder radius is $\sigma_c < 22$ nm in this simulation, while the standard deviation of the torus radius is $\sigma_w < 2.87$ µm. The random deviations of the torus are 2 orders of magnitude higher than those of the cylinder radius, which is assumed to result from the approximation of only a part of the geometric torus object and agrees with earlier findings, e.g., the error of a spherical center approximation depending on the size of the measured spherical cap [Bou93], or the increased diameter [Fla01] or center uncertainty [McC79] with decreasing arcs of a circle. A second reason for the increased standard deviation of the torus radius might be the number of evaluated points. The torus was simulated with approximately 100,000 points, which is only a third of the number of points on the cylinder. Nevertheless, for both parameters the standard deviation is only a fraction of the initial amplitude of noise due to the high number of data points available.

In order to demonstrate the holistic approximation, the data acquired with DHM were evaluated based on the geometric model in Fig. 5.7. The three-dimensional approximation results with a measured cylinder radius of 497 µm and a torus radius of 234 µm are illustrated in Fig. 5.10 for a cross-section through the symmetry axis of the micro cup. Systematic deviations between the measured and approximated surface points occur, which are a result of real deviations of the cylindrical part of the three-dimensional micro cup from the desired geometry. Hence, the holistic approximation allows the identification of geometric deviations and, thus, the automatic quality inspection of geometric features in micro production. However, a crucial point is a tailored geometric model. Depending on the inspection task, the degrees of freedom of the geometric model can include the desired geometric parameters only (workpiece quality) or additional parameters for quantifying typical manufacturing errors (manufacturing process).

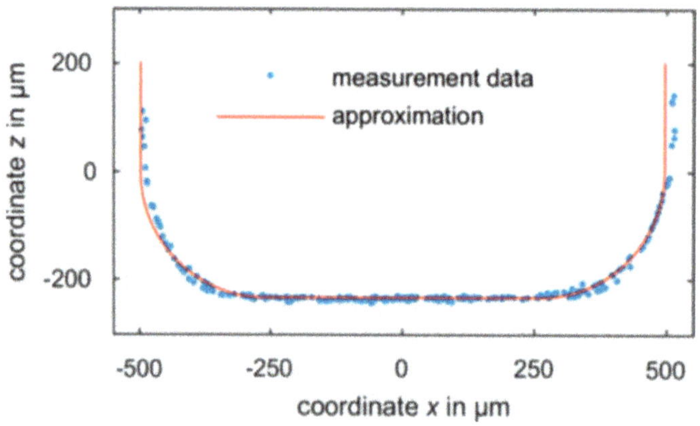

Fig. 5.10 Cross-section of the acquired data (measurement data, see Fig. 5.6) with the result of the holistic approximation. Note that shape deviations of the cylindrical part of the three-dimensional micro cup are responsible for the systematic deviations between the measured and approximated data points in the cross-section shown

5.2.4 Detection of Surface Defects

Surface defects such as scratches or dirt might be too small to cause a detectable change in the measured phase distribution. Hence, reliable detection necessitates additional methods, which incorporate the measured amplitude image.

5.2.4.1 State of the Art

Currently, algorithms for automatic surface inspection are to a large extent based on manually engineered features [Xie08], most commonly statistical and filter-based [Neo14]. While the introduction of expert knowledge often allows for the creation of powerful features, this process is laborious and might be necessary for each new product. General solutions that can automatically adapt to new problem sets could thus yield significant time and cost advantages. One such solution is convolutional neural networks (CNN). These have become the driving factor behind many recent innovations in the field of computer vision and allowed significant advances in various applications, such as object classification [Kri12] or semantic image segmentation [Yu15]. CNNs have also recently been successfully applied for industrial surface inspection [Wei16].

5.2.4.2 Methods

The core building block of CNNs is the convolutional layer. Instead of processing, e.g. an image all at once, it is divided into small (usually) overlapping windows and fed piece-wise into a neural network. Each window is thus mapped to a vector of activations of a shared neural network. Convolutional layers hence automatically learn a set of filters in the form of the networks weights.

In the most common framework, convolutional layers are combined with pooling layers, usually max-pooling layers. Max-pooling layers summarize the extracted features by taking the maximum activation for each unit over a small area. Deep CNNs are built by stringing together multiple convolutional and pooling layers. With increasing depth, the network thus extracts increasingly complex features for increasingly large image areas or receptive fields. The application of max-pooling thereby yields multiple advantages. By scaling down the input, the number of parameters is decreased, which increases the computational efficiency. At the same time, the receptive field sizes are increased and hence the amount of context that can be integrated by each unit of the neural network. Additionally, the use of max-pooling yields a small degree of translation invariance, which increases the network's robustness towards these operations. One disadvantage, however, is that, by scaling down the input, spatial resolution gets lost. This becomes an issue when the goal is spatially precise defect detection. To solve this problem, multiple solutions have been proposed in the field of semantic image segmentation, e.g. the use of dilated convolutions [Yu15], the U-Net architecture [Ron15] and the LinkNet architecture [Cha17].

One solution is to augment the classical CNN architecture with a second network for upscaling the spatial resolution. High-level, low-resolution features are thereby up-sampled and merged with the corresponding low-level, high-resolution features. This architecture is largely known as U-Net [Ron15]. The advantage is that it harnesses the benefits of max-pooling while still being able to give precise defect labels.

Here we implemented a modified version of the U-Net. Aside from accurate defect detection, our objective was thereby to keep the hardware requirements and computing time of the network as low as possible. To achieve this goal, we employed three recently developed methods. Firstly, our architecture is heavily inspired by densely connected CNNs [Hua17]. Secondly, we opted for depth-wise separable convolutions [Cho16]. Finally, our network takes inspiration from the LinkNet architecture [Cha17].

The basic idea behind densely connected CNNs is to feed into each new layer the activation of each previous layer. This allows a significant decrease of the number of connections in each layer as all the information from previous layers can be directly accessed instead of having to be repeated over each layer. Figure 5.11a illustrates the blocks of densely connected layers used in this work. Each layer propagates its activations to all the successive layers within a block. Filter sizes were chosen to be 5×5 for layers one and three in each block and 1×1 for layers two and four as well as the final output layer. Each layer thereby uses h units. In our experiments h was set equal to 12. As a bottleneck, we only concatenate the activations of all layers within the block to produce the blocks output, i.e. the input to a block is not propagated after the block. Implementation details are given in Table 5.1.

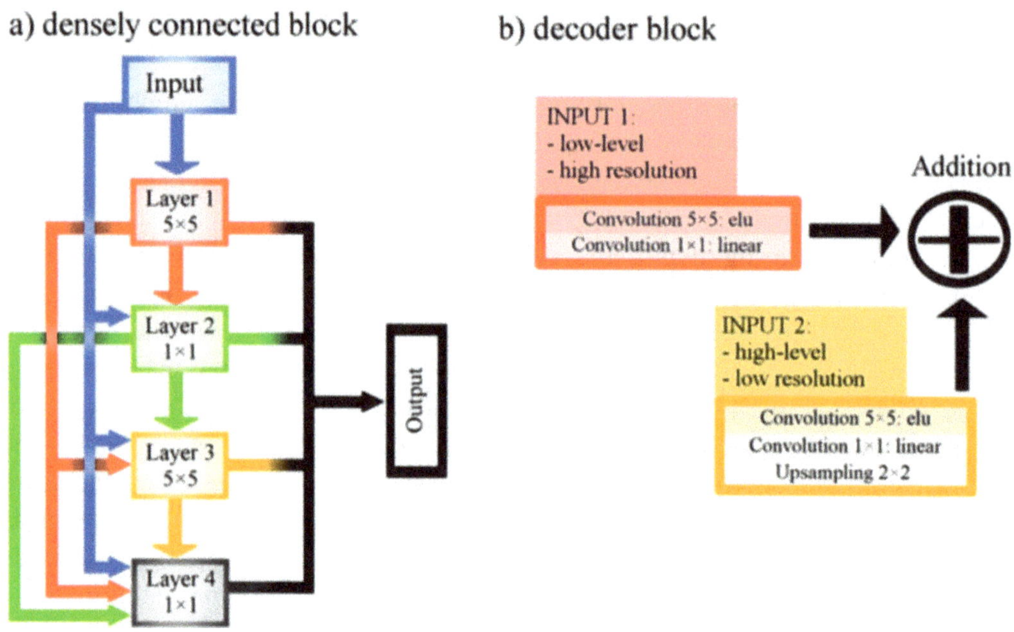

Fig. 5.11 **a** Densely connected block as used in this work. The Input (blue) is fed into all successive layers of the block. The activations of each layer (red, green, orange, gray) are fed into all successive layers. The output is constructed by concatenating the activations of all layers of the block (red, green, orange, gray), but not the input. **b** Decoder block to integrate low- and high-level features. Input 1 (upper left): Low-level features are subjected to one non-linear (exponential linear unit: elu) and one linear 1 × 1 convolution. The number of features is thereby reduced by a factor of 0.25. Input 2 (lower right): High-level features are subjected to one non-linear and one linear 1 × 1 convolution and then sampled up to match the resolution of the low-level input. The number of features is thereby also reduced to match the reduced low-level features

Table 5.1 Layer types, filter sizes, number of units and input/output dimensions for each layer within a block (W: width, H: height, SC: separable convolution)

Layer type	Filter size	# units	Input resolution	Output resolution
Input		d	–	W × H × d
SC	5×5	h	W × H × d	W × H × h
SC	1×1	h	W × H × h+d	W × H × h
SC	5×5	h	W× H × 2•h+d	W × H × h
SC	1×1	h	W × H × 3•h+d	W × H × h
Output			–	W × H × 4•h

The second method for increasing the model efficiency is the use of depth-wise separable convolutions [Cho16]. The principle behind depth-wise separable convolutions is that, instead of performing convolutions over all channels within a spatial window simultaneously, the spatial and the depth/channel-wise convolutions are performed separately. This allows for a significant decrease in the amount of network connections.

The third method for increasing the model efficiency takes inspiration from the LinkNet architecture [Cha17]. There are two ways in which LinkNet increases the efficiency of the standard U-Net architecture. Firstly, high- and low-level features are merged via addition instead of concatenation. Secondly, the number of features is also reduced before the summation. Our implementation of this procedure is shown in Fig. 5.11b.

The final network consists of a down-sampling and up-sampling part. The spatial resolution of the input image is thereby down-sampled from 512 × 512 to 8 × 8 and then up-sampled again to 512 × 512. The down-sampling part consists of densely connected blocks followed by a 4 × 4 max-pooling operation and batch normalization [Iof15]. In the up-sampling part, we use decoder blocks as illustrated in Fig. 5.12 to efficiently integrate low- and high-level features.

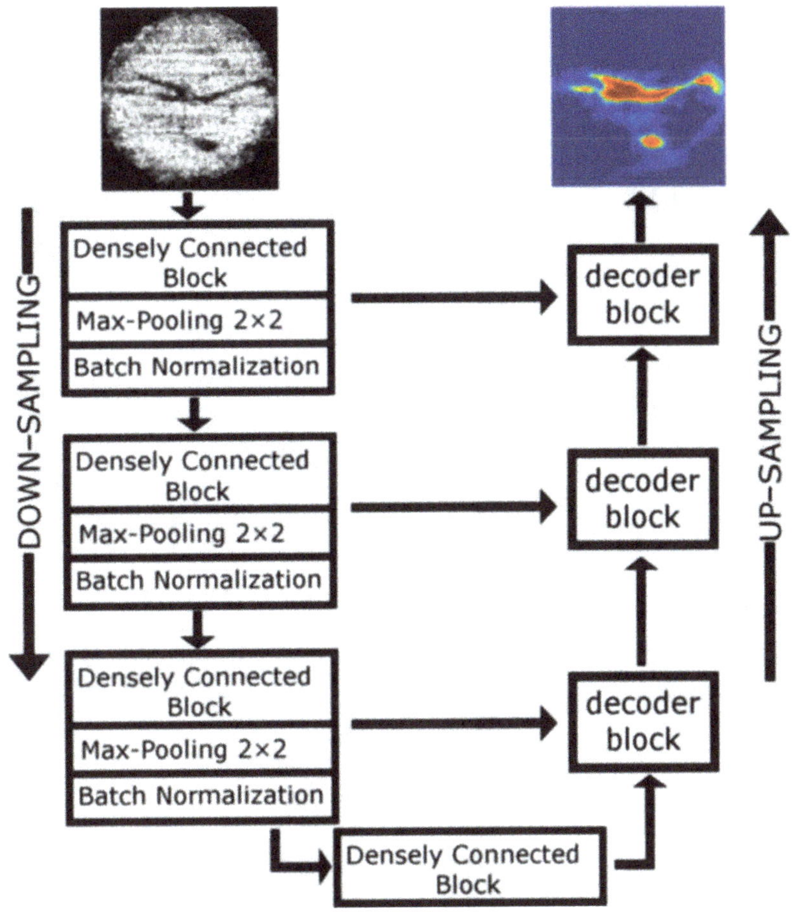

Fig. 5.12 Dense U-Net as implemented for this work. The picture on the upper left shows the input image of the test part to be inspected, while the result on the upper right shows the predicted defect position. In the down-sampling part, the input is fed into blocks of densely connected convolutional layers (see Fig. 5.11a), followed by 4 × 4 max-pooling and batch normalization. The up-sampling part uses decoder blocks as described in Fig. 5.11b. The spatial resolution is restored via up-sampling and concatenation with the corresponding layer of the down-sampling part as well as another densely connected block

As activation functions, we used exponential linear units (elu) [Cle15] in all but the output layers. For the output layer, we used sigmoid units to constrain the output to the interval [0, 1]. The network was trained by minimizing binary cross-entropy (also known as log-loss) using the Adam optimizer [Kin14]. The learning rate was initialized at 0.001 and automatically reduced by a factor of 0.1 when no decrease in loss was observed for more than 10 epochs. The mini-batch size was set to eight. All experiments were conducted using the keras library for Python.

To improve the defect labeling, we automatically created a mask for background subtraction for each measurement. The steps are shown in Fig. 5.13. First we applied a strong low-pass filter to the measurement (i.e. convolution with a 55×55 matrix of ones). We binarized the resulting image using its mean as a threshold. In the resulting binary matrix, the largest contour was detected using methods provided by the OpenCV software library [Bra00]. The background mask was directly fed into the CNN and multiplied with its output layer. This allowed the network to learn features only for the relevant parts of the image.

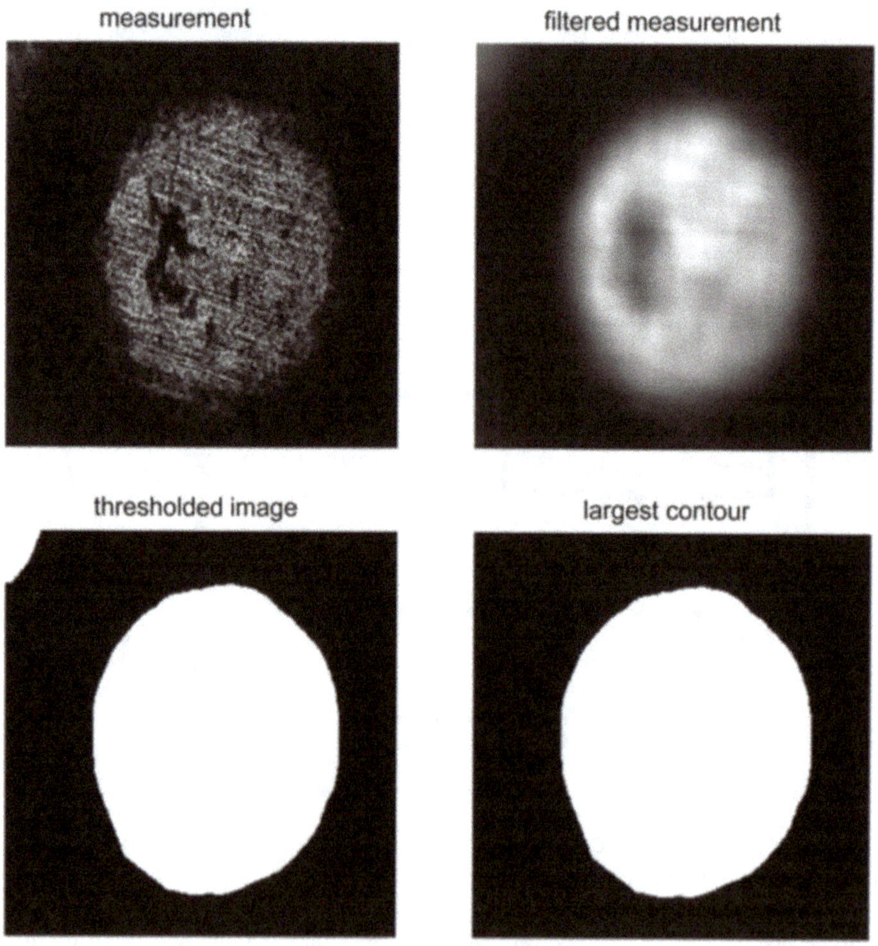

Fig. 5.13 Background removal: In order to improve defect detection, the background was removed from the final classification result. The measurement (upper left) was low-pass filtered. The resulting image (upper right) was thresholded by its mean value. In the resulting image (lower left) the largest contour was detected and used as a mask for background removal (lower right)

We evaluated our method by using 64 samples for training and the remaining 5 samples for evaluation. To increase the amount of training data artificially, we used the following operations: horizontal flipping, vertical flipping, random rotations, and scaling the size by a factor between 0.9 and 1.1 (cropping or adding the additional/missing pixels at the boundaries).

5.2.4.3 Validation

Figure 5.14 shows the resulting defect maps for the testing data. All defects are detected and marked correctly. However, it should also be noted that the defect labels are still rather coarse, especially around the borders.

Defect detection for a single input image takes <130 ms on our test system (AMD Ryzen Threadripper 1900X 8-Core Processor $\times$ 16, 64 GB RAM, GeForce GTX 1080 Ti). Additional speed gains can be achieved by processing multiple images at once, as this would decrease the amount of data transfer towards and from the Graphics Processing Unit (GPU).

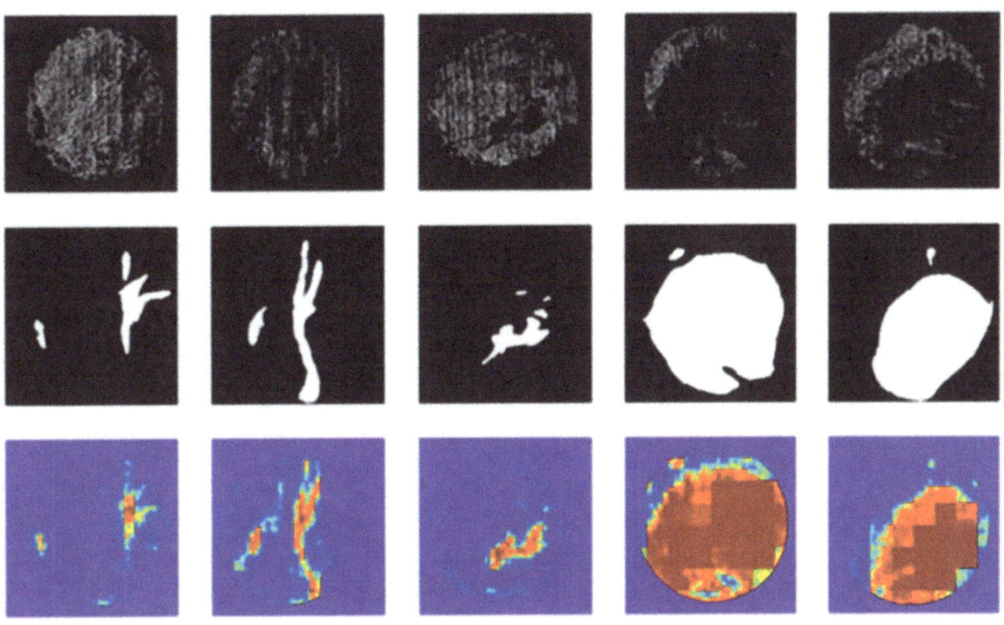

Fig. 5.14 Network prediction for all five test measurements. Top row: input measurement after background subtraction. Middle row: defect masks. Bottom row: defect predictions. All defects are marked correctly

5.3 Inspection of Functional Surfaces on Micro Components in the Interior of Cavities

Aleksandar Simic*, Benjamin Staar, Claas Falldorf, Michael Lütjen, Michael Freitag and Ralf B. Bergmann

Abstract A fast and precise solution for the inspection of the interior of micro parts using digital holography is presented in this chapter. The system presented here is capable of operating in an industrial environment. For this purpose, a compact Michelson setup in front of the imaging optics is used, so that the light paths of the object- and reference arm are almost identical. This makes the system less vulnerable to mechanical vibrations. A further improvement is obtained using the two-frame phase-shifting method for the recording of a complex wave field. This enables the usage of two cameras in order to allow the recording of a complex wave field in a single exposure. With the help of two-wavelength contouring, optically rough objects with a synthetic wavelength of approximately 93 µm are investigated. The measurement results make it possible to determine the shape of the interior surface and faults such as scratches with a resolution of approximately 5 µm. In order to fully utilize the measurement speed of the setup, a fast and reliable solution for automatic defect detection is required. For a profitable industrial application, it is therefore crucial to reliably detect all defective parts while producing little to no false positives (i.e. pseudo-rejections). This is realized by utilizing prior knowledge about the object shapes to implement fast phase unwrapping for defect detection. Defects are then reliably detected by identifying consecutive areas of deviation in relative depth. The evaluation of measurements taken in an industrial environment shows that this approach reliably detects all defects with a false-positive rate of less than one percent.

Keywords Quality control · Optical monitoring · Digital holography

5.3.1 Introduction

Micro cold formed parts are produced in high quantities, as many of such are incorporated inside a complete system. The mass production of micro parts can only be efficient if the quality inspection of these parts is incorporated within the production line. Optical metrology offers the opportunity to determine the shape of the structures of such parts and allows for quality control. Automated quality control with the help of an automated optical system within the industrial production line reduces the costs and time that would otherwise be required for a sophisticated manual inspection. Up to now, tactile methods have been used to inspect components on a sample basis, but these are not suitable for fast quality inspection in the production line as they are too slow and might alter the sample.

Among the non-tactile methods, confocal microscopy is commonly used for inspection but is clearly too slow for an automated 100% inspection. An overview of such methods can be found in [Ber12, Kop13]. Alternatively, white-light interferometry (WLI) is suitable, as it measures with high speed and is highly precise. For a review of WLI, see [Gro15]. However, WLI requires a comparatively large number of recordings, commonly by depth scanning, to capture depth information.

Digital holography (DH) is precise and only requires a small number of recordings to obtain the object shape, as shown in [Fal15]. This makes it a good candidate for the fast three-dimensional inspection of micro parts. Usually DH uses the method of temporal phase-shifting for phase evaluation, which is generally realized with a piezoelectric device. To realize a system which exhibits an even higher robustness, the method of two-frame phase-shifting is used to measure the object shape in two consecutive exposures.

5.3.1.1 Digital Holography

Historically, an interference pattern of an object- and reference wave was recorded on holographic plates and evaluated afterwards. However, such holographic plates can only be used a single time and require wet chemical processing. With the rise of computational methods and image-processing methods with the use of CCD and CMOS cameras, classical holography was replaced by digital holography and digital holographic microscopy (DHM) for the inspection of microscopic objects. An interference pattern arising from the light scattered by the investigated object and a reference wave is recorded in the CCD plane to extract the complex wave field of the object under investigation (see Fig. 5.15).

The intensity distribution of the interference pattern is given by

$$\begin{aligned} \mathrm{I}(\vec{\mathrm{r}}) &= |\mathrm{U_O}(\vec{\mathrm{r}}) + \mathrm{U_R}(\vec{\mathrm{r}})|^2 \\ &= |\mathrm{U_O}(\vec{\mathrm{r}})|^2 + |\mathrm{U_R}(\vec{\mathrm{r}})|^2 + \mathrm{U_O^*}(\vec{\mathrm{r}}) U_{R(\vec{r})} + U_R^*(\vec{\mathrm{r}}) \end{aligned} \tag{5.10}$$

where $\vec{\mathrm{r}}$ is the position vector and $\mathrm{U_O}(\vec{\mathrm{r}})$ and $\mathrm{U_R}(\vec{\mathrm{r}})$ are the object- and reference-wave fields respectively, with $\mathrm{U_O^*}(\vec{\mathrm{r}})$ and $\mathrm{U_R^*}(\vec{\mathrm{r}})$ being the particular conjugated wave fields. When multiplying this interference pattern with the complex amplitude of the reference wave, the following fundamental equation is obtained:

$$\begin{aligned} \mathrm{I}(\vec{\mathrm{r}}) \cdot \mathrm{U_R}(\vec{\mathrm{r}}) = &\left(|\mathrm{U_O}(\vec{\mathrm{r}})|^2 + |\mathrm{U_R}(\vec{\mathrm{r}})|^2\right) \cdot \mathrm{U_R}(\vec{\mathrm{r}}) + \mathrm{U_O^*}(\vec{\mathrm{r}}) \cdot \mathrm{U_R}(\vec{\mathrm{r}})^2 \\ &+ \mathrm{U_O}(\vec{\mathrm{r}})|\mathrm{U_R}(\vec{\mathrm{r}})|^2. \end{aligned} \tag{5.11}$$

The first term of Eq. 5.11 represents the DC term, which can be observed in the center of the recorded picture as the brightness of the image. The second term

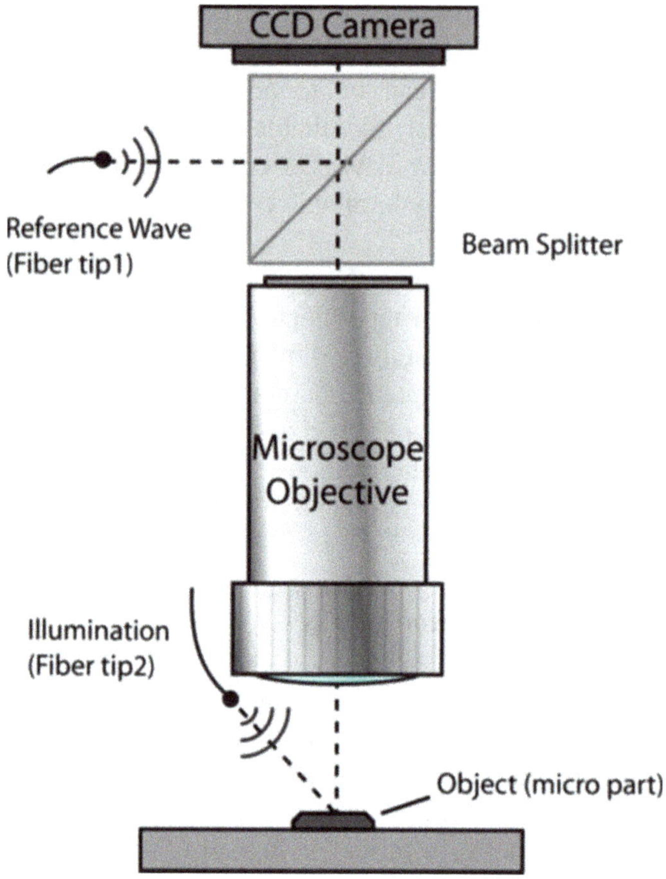

Fig. 5.15 Sketch of a conventional setup in digital holography. The object (micro part) is illuminated and the image is magnified with the help of a microscope objective and projected on a CCD camera. At the same time, the CCD is illuminated with a reference wave. The arising interference pattern gives the opportunity to extract the complex wave field [Ago17]

describes an inverted image $U_O^*(\vec{r}) \cdot U_R(\vec{r})^2$ of the object. The last term represents a virtual image of the object. The phase Φ of the wave field $U_O(\vec{r})$ contains information on the form of the object. The height h of the observed object can be calculated with the phase Φ from

$$\Phi = \frac{2\pi}{\lambda} \cdot 2h \quad (5.12)$$

5.3.1.2 Two-Wavelength Contouring

Two-wavelength contouring is an established method for the form recognition of diffusely reflecting objects. Falldorf et al. managed to significantly enhance the signal-to-noise ratio of the holographic measurement of optically rough objects with the help of this method [Fal12]. For objects that have a roughness larger than half of the wavelength of the illuminating light, there will be ambiguous results according to Eq. 5.12, as the phase only varies in the range of $-\pi$ to π. In two-wavelength contouring, two light sources are used to perform the measurement with slightly different wavelengths, λ_1 and λ_2. The corresponding phase distributions are subtracted and the resulting phase difference can be interpreted as a single measurement with a synthetic wavelength [Fal15] of

$$\Lambda = \frac{\lambda_1 \lambda_2}{|\lambda_1 - \lambda_2|} \tag{5.13}$$

The synthetic wavelength can be chosen to be much larger than the surface roughness by adjusting λ_1 and λ_2 to resolve the ambiguity problem.

5.3.1.3 Two-Frame Phase-Shifting

The recorded digital hologram from Eq. 5.11 only contains object information in the virtual image, and the remaining inverted image and DC term are generally not of interest. Therefore, only a small part of the camera resolution can be used with this method. To use the complete spatial bandwidth, the method of temporal phase-shifting is used, where the phase is shifted by a fractional amount of the wavelength to generate several equations and extract detailed phase information.

The recorded interference pattern from Eq. 5.10 can be expressed with the help of the phase difference $\Delta\Phi$ of the involved wave fields, to extract 3D-information on the considered object by using

$$\begin{aligned} I(\vec{r}) &= |U_O(\vec{r}) + U_R(\vec{r})|^2 \\ &= I_O(\vec{r}) + I_R(\vec{r}) + 2\sqrt{I_O(\vec{r}) I_R(\vec{r})} \cos(\Delta\Phi) \end{aligned} \tag{5.14}$$

This equation contains the three unknown variables $I_O(\vec{r}), I_R(\vec{r}), \Delta\Phi$, which makes it impossible to extract the phase difference $\Delta\Phi$. To solve this problem the phase difference $\Delta\Phi$ of the interference pattern produced is shifted with a known factor Δ to obtain a system of at least three equations from the corresponding recorded intensities. This process is generally accomplished with the help of a piezoelectric device in the setup. The resulting phase distribution is wrapped in the bounded interval $[0, 2\pi]$ and has to be unwrapped to determine the continuous behavior.

To incorporate this method in an industrial environment, its robustness is improved by replacing the piezoelectric device. To maintain the complete space–bandwidth product of the detected signal, it is vital to find a way of using temporal phase-shifting in a single camera exposure. For this purpose, the temporal phase-shifting method is used with the help of only two recorded interference patterns. This is accomplished by using circular polarized light from the object and linear polarized light from the reference mirror. Nozawa et al. have already used this system for single-shot and highly accurate measurements of complex amplitude fields with a simple optical setup [Noz15].

For the recorded interference patterns in the CCD plane, the wave field $U_G(\vec{r} = z_0)$ in the CCD plane can be written as

$$U_G = (I_1 - I_0) + i(I_2 - I_0), \tag{5.15}$$

with I_1 and I_2 being the phase-shifted single recorded interference patterns shifted by $\pi/2$, respectively. Liu et al. showed that the DC term I_0 is given by [Liu09]

$$I_0 = \frac{2R^2 + I_1 + I_2}{2} - \sqrt{\frac{(2R^2 + I_1 + I_2)^2 - 2(I_1^2 + I_2^2 + 4R^4)}{2}}, \tag{5.16}$$

with the amplitude of the reference wave R.

5.3.2 Experimental Alignment

For the inspection of the interior of a micro deep drawing component with DHM, a Michelson interferometer based setup, shown in Fig. 5.16, is constructed with the object in one arm and a plane mirror in the other arm to provide a reference wave. The Michelson interferometer has been placed as close to the object as possible in order to minimize the lengths of the separated light paths. The object is introduced to the setup with the help of a positioning apparatus, constructed by the industrial partner Stüken Corp. Furthermore, the usage of polarization optics and two cameras enables the simultaneous recording of two phase-shifted images. In this configuration, the setup is more stable with respect to exterior disturbances compared to a setup using a piezoelectric device.

After leaving the fiber switch, the light is parallelized with the help of a collimating lens and is linearly polarized. A $\lambda/4$ plate then polarizes the light circularly and illuminates the object through a beam splitter. At the same time, half of the intensity is redirected and linearly polarized again to illuminate the reference mirror. After traveling through a microscope with 5x magnification, the light is again divided with the help of a polarization-sensitive beam splitter to lead it to two camera targets at the same time. By using such a beam splitter, two interference patterns are projected on the camera targets, shifted by 90°.

As light sources, two diode lasers with output powers of $P_1 = 70$ mW and $P_2 = 20$ mW and with wavelengths $\lambda_1 = 642.2$ nm and $\lambda_2 = 637.8$ nm are used. With that, the camera exposure times were set to 8 ms to fully illuminate the camera targets. With the employed wavelengths, the synthetic wavelength amounts to $\Lambda =$ 93.1 μm. To avoid coherent amplification and to minimize speckle noise, laser light with coherence lengths of less than 1 mm is used.

5.3.2.1 Experimental Results

Figure 5.17 shows a sketch of the investigated parts, which have a functionality area that is vital for the component. This area must not exhibit scratches or

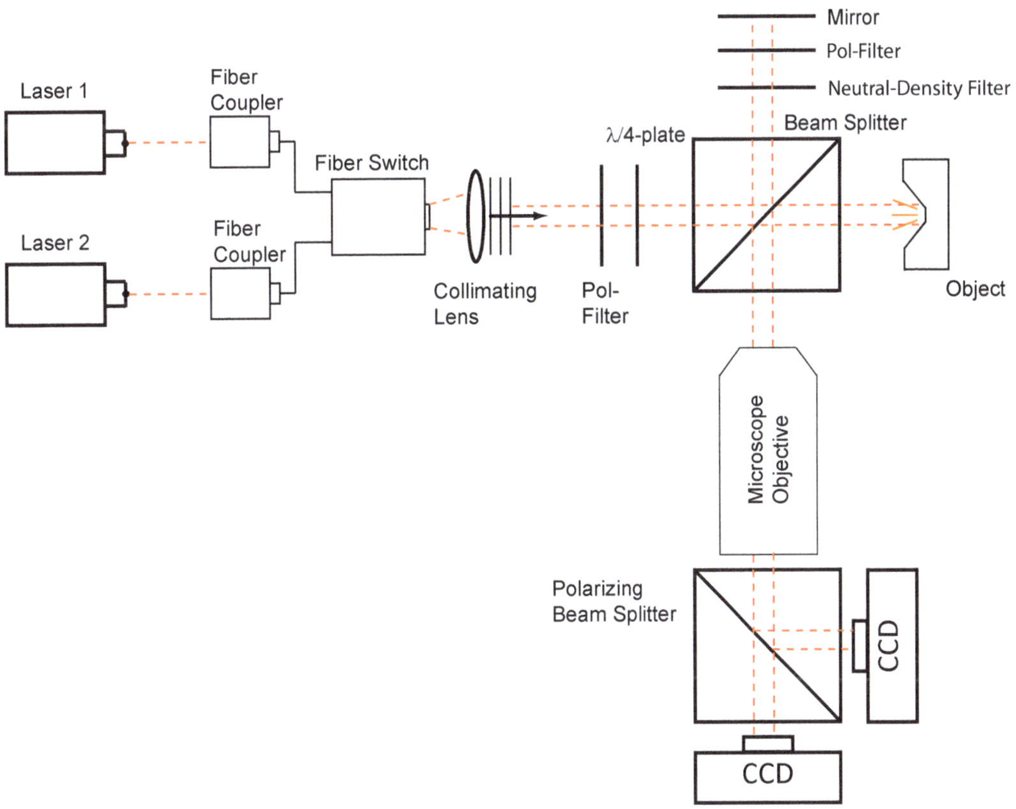

Fig. 5.16 Setup for the internal inspection of micro deep-drawing parts. Before entering the interferometer shown in the right part of the drawing, the light is polarized circularly with a λ/4-plate. The interferometer consists of a mirror in one arm and an object in the other arm. With the help of a polarizing filter, the reference light is polarized linearly while the object wave has a circular polarization. The image is magnified 5x using the microscope objective. By using a polarizing beam splitter, two interference patterns are projected on two cameras at the same time, being shifted by 90°

imperfections and therefore has to be inspected. 247 parts were inspected, of which 230 were defect-free and 17 were bad parts which were identified prior to the measurement.

The setup was implemented in the department for quality assurance of Stüken Corp., a producer of micro deep drawing parts. The recorded phase distribution displays the inner form of the object and allows for a classification of the functionality. Figure 5.18a shows the phase of the functional area of the component presented above. Two consecutive fringes indicate a height difference of Λ/2 on the object surface. Any deviation of the concentric inner form shows potential faults in the functional area, which can be seen in Fig. 5.18b, for example.

The phase can be unwrapped to obtain metric data with the help of Eq. 5.12. The unwrapped area around the defect can be seen in Fig. 5.19. Eight points on the dashed red line inside the defect were considered and compared with eight points on the parallel dashed red line outside the defect to calculate the mean depth of the error. Figure 5.19 contains measured values for the phase difference of the

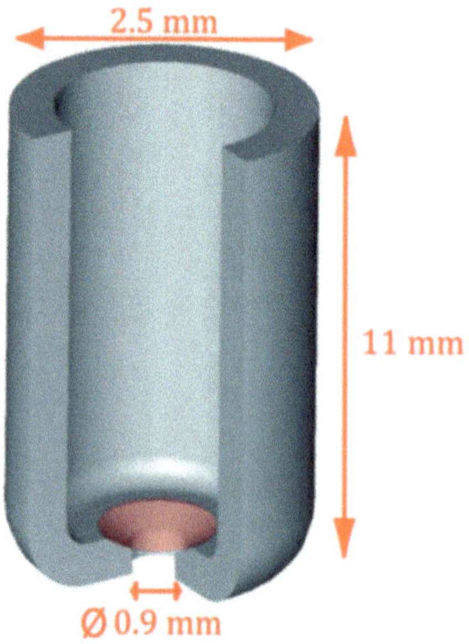

Fig. 5.17 Sketch of the investigated micro part. The area marked in light red around the lower hole serves as a functional area and has to be inspected. Taken from [Sim17]

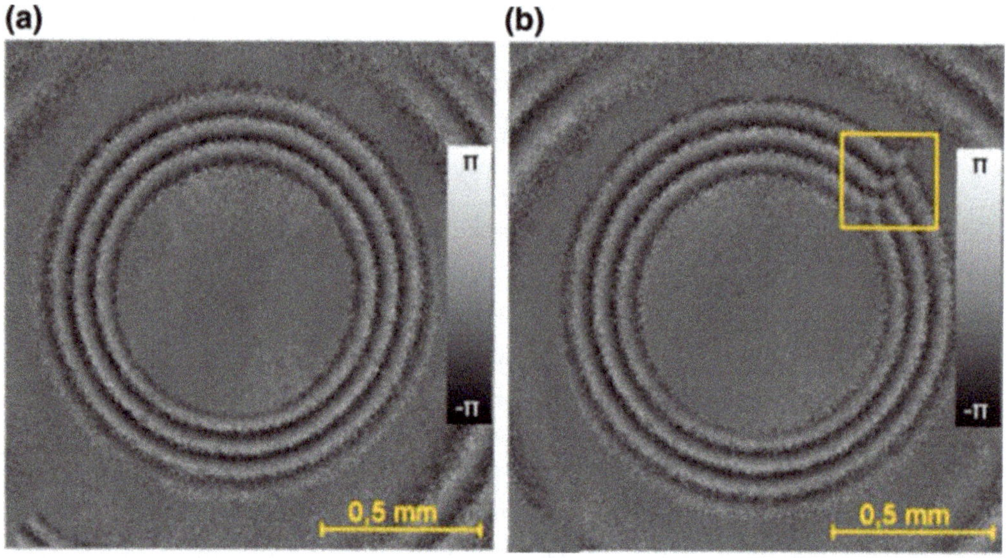

Fig. 5.18 **a** Phase distribution of the recorded functional area of the inspected micro part, which is an acceptable part. In **b** one can see the function area with a potential fault. Taken from [Sim17]

measured wave field. To convert the measured values into metric data, Eq. 5.12 is used with respect to a doubled light path as the light is reflected from the object. Then the mean depth d of the error can be calculated using the averages of the phase difference $\Delta\Phi_1, \Delta\Phi_2$ at the dotted lines of

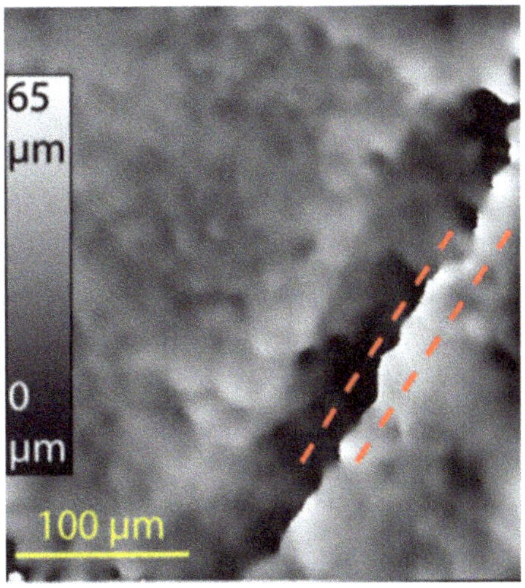

Fig. 5.19 Defect detected in the functional area after unwrapping the detected phase. The mean depth of the scratch is calculated on the dotted lines and amounts to d (x, y) = (20.2 ± 1.5) µm. Taken from [Sim17]

$$d(x,y) = \frac{\Delta\bar{\Phi}_1(x,y) - \Delta\bar{\Phi}_2(x,y)}{4\pi} \cdot \Lambda = (20.2 \pm 1.5)\mu m. \tag{5.17}$$

With this system, a lateral as well as a depth resolution of 5 µm can be achieved.

5.3.2.2 Comparison with X-Ray Tomography

For validation, the functionality area of the bad part from the last section was inspected using X-ray tomography. Figure 5.20 shows the measurement result, which depicts a scratch with a depth of 23 µm at the marked spot. With that, the result from Eq. 5.16 can be validated.

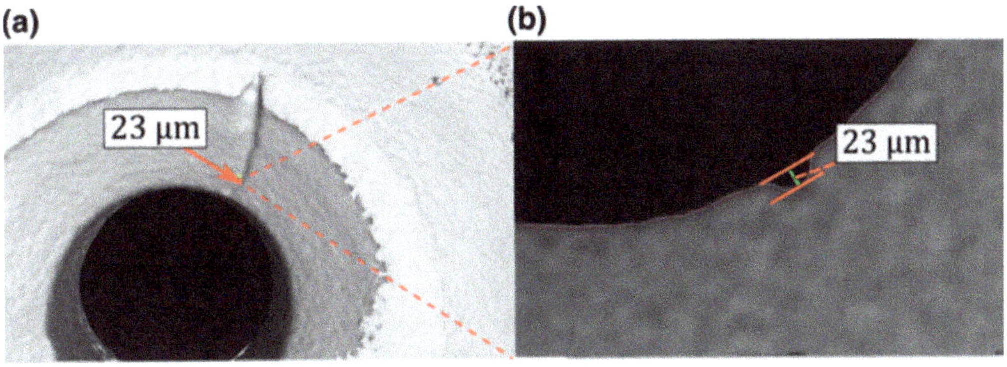

Fig. 5.20 **a** Measured spot on the functional surface with X-ray tomography. **b** Cross-section with the result of the depth measurement

5.3.2.3 Different Batches of Material

Different batches of materials were investigated with the presented system to evaluate the method. These were glossy parts, oily parts and heat-treated parts. A comparison of those three material types is shown in Fig. 5.21. Especially glossy parts do not scatter the incoming light and directly reflect most of it instead. Therefore, the signal-to-noise ratio decreases especially for the glossy parts, making further evaluation therefore impossible.

5.3.3 Automatic Defect Detection

For the effective utilization of the setup's measurement speed, manual evaluation of the measurements is not feasible. Hence a solution for automatic defect detection was developed. Thus the challenge was threefold: Firstly, the method had to be fast, as a slow algorithm would be detrimental to the fast measurement system. Secondly, defect detection had to be very accurate, with zero false negatives (undetected defects) and less than 4% false positives (intact parts falsely labeled as defective). Thirdly, due to the well optimized process, the number of defective samples was very small. Consequently, the application of state-of-the-art machine learning methods, like e.g. convolutional neural networks, which have been applied by Ronneberger et al. [Ron15] and Weimar et al. [Wei16] for example, was not feasible.

The detection of the measured part is achieved by detecting circles in the phase distribution $\Delta\Phi$. Potential defects are then filtered out by applying a low-pass filter (LPF) in a circular motion. The resulting prototype is then subtracted from the measured $\Delta\Phi$ to identify deviations via the application of a threshold. The whole defect detection pipeline is schematically shown in Fig. 5.22.

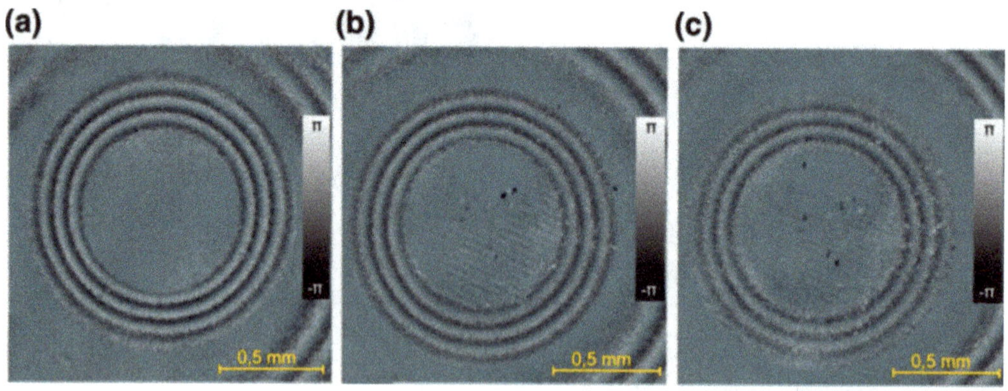

Fig. 5.21 Measurements of the same part of different batches of **a** heat-treated, **b** oily and **c** glossy material. The signal-to-noise ratio decreases for oily and glossy parts and does not allow a precise evaluation

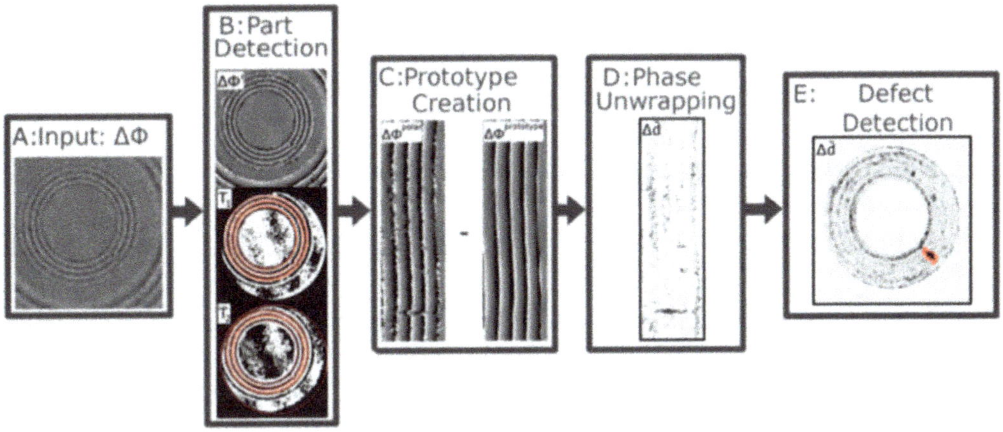

Fig. 5.22 Defect detection pipeline. **a**: phase distribution image $\boldsymbol{\Delta\Phi}$. **b**: Circles are detected in the phase distribution image $\Delta\Phi$ via thresholding and contour detection. **c**: A defect-free prototype, $\Delta\Phi^{prototype}$, is created by mapping $\Delta\Phi$ to polar coordinates, yielding $\Delta\Phi^{polar}$ and applying a LPF in the angular direction. **d**: Phase unwrapping is realized by subtracting the prototype $\Delta\Phi^{prototype}$ from $\Delta\Phi^{polar}$, yielding the depth deviation image $\Delta\bar{\mathbf{d}}$. **e**: Defects are identified from $\Delta\bar{\mathbf{d}}$ and marked accordingly (mapped back to Cartesian coordinates for better visualization)

5.3.3.1 Preprocessing

In order to decrease noise, $\Delta\Phi$ was filtered by applying a sin/cos LPF. Since $\Delta\Phi$ is a phase distribution, low-pass filtering in the complex plane may be applied, i.e. to the complex phasor $f_c(x) = e^{if(x)}$ which exhibits unit amplitude and $f(x)$ as the phase. This approach prevents filter artifacts at the phase transitions.

5.3.3.2 Part Detection

While objects are positioned close to the center of the measured image, there are significant deviations that do not allow the assumption of a fixed center without notable loss of accuracy. For the problem at hand, the measured part of the object is circular and the measurement is taken orthogonally. This is manifested in the resulting phase distributions as nearly concentric circles. The measured part's center can hence be located by the detection of these circles. This was implemented by first applying a strong sin/cos LPF to $\Delta\Phi$, yielding $\Delta\Phi'$ and subsequently applying a binary threshold to $\Delta\Phi'$, resulting in two images T_1 and T_2 with:

$$T_1 = \begin{cases} 1, & \Delta\Phi'_{i,j} \le 0 \\ 0, & \Delta\Phi'_{i,j} > 0 \end{cases} \quad T_2 = \begin{cases} 0, & \Delta\Phi'_{i,j} \le 0 \\ 1, & \Delta\Phi'_{i,j} > 0 \end{cases} \tag{5.18}$$

An example is shown in Fig. 5.22b. Subsequently, contours in T_1 and T_2 are detected and sorted by area. The largest contours are then fitted by their minimal enclosing circle. As a robust estimate of the object's center, the median of the resulting centers is taken, yielding the center (c_x, c_y).

5.3.3.3 Prototype Creation and Phase Unwrapping

In order to detect defects, the measured phase distribution $\Delta\Phi$ is compared to an ideal prototype of that measurement. Expecting the measured part to have a smooth surface, prototype creation is achieved by the application of a sin/cos low-pass filter in a circular motion. One could think of this as virtually regrinding the object to smooth out defects. The detailed steps for this process are as follows:

First $\Delta\Phi$ is mapped to polar coordinates with respect to the object's center detected in the previous step. The result is an $\beta \times r$ image $\Delta\Phi^{polar}$ whereby β marks the angular resolution and r the radius.

$$\begin{aligned} \Delta\Phi^{\text{polar}}[x, y] = \Delta\Phi\left[y \cdot \cos\left(\tfrac{x2\pi}{\beta}\right) + c_x, y \cdot \sin\left(\tfrac{x2\pi}{\beta}\right) + c_y\right] \\ x = 1, \ldots, \beta \in \mathbb{N} \\ y = 1, \ldots, r \in \mathbb{N} \end{aligned} \tag{5.19}$$

Low-pass filtering in a circular motion can thus be achieved by applying a sin/cos LPF to $\Delta\Phi^{polar}$, yielding $\Delta\Phi^{prototype}$ (for an example see Fig. 5.22c). The $m \times n$ filter matrix was thereby chosen to be much larger in angular direction m than in radial direction n. The reasoning is that, due to phase transitions, high-frequency components are expected in the radial direction even for smooth surfaces. In the angular direction, however, a smooth surface should only exhibit low-frequency components, as there should not be any phase transitions. Deviations in depth can thus be calculated via (Fig. 5.22d)

$$\Delta\bar{d} = \left|\Delta\Phi^{\text{prototype}} - \Delta\Phi^{\text{polar}}\right|. \tag{5.20}$$

5.3.3.4 Defect Detection

Potential defects are marked by deviations from zero in $\Delta\bar{d}$. Due to roughness of the measured part's surface that lies within tolerance, there might exist multiple such areas, even for intact parts. To differentiate between this background noise and actual defects, two different features are used. Firstly, errors are assumed to be marked by larger connected areas of deviations from zero, i.e., the area of an actual defect exceeds a certain threshold. Secondly, it is assumed that for defective areas the mean deviation from the background exceeds a certain threshold.

Accordingly, the defect detection routine searches for connected areas of deviations from zero in $\Delta\bar{d}$ with areas above an area threshold t_A where the mean deviation exceeds a depth threshold t_D. An example is shown in Fig. 5.22e.

5.3.3.5 Detecting Loss of Focus

Larger defects in the measured part's geometry as well as environmental influences might cause the measured phase distribution to be out of focus. However, the defect detection routine does not necessarily capture this case. Therefore, an additional method for detecting loss of focus was implemented. As a marker for focus, the orientation of gradients in $\Delta\Phi^{polar}$ is employed. The underlying assumption is that well focused parts of $\Delta\Phi^{polar}$ show homogeneous orientation of gradients, while unfocused areas show gradient orientations that are more or less random. Focused areas are hence marked by low standard deviation in the gradient orientations, while unfocused areas are marked by large standard deviation. The gradient orientation in $\Delta\Phi^{polar}$ is calculated in the following way. First, the Sobel derivatives [Sob90] are calculated by convolution with Sobel operators S_x and S_y, resulting in

$$G_x = S_x * \Delta\Phi^{polar} = \begin{bmatrix} 1 & 0 & -1 \\ 2 & 0 & -2 \\ 1 & 0 & -1 \end{bmatrix} * \Delta\Phi^{polar} \tag{5.21a}$$

and

$$G_y = S_y * \Delta\Phi^{polar} = \begin{bmatrix} 1 & 2 & 1 \\ 0 & 0 & 0 \\ -1 & -2 & -1 \end{bmatrix} * \Delta\Phi^{polar}. \tag{5.21b}$$

Large derivatives caused by phase transitions (Fig. 5.23b) are then removed by looking for entries in G_x and G_y with absolute values above five times the median absolute deviation (mad), i.e. $|G_x| > 5mad(|G_x|)$ and $|G_y| > 5mad(|G_y|)$ respectively. The identified values are then replaced by the respective median value (Fig. 5.23c).

Subsequently, the gradient orientation (Fig. 5.23d) is calculated via

$$\Theta = \arctan(G_y, G_x). \tag{5.22}$$

Then the local standard deviation $std(\Theta)$ of Θ over windows of 25×25 pixels is calculated

$$\mathrm{std}(\Theta) = (\Theta - (\Theta * K)) * K \tag{5.23}$$

with a 25×25 unit matrix K as the convolution kernel. The normalized sum f of the values in $\mathrm{std}(\Theta)$ then serves as an indicator value on how focused the image is (Fig. 5.23e):

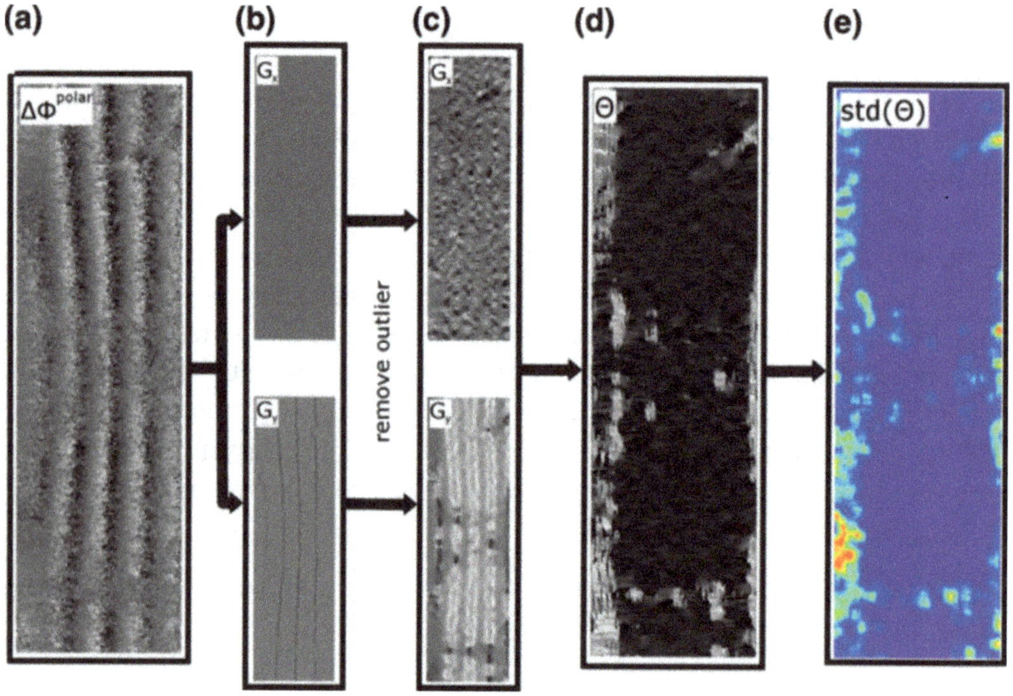

Fig. 5.23 **a**: Polar coordinate image $\Delta\Phi^{polar}$ of the phase distribution $\Delta\Phi$. **b**: Result of calculating Sobel derivative of $\Delta\Phi^{polar}$ in x- and y-direction ($\mathbf{G_x}$ and $\mathbf{G_y}$ respectively). **c**: Result of removing outlier derivatives caused by phase transitions. **d**: The gradient orientation image $\mathbf{\Theta}$. **e**: Local standard deviation of gradient orientation image $\mathbf{\Theta}$. Blue marks low values while red marks large values

$$f = \frac{1}{r\beta}\sum_{i=1}^{r}\sum_{j=1}^{\beta} \mathrm{std}(\Theta)_{ij} \tag{5.24}$$

If f exceeds a certain threshold t_{focus}, the measurement is said to be out of focus.

5.3.3.6 *Results*

In total, this defect detection routine has the four parameters shown in Table 5.2 with the thresholds for area t_A, depth t_D and focus t_{focus}. After evaluating different settings manually, the set of parameters shown in Table 5.2 was used for further work

Table 5.2 Parameter set for defect routine

Parameter	Variable	Value
Angular resolution	β	1800 pixels
Threshold: area	t_A	1000 pixels
Threshold: mean depth	t_D	0.1 (Range: 0–2 π)
Threshold: focus	t_{focus}	0.07

Defect detection was evaluated on 296 measurements of 247 parts (230 parts of those were previously inspected and found to be acceptable and 17 parts were identified as bad parts) in the department of quality assurance of Stüken Corp. with a measurement speed of approximately one part per second. Defective parts were all measured *at least* 3 times with different orientations to verify reproducible defect detection. Out of the 296 measurements, all the defective parts were reliably detected (true positive), while eleven intact parts were sorted out. Of these eleven parts, nine were correctly sorted out due to the measurement being out of focus, leaving two false positive detections.

5.4 In Situ Geometry Measurement Using Confocal Fluorescence Microscopy

Merlin Mikulewitsch* and Andreas Fischer

Abstract Due to the challenging environment of micro manufacturing processes such as laser chemical machining (LCM) where the workpiece is submerged in a fluid, a contactless in situ capable measurement is required for quality control. However, the in situ geometry measurement has several challenges for optical measurement systems because the high surface gradients of the micro geometries and the fluid environment complicate the use of conventional metrology. Confocal fluorescence microscopy allows for the determination of the surface position by adding an isotropically scattering fluorophore to the fluid and detecting the signal drop at the boundary layer between the measured object and the fluid. This technique, capable of improving the measurability of metallic surfaces with strong curvatures, is evaluated for suitability as an in situ measurement method for the LCM process. Unlike in thinner layers, however, the signal with fluid layers ≥ 1 mm, as needed for LCM in situ applications, shows strong dependencies on the fluorophore concentration and fluid depth. Thus, a physical model of the fluorescence intensity signal was developed for the evaluation of the surface position. To validate the method for the in situ measurement of geometry parameters, the step height of a submerged reference step was determined by measuring the surface positions along a line over the step. The step height measurement results in an uncertainty of 8.8 μm that is verified by deriving the potential measurement uncertainty of the model-based measurement approach. Further investigation of the uncertainty budget will allow a reduction of the measurement uncertainty and enable in situ monitoring and control of the LCM process.

Keywords In situ measurement · Confocal microscopy · Signal modeling

5.4.1 Challenges of Optical Metrology for In-Process and in situ Measurements

Laser chemical machining (LCM) is a promising alternative process that allows for inexpensive manufacturing of micro geometries in hard metals, such as dies for micro forming, without heat damage or structural alterations to the material [Mik17]. Laser chemically machined geometries can reach structure sizes between 10 μm and 400 μm, with steep slopes and a surface roughness of up to 0.3 μm [Ste10]. However, factors in the process environment, such as chaotic thermal interactions between the fluid and workpiece geometry, complicate the manufacture of a desired geometry, necessitating a closed-loop quality control [Zha17] with an in situ measurement feedback to improve the manufacturing quality (see Sect. 4.3). The challenging conditions of the LCM process, such as the requirement that the workpiece needs to be submerged by a fluid layer (typically 1–40 mm thick), hinder the in situ application of many measurement methods. The general lack of accessibility to the workpiece, for instance, requires the use of contactless measurement methods based on optical acquisition.

Conventional micro-topography measurement techniques can be separated into interferometric methods (e.g. displacement interferometry, digital holography [Kop13]) and other techniques, such as conventional laser-scanning confocal microscopy [Han06]. Conventional confocal microscopy is hindered, however, by the in situ conditions of high surface angles of the specimen [Liu16]. Interferometric methods were also investigated as a means of control feedback [Zha13], but were found to be unsuitable: The tested measurement systems integrated the interferometer directly into the machining head of the laser jet system as a two-beam interferometer according to the Michelson principle in order to increase the signal strength. The measuring beam was guided coaxially to the etchant and processing beam onto the surface of the workpiece. To obtain the path difference from the interferogram, the number of interference fringes was determined with phototransistors. If the measuring and reference arms are in different ambient media, a correction with the refractive indices of the media is also required. Evaluating the interferometer with samples of different surface roughness, it was determined that the measuring signal strength decreases with increasing surface roughness [Ger10]. In the end however, successful in situ measurement application proved to be unfeasible due to the formation of thermal gradients and gas bubbles that act as moving micro lenses and cause strong disruptions of the measuring beam [Ger10]. Thus, a suitable in situ measurement method capable of dealing with the process-induced currents, thermal gradients, and refractive index fluctuations is needed to improve the feasibility and acceptance of laser chemical machining as a competitive manufacturing process. A method based on the confocal detection of the fluorescence emitted by the fluid shows promise for in situ measurement application. The measurement is based on detecting the boundary position of the specimen surface and the fluid through the change in fluorescence signal while the confocal detection volume is scanned vertically through the fluid [Mic14].

5.4.2 Principle of Confocal Microscopy Based Measurement

The principle of measurement as shown in Fig. 5.24 is based on the detection of the fluorescence intensity emitted from the fluid covering the specimen using a confocal microscopy setup [Mik18]. The light of a green diode laser (λ = 532 nm) is expanded by a Keplerian beam expander and redirected by a beam splitter to the objective lens (NA = 0.42, WD = 20 mm), exciting the fluorescent fluid (aqueous solution of Rhodamine B) submerging the specimen. The specimen container is positioned using a 3-axis linear stage to enable the scanning of the focus position through the fluid. Only the fluorescence light emitted by the fluid (λ = 565 nm) is collected by a charge-coupled device (CCD). The confocal principle causes light not originating from a volume around the focus of the objective (*confocal volume*) to contribute less to the detected fluorescence signal. Scanning the confocal volume

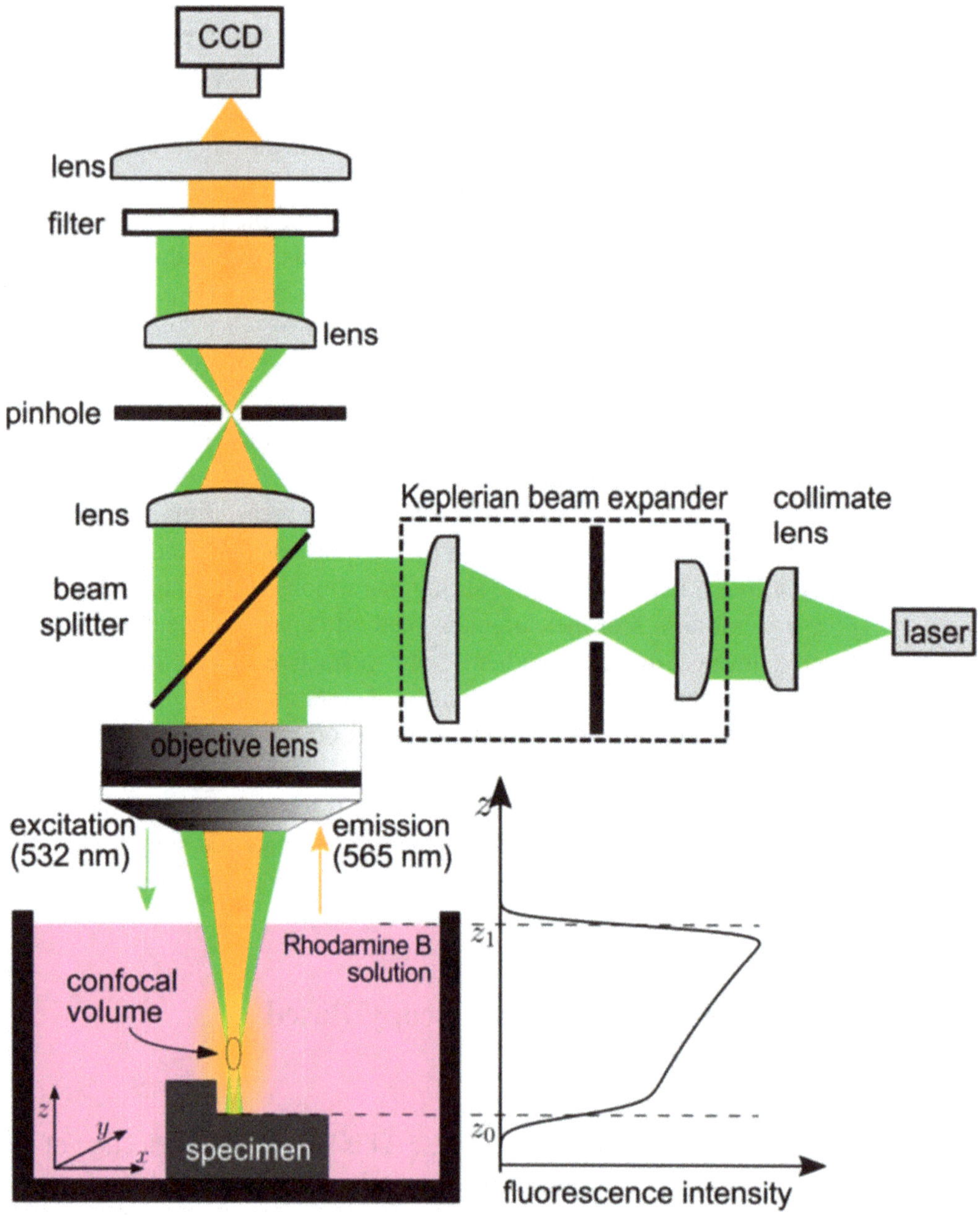

Fig. 5.24 Schematic diagram of the experimental setup and measurement principle of the confocal fluorescence microscopy system. Moving the confocal volume vertically through the fluid, a characteristic fluorescence intensity signal is generated (bottom right) [Mik18]

of the excitation laser vertically (in the z-direction) through the fluorescent fluid produces a characteristic fluorescence intensity signal (see Fig. 5.24, bottom right).

Since the excitation light is filtered out, only light that is emitted by the fluid inside the confocal volume is detected. Only for values of z inside the boundaries of the fluid ($z_0 < z < z_1$, see Fig. 5.24) will a significant signal be produced, since no fluorescent fluid is present to generate light when the confocal volume is fully located in either air or the specimen. The signal does not decay abruptly at the boundary but gradually, depending on the vertical extent of the confocal volume. The exact determination of the surface position z_0 is not trivial, as opposed to the case of very thin fluid layers, where the depth response is more similar to that of conventional confocal microscopy where the intensity peak corresponds directly to the surface position. For the case of thicker fluid layers, the properties of the fluorescence signal depend strongly on the fluorophore concentration and the fluid depth. With high fluorophore concentrations or thick fluid layers, the Lambert–Beer law of absorption causes less excitation light to reach far into the fluid, resulting in the decay of the fluorescence signal before the confocal volume reaches the specimen surface. This effect is negligible for thin fluid layers, but needs to be taken into consideration when choosing the fluorophore concentration for measurements in thicker layers. For the purpose of determining the surface position of the specimen from the fluorescence signal, a physical model of the fluorescence signal is used.

5.4.2.1 Model Assumptions

In order to solve the inverse problem of how to determine the surface position of the specimen from the acquired fluorescence intensity signal, a model of the fluorescence signal formation was developed [Mik18]. The model is based on several assumptions:

1. The detected fluorescence intensity is only generated in the confocal volume
2. The shape of the confocal volume is simplified to a 3D Gaussian function
3. The shape of the confocal volume is not affected by refraction
4. The specimen surface is non-reflective
5. The fluid surface does not move
6. A constant and uniform fluorophore concentration
7. A constant excitation power
8. The confocal volume is cut off by the horizontal surface element

The model assumptions are the source of model uncertainties that propagate into the uncertainty of the geometry parameter determination. However, it could be shown [Mik18] that even this simplified model is capable of enabling the surface position to be determined within thick fluid layers. The advantage of these simplifications is the existence of a closed mathematical formula to describe the fluorescence intensity signal (see Eq. 5.27).

5.4.2.2 *Model Description*

To model the fluorescence signal, the confocal volume in which it is generated needs to be described first. Since the confocal microscope only detects light from inside the confocal volume, the signal rapidly decreases if this volume moves outside the fluid that generates the fluorescent light. The spatially distributed contribution of each infinitesimal volume element to the detected fluorescence light power can be characterized in a first approximation by the three-dimensional Gaussian function [Rüt08]

$$I(\boldsymbol{r}, z) = I_0 \cdot \exp\left(-\frac{2}{w_0^2}\left(\boldsymbol{r}^2 + \frac{z^2}{\kappa^2}\right)\right), \tag{5.25}$$

$$\text{with}\, \boldsymbol{r} = \begin{pmatrix} x \\ y \end{pmatrix}.$$

This confocal volume function has a width of w_0 in the xy-direction and κw_0 in the z-direction, where $\kappa > 1$ is a constant factor dependent on the confocal setup. The parameter I_0 describes the maximum fluorescence light power determined by the excitation power and fluorophore concentration. Because the signal is generated by scanning the confocal volume through the fluid, a weighting factor, which is zero outside the boundaries of the fluid and follows the Lambert–Beer law of absorption inside the fluorophore, needs to be considered. The fluorescence intensity signal $I_F(z)$ detected at position z (see Fig. 5.25) is obtained by integrating the total contribution described by weighting the confocal volume function over all dimensions. The integral over the confocal volume function can be thought of as a vertical (z) convolution of the horizontal (x, y) integral $\int_{-\infty}^{\infty} I(\boldsymbol{r}, z)\mathrm{d}\boldsymbol{r}$ with the weighting function $\eta(z)$ of the fluid

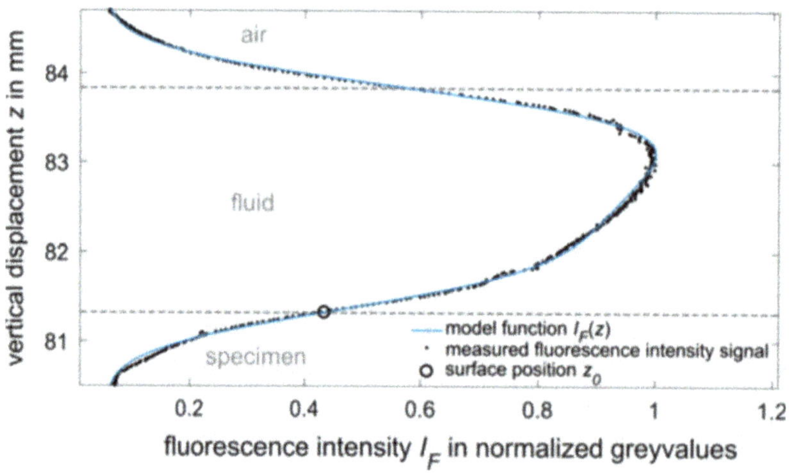

Fig. 5.25 Measured fluorescence intensity signal and fitted model function $I_F(z)$, (see Eq. 5.27). The surface position parameter z_0 (see diagram in Fig. 5.24) resulting from the non-linear least-squares fit is marked with a circle

$$I_F(z) = \eta(z) * \int_{-\infty}^{\infty} I(\boldsymbol{r}, z)\mathrm{d}\boldsymbol{r}, \tag{5.26}$$

$$\text{with } \eta(z) = \begin{cases} \exp(\varepsilon \cdot (z - z_1)) & z_0 \leq z \leq z_1 \\ 0 & \text{otherwise} \end{cases}.$$

Evaluating the convolution integral from Eq. 5.26 with the confocal volume function $I(\boldsymbol{r}, z)$ from Eq. 5.25 gives the model function of the fluorescence intensity signal $I_F(z)$ as

$$I_F(z) = \tilde{I}_0 \cdot \left(\operatorname{erf}\left(\frac{z - z_0}{2\xi} + \varepsilon\xi\right) - \operatorname{erf}\left(\frac{z - z_1}{2\xi} + \varepsilon\xi\right)\right) \cdot \exp(\varepsilon \cdot (z - z_1)) + C, \tag{5.27}$$

$$\text{with } \xi = \frac{\kappa w_0}{2\sqrt{2}}, \tilde{I}_0 = \frac{I_0 w_0^2 \xi \pi^{\frac{3}{2}}}{4} \cdot \exp\left(\varepsilon^2 \xi^2\right).$$

The surface position z_0 is then determined by a non-linear regression of the measured fluorescence intensity signal with the model function $I_F(z)$ using a least squares method. The approximation parameters are the amplitude $\tilde{I}_0$, the offset C, the fluorophore concentration-dependent attenuation coefficient ε, the confocal volume shape parameter ξ and the position parameters z_1 (fluid surface) and z_0 (specimen surface).

5.4.3 Experimental Validation

The in situ measurement technique was validated by measuring the geometry parameter step height of a referenced step object [Mik18]. The fluorescence intensity signal resulting from the measurement of a single point on the step-specimen is shown in Fig. 5.25.

In order to obtain the geometry parameter step height, the fluorescence intensity signal (see Fig. 5.25 for the signal of a single point) of 23 points along a line perpendicular to the edge of the step-specimen was acquired (i.e. performing a z-scan for each (x, y)-point). The surface position z_0 was determined from each measured intensity signal, with a least-squares approximation using Eq. 5.27. The resulting specimen surface positons z_0 are shown in Fig. 5.26. After correcting the tilt of the step and the focus shift due to refraction at the fluid surface, the step height h was determined by the difference of the mean surface positions of each step surface

$$h = h_{\text{upper}} - h_{\text{lower}}. \tag{5.28}$$

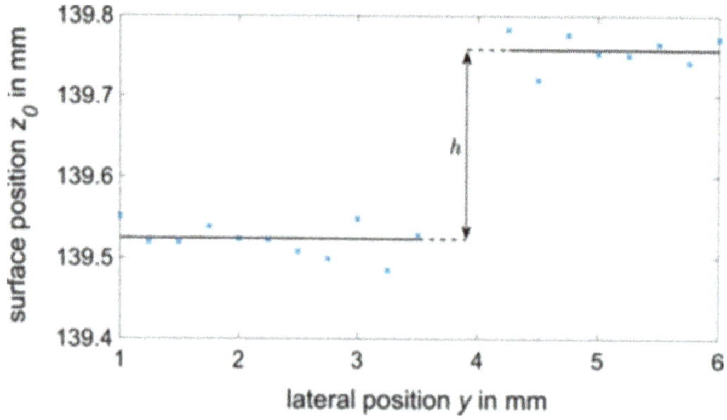

Fig. 5.26 Result of the surface position measurements for the step object. The geometry parameter step height h was determined by the difference of the two mean surface positions of each step surface, resulting in $h = (258.2 \pm 8.8)$ μm [Mik18]

Comparing the step height result of $h = (258.2 \pm 8.8)$ μm with the tactile reference measurement of (253.5 ± 0.2) μm shows no significant systematic deviations. Since the positions on each step surface show a relatively large stochastic scattering (up to 20%), the uncertainties are most likely caused by the general surface condition or uncertainties in the fitting model. The measurement technique based on confocal fluorescence microscopy was thus shown to be capable of determining the geometry parameter step height for microstructures submerged in thick fluid layers >100 μm, which demonstrates the suitability of the model-based approach for in situ application. However, the sources of the uncertainty of 8.8 μm need to be further characterized in order to reduce it to the desired 1 μm. In order to find the lower boundary of uncertainty for the confocal microscopy-based geometry measurement of submerged micro-structures, a determination of the measurement uncertainty with the approximation of the signal model is necessary.

5.4.4 Uncertainty Characterization

To obtain the fundamental uncertainty of the surface position $\sigma_{z_0} = \sqrt{\mathrm{Var}(z_0)}$ from the measurement with the non-linear least squares approximation method, the covariance matrix of the estimator

$$\hat{\theta}_{I_F} = \left[\hat{\mathrm{I}}_0, \hat{\mathrm{C}}, \hat{\epsilon}, \hat{\xi}, \hat{z}_0, \hat{z}_1\right]^T, \tag{5.29}$$

based on the fit function $I_F(z)$ (see Eq. 5.27) needs to be calculated. Applying an uncertainty propagation calculation to the least squares estimator gives the following relation for the estimator's covariance matrix [Kay93]:

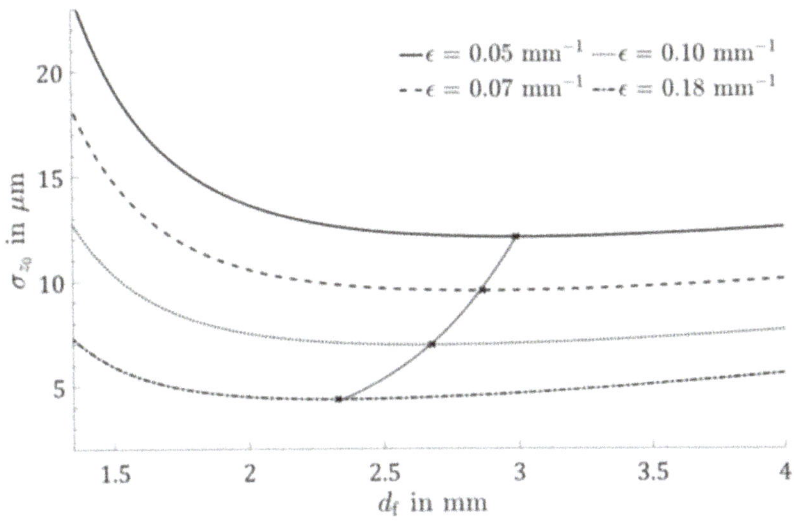

Fig. 5.27 Calculated uncertainty of the surface position z_0 as a function of the fluid depth d_f. The minimum uncertainty for each different attenuation coefficient ϵ is shown as a black cross

$$\mathrm{Cov}\left(\underline{\hat{\theta}}_{I_F}, \underline{\hat{\theta}}_{I_F}\right) = \left(H_{I_F}^{\mathrm{T}} C_{\tilde{I}_F}^{-1} H_{I_F}\right)^{-1}, \tag{5.30}$$

where H_{I_F} denotes the Jacobian matrix with the partial derivatives of the approximation function with respect to $\hat{\theta}_{I_F}$ at each position z_i of the measured fluorescence intensity signal $\tilde{I}_F(z_i)$, and $\mathrm{C}_{\tilde{I}_F}$ the covariance matrix whose main diagonal contains the variance $\sigma^2_{\tilde{I}_F}$ of each z_i of the fluorescence signal $\tilde{I}_F(z_i)$, since the covariance between the individual values is assumed to be zero.

To calculate the surface position uncertainty σ_{z_0} of the measurement using the non-linear regression of the model function $I_F(z)$ (see. Eq. 5.27), the average variance of the measured fluorescence signal around the model curve, i.e. $\sigma^2_{\tilde{I}_F} = 1.26 \cdot 10^9$ (in units of detected photons), is used. The calculation results in an uncertainty of 8.56 μm for the surface position z_0 and 3.75 μm if propagated into an uncertainty for the step height (based on 23 surface position measurements, according to the results from Fig. 5.26). The uncertainty of the experimental result of 8.8 μm is still larger than the calculated uncertainty by a factor of 2.3, suggesting varying conditions during the measurement, such as the form of the fluid surface or the micro topography of the specimen. To analyze the effects of the parameters fluid depth $d_f = z_1 - z_0$ and concentration-dependent attenuation coefficient ϵ, the uncertainty σ_{z_0} is determined as a function of the fluid depth d_f for different attenuation coefficients ϵ, shown in Fig. 5.27. The calculation reveals, that for each ϵ, a fluid depth exists where the uncertainty of the surface position σ_{z_0} is at a minimum (marked as a black cross).

The minimum decreases with higher attenuation coefficients ϵ and lower fluid depths d_f, which allows a suitable concentration (i.e. ϵ) to be chosen for any particular application-dependent fluid depth to achieve optimal uncertainty.

A Monte-Carlo (MC) simulation of the surface position determination (fitting an artificial signal with equivalent noise 10,000 times) was used to verify the results of the uncertainty calculations. To determine the theoretical limit of the achievable uncertainty, the variance $\sigma_{I_F}^2$ of the fluorescence intensity is decreased to its theoretical minimum. For this, an ideal, shot-noise limited signal with a Poisson distributed variance is considered, resulting in a surface position uncertainty of $\sigma_{z_0} = 0.07\,\mu\text{m}$ for the 23 position measurements of the experiment (see Fig. 5.26). As a result, the shot-noise limited uncertainty is two orders of magnitude lower than the currently achieved measurement uncertainty. However, if the variance of the fit residuum is considered, the theoretically achievable uncertainty of the surface position for a fluid thickness of 2.6 mm amounts to $\sigma_{z_0} = 8.56\,\mu\text{m}$, which corresponds to a step height uncertainty of $\sigma_h = 3.75\,\mu\text{m}$. These results demonstrate the untapped potential of the measurement technique and suggest that the step height uncertainty of the model-based measurement approach is only limited by the natural variation of the surface or effects not considered in the model (model uncertainties). Hence, a sub-micrometer precision seems feasible with the proposed confocal fluorescence microscopy technique, if the model uncertainties are decreased and the signal-to-noise ratio is maximized in future investigations.

5.5 Characterization of Semi-finished Micro Products and Micro Components

Bernd Köhler, Brigitte Clausen* and Hans-Werner Zoch

Abstract The mechanical properties of semi-finished micro products and micro components cannot easily be extrapolated from macro material. The effects of microstructure and surface conditions have a strong influence on the reliability of measuring results and on the material properties. To evaluate new approaches in micro wire and foil production new testing techniques had to be applied. Furthermore, the suitability of bonding techniques in production processes had to be tested dynamically. Finally, the mechanical properties of micro components produced with the newly developed techniques had to be validated.

Keywords Microstructure · Tensile strength · Fatigue

5.5.1 Introduction

Acquiring the mechanical properties of micro samples demands additional awareness concerning the effects of microstructure, surface, and alignment influences. In both tensile and fatigue tests of thin sheets, the edges of samples have to be very smooth and without micro notches to prevent them from early failure [Köh10b]. It is even more important to observe a correct alignment of the samples than for common samples, since eccentricity in the driving direction can lead to wrinkling and accelerate failure [Hon03]. Frames to place them into the testing device [Hon03] can additionally support especially sensitive samples. The ratio between the sample dimensions and the spatial extent of micro structural features causes size effects on, for example, the yield strength.

The strongest influence is caused by the inhibition of dislocation movement due to surface effects and a lack of dislocation sources. For sample sizes between 1 and 10 μm, this effect causes a strong increase of the yield strength. Furthermore, strain gradient plasticity effects influence the yield strength, due to additionally generated geometrically necessary dislocations (GND). Assuming a length scale of possible interactions with mobile dislocations between 10 nm and 100 μm, the plastic deformation in micro samples can be described satisfactorily [Vol09]. The surface grain model [Kal96] takes into account the difference between the influence of the surface grains and the influence of core grains on the yield strength. The plastic deformation of surface grains occurs more easily than for core grains due to a smaller contribution to strain hardening. The effect is measureable in sample sizes less than 20 times the average grain size [Kal96]. Among the different approaches to calculation, Köhler et al. [Köh13] summarized the yield strength dependencies from structural features. They show that for micro metal forming applications, the temperature and strain rate dependence on the yield strength can be neglected. For samples with less than 20 grains in the cross-section, an increasing yield strength effect in comparison to single crystals occurs due to the different orientation of grains towards the applied stress and the lack of strain continuity across grain boundaries. The increase can be calculated by a crystal orientation factor m_{or} [Tay56]. For samples with more than 20 grains in the cross-section, the Hall–Petch relation works well to calculate the grain size influence. An additional term taking the grain boundary resistance (GBR) into account effects a further increase of yield strength. Kim et al. [Kim07], using an extended Hall–Petch relation, provided a combination of both.

Since the local microstructure gains a higher influence on the mechanical properties, the natural scatter in the local microstructure results in larger scatter in the mechanical properties (see Sect. 1.4.2.2 Scatter).

5.5.2 Equipment for Testing Micro Samples

5.5.2.1 Mechanical Testing

The results depicted in the following were produced on an electrodynamic testing machine Instron® ElectroPuls™ E1000 type powered by a linear motor, described thoroughly by Köhler et al. [Köh10b]. Its control allows a wide range of specimens and specimen stiffnesses to be tested in dynamic mode with loads up to ±1000 N as well as in static mode with loads up to ±710 N. The cyclic tensile tests were carried out under sinusoidal load at a stress ratio $R = 0.1$ and a frequency $f = 20$ Hz. Spark erosion and an additional barrel finishing process could achieve the required high surface quality of dynamically tested sheets with a thickness of about 10 μm. The parameters of this finishing process are the results of internal investigations. The testing of micro wires required the development of new clamping systems, since a conventional clamping of wires would provoke an early failure caused by the notch effect. Quasi-static testing of soft samples allows the application of a reinforcement at the clamping ends to avoid breakage at the clamps. For samples with higher strength and dynamic tests, the demand was solved by a concept known from testing fibers. The incrementally formed wire is guided two turns around a roll at both ends and only mechanically clamped at the end (Fig. 5.28). In this way, the force is dissipated via the friction of the wire on the rollers and the notch effect at the end of the clamping can be neglected.

Ultra-micro hardness measurements on the rotary swaged wires made of 304 stainless steel showed that they undergo a strong work hardening due to the forming process and have a surface hardness up to 550 HV 0.03 [Köh17]. To ensure a satisfactory endurance of the rollers, a nitriding steel (31CrMoV9) was chosen as material. After manufacturing, the rolls were quenched and tempered (870 °C 2 h/oil/550 °C 2 h) and nitrided (510 °C 24 h) up to a nitriding depth of about 0.3 mm. A surface hardness of 900 HV 0.1 or 65 HRC, respectively, was achieved.

Fig. 5.28 Clamping device used for the quasi-static and cyclic testing of wires

5.5.2.2 Metallographic Investigations

The change of microstructure due to deformation and heat treatment has to be documented and interpreted to understand the mechanisms occurring in micro cold forming. Besides classic metallographic methods, an electron backscatter scanning diffractometer (EBSD) was used to obtain the grain sizes in aluminum sheets, recognize the anisotropic microstructure, and scan the phase distribution. For this purpose, a Philips XL30 scanning electron microscope (SEM) with an EBSD detector EDAX DigiView IV was available. The sample preparation for optimized EBSD imaging was dependent on the material to be analyzed. To avoid misinterpretation due to preparation failures, an investigation of the influence of different preparation techniques on the results of the EBSD micrographs was started. The results were published by Köhler et al. [Köh18].

5.5.3 Tensile Tests on Micro Samples

The first results from tensile tests on thin DC01 sheets with a chemical composition Fe-0.027C-0.047Al-0.005Si-0.190Mn-0.007P-0.009S (mass%) and a thickness of about 50 μm demanded an appropriate model to explain the unexpected dependence on the grain size [Köh10a]. The first received results are displayed in Fig. 5.29. The high yield strength in the cold worked state can easily be explained by the high dislocation density due to cold working in the forming process. The dislocation density was reduced by heat treatment for 15 min at 850 °C in a salt bath from 10^{10} to about $2 \times 10^8\ m^{-2}$. An elongation of the holding time at the heat treatment temperature caused no further reduction of the dislocation density.

To explain the deviance towards the Hall–Petch relation due to the small number of grains in the cross-section of the tensile test samples, a model inspired by Janssen [Jan07] was applied successfully [Köh11a]. A simple Taylor approach assigning different yield strengths to the core and surface provides a satisfactory description of the yield strength of the sheets:

$$\sigma_{p0.2}^{sheet} = \alpha \cdot \sigma_{p0.2}^{surface} + (1 + \alpha) \cdot \sigma_{p0.2}^{core} \tag{5.31}$$

with α representing the volume fraction of the surface grains. The yield strengths $\sigma_{p0.2}^{surface}$ and $\sigma_{p0.2}^{core}$ in the Hall–Petch relation are extended by the hardness-dependent contribution $c \cdot (HV - 150)$ (see Eqs. 5.32 and 5.33), wherein c is a material-dependent parameter. A hardness of 150 HV is assumed for the soft annealed state of the mild steel. The differing grain sizes d and grain boundaries resistances of surface and core grains, $k^{surface}$ and k^{core}, as well as the friction stress σ_0 are explicitly taken into account:

$$\sigma_{p0.2}^{surface} = \sigma_0 + c(HV - 150) + \frac{k^{surface}}{\sqrt{d^{surface}}} \tag{5.32}$$

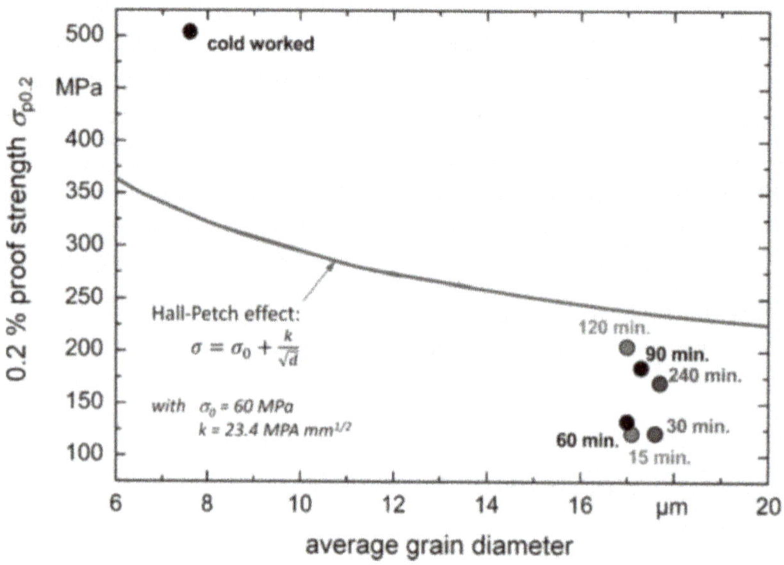

Fig. 5.29 0.2% yield strength of DC01 sheets in dependence on heat treatment and grain diameter compared to the theoretical yield strength due to Hall–Petch relation. The cold worked sheets were annealed at 850 °C in a salt bath for the given time [Köh10a]

$$\sigma_{p0.2}^{core} = \sigma_0 + c(HV - 150) + \frac{k^{core}}{\sqrt{d^{core}}} \tag{5.33}$$

Using the measured values and values from the literature [Hut63], the relations could be completed for DC01sheets to:

$$\sigma_{p0.2}^{surface} = 56\,\text{MPa} + 2.7\frac{\text{MPa}}{\text{HV}}(HV - 150) + \frac{382\,\text{MPa} \cdot \sqrt{\mu\text{m}}}{\sqrt{d^{surface}}} \tag{5.34}$$

$$\sigma_{p0.2}^{core} = 56\text{MPa} + 2.7\frac{\text{MPa}}{\text{HV}}(HV - 150) + \frac{696\text{MPa} \cdot \sqrt{\mu\text{m}}}{\sqrt{d^{core}}} \tag{5.35}$$

With this newly developed relation, the dependence of the 0.2% yield strength of micro sheets on the grain size could be predicted with a satisfactory accuracy [Köh11a].

The characterization of the first thin films generated by physical vapor deposition required the simultaneous consideration of the test result and the microstructure [Sto10]. Typical deposition effects, like hillocks, explained the unusually high differences in the results of nominally equal batches. The enhancement of the process quality resulted in more homogeneous tensile test results, suitable to adjust the process parameters of the sputter process [Kov17].

A customized clamping technique for micro wires was successfully applied on the incrementally forged samples. As expected, the specimen failure occurred predominantly within the gauge length. In the tensile test results, the scatter was tolerable, though the results were sometimes unexpected.

Rotary swaging of micro wires produced from 304 stainless steel causes the transformation-induced formation of martensite [Kuh15]. The fraction of martensite increases with decreasing feed rate in the forming process. Though the hardness increase is in accordance with the martensite fraction, the tensile test results show an almost diametrical tendency [Köh17]. The ultimate strength and the yield strength increase slightly with the increasing feed rate (Fig. 5.30). The increase can be explained by the strength hardening of the samples due to the forming process. It should be mentioned that the scatter of the strength values of these samples is comparatively small.

5.5.4 Endurance Tests on Micro Samples

In contrast to the tensile test results, the results of the fatigue test on rotary swaged micro wires show a considerable influence of the feed rate [Köh17]. With 1 and 3 mm/s feed rates, the results show an increase of the endurance limit σ_e with an enormous scatter, which is represented by the scattering parameter $T = \sigma_{e,10\%}/\sigma_{e,90\%}$. With 5 mm/s, the scatter and the endurance limit decrease significantly in

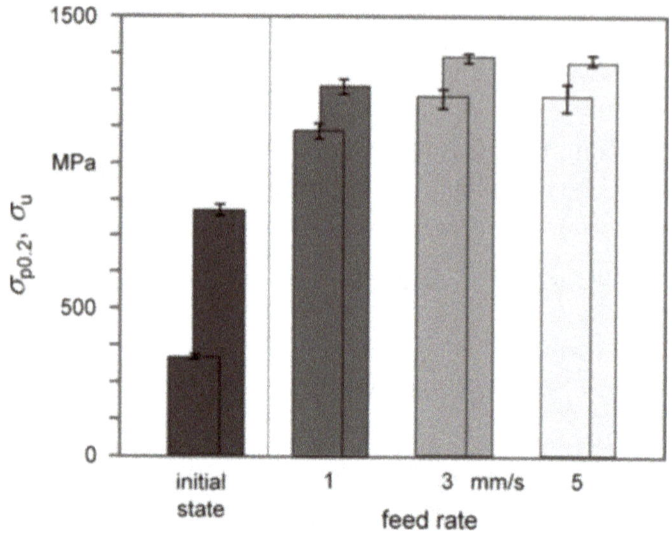

Fig. 5.30 0.2% yield strength $\sigma_{p0.2}$ and ultimate tensile strength σ_u of rotary swaged micro wires produced from 304 stainless steel as a function of feed rate [Köh17]

accordance with the decrease in hardness (Fig. 5.31). The scatter of the results is significantly higher than in the initial material. A comparison of these results with results gained on macro swaged samples showed that, though the scatter of the results is much lower, the mean value of the macro swaged samples lies well below the results for the micro swaged samples [Gan96]. This result can be explained by the statistical influence of the appearance of failure on the fatigue values. The less commonly a critical failure appears in the material, the more the scatter and the

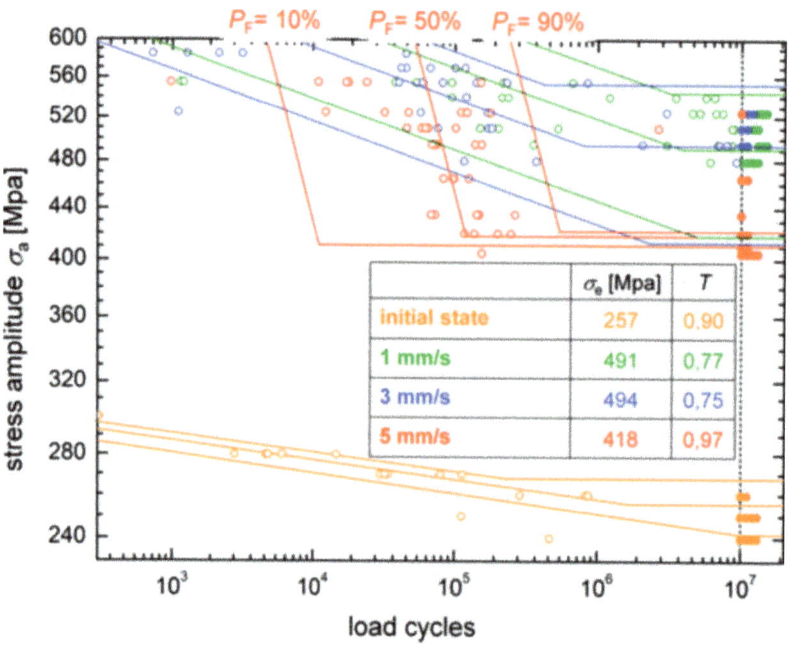

Fig. 5.31 Woehler diagram of rotary swaged micro wires produced from 304 stainless steel for varied feed rates supplemented by calculated failure probabilities P_F (● = run out, ○ = failure) [Köh17]

Table 5.3 Fraction of failure occurrence in surface, volume and at non-metallic inclusions of the dynamically tested rotary swaged micro wires produced from 304 stainless steel

	Initial state (%)	1 mm/s (%)	3 mm/s (%)	5 mm/s (%)
Surface	0	20	27	22
Inclusion	12	10	15	0
Volume	88	70	58	78

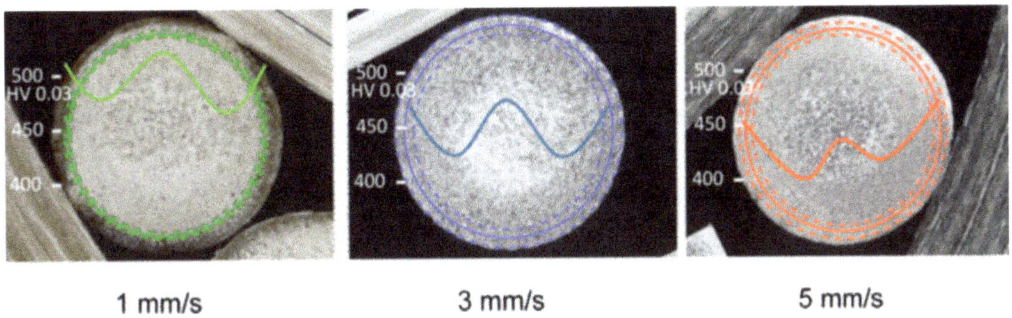

Fig. 5.32 Superposition of micrographs (with martensite proof etchant), position of failures with highest probability, and hardness curves of rotary swaged micro wires produced from 304 stainless steel for different feed rates (feed rate from left to right: 1, 3, and 5 mm/s) [Köh17]

mean value for the endurance limit σ_e increase, and the more samples have to be tested to receive a loadable result.

An additional interesting point is discovered in the origin of the failure of the fracture (Table 5.3). Stereo microscopy and additional scanning electron microscopic (SEM) investigations of the fracture surface showed that the fracture mainly does not start from a non-metallic inclusion or from the surface, as normally expected, but from under the surface.

A comparison of the results of the hardness measurements, metallographic investigations and the failure origins position in the fractured samples showed that the fracture starts predominantly in areas with an increased martensite fraction and the highest hardness values. In Fig. 5.32 the micrographs of the wires are superimposed with the position of the highest probability of failure (colored rings = mean value of failure probability; surrounding dotted rings = mean value +/− standard deviation) and the hardness measured across the cross-section. Residual stress measurements revealed additionally an increase of residual stresses in the martensitic phase below the surface. The position of the maximum residual stresses correlates with the fracture initiation position of the dynamically tested samples (Fig. 5.33).

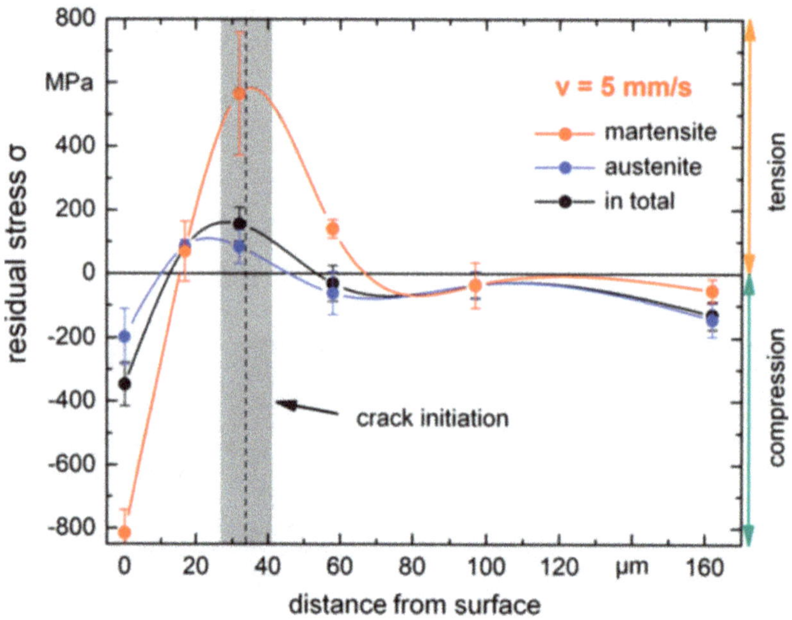

Fig. 5.33 Residual stresses in the surface of rotary swaged micro wires produced from 304 stainless steel with a feed rate of 5 mm/s

5.5.5 Microstructure Analysis with EBSD on Rotary Swaged Samples

Köhler et al. [Köh18] published the results of the comparison of preparation techniques for EBSD measurements. The most interesting result was that the indexability of grains depends not only on the quality of preparation but also on the degree of deformation of the grains due to the forming process. Any distortion to the crystal lattice within the diffracting volume produces a lower quality due to more diffuse diffraction patterns. This enables the parameter to provide a qualitative description of the strain distribution in the microstructure, if the underlying structures can still be dissolved in principle [War94]. However, severe deformation can also cause ultra-fine grains, which can also be the reason for insufficient resolution of the diffraction patterns. An example of the effect of severe deformation on the image quality is shown in Figs. 5.34 and 5.35. Figure 5.34 shows the image quality and the EBSD image of the microstructure of a 304 stainless steel wire in the initial state before forming. The image quality is very good and there are only a few dots along the grain boundary that cannot be indexed correctly. Figure 5.35 shows the same steel wire after rotary swaging with a degree of deformation $\varphi = -1.46$. The grains have been deformed differently according to their alignment to the main shear stresses. Between the deformed grains, large black areas appear, which are not indexable and which indicate a severe deformation or the formation of nano-grained structures.

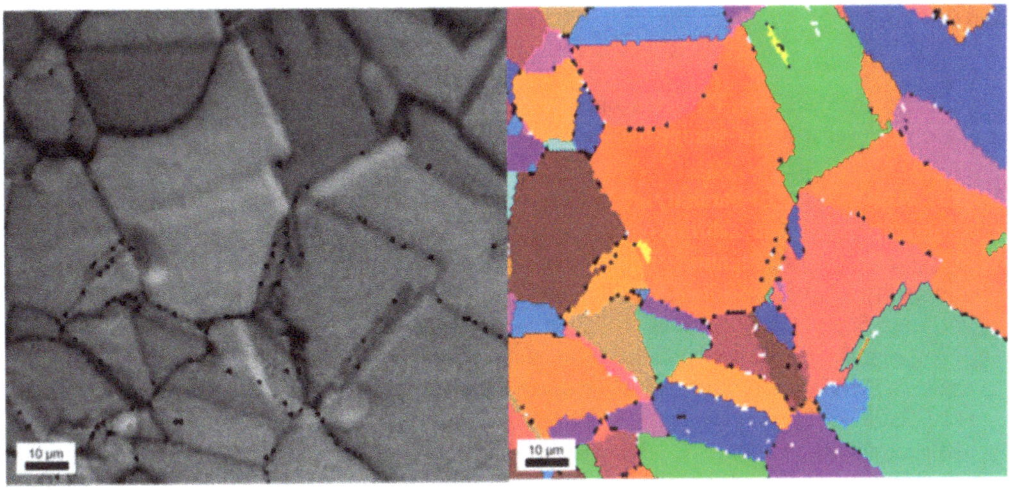

Fig. 5.34 Image quality (left) and EBSD image (right) of 304 stainless steel wire microstructure in the initial state

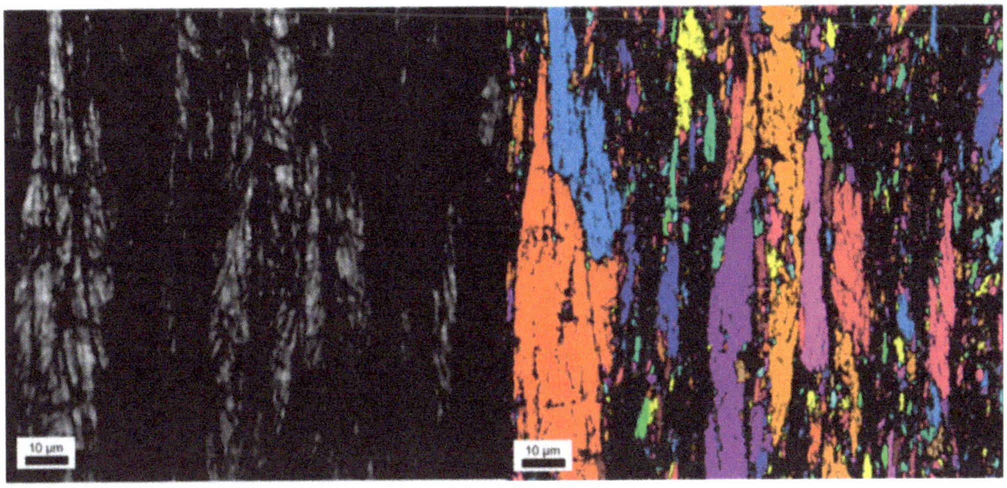

Fig. 5.35 Image quality (left) and EBSD image (right) of 304 stainless steel wires microstructure in rotary swaged state with degree of deformation $\varphi = -1.46$

References

[Ago10] Agour, M., Huke, P., von Kopylow, C., Falldorf, C.: Shape measurement by means of phase retrieval using a spatial light modulator. In: AIP Conference Proceedings, vol. 1236, pp. 265–270 (2010)

[Ago15] Agour, M., El-Farahaty, K., Seisa, E., Omar, E., Sokkar, T.: Single-shot digital holography for fast measuring optical properties of fibers. Appl. Opt. **54**, E188–E195 (2015)

[Ago17] Agour, M., Klattenhoff, R., Falldorf, C., Bergmann, R.B.: Spatial multiplexing digital holography for speckle noise reduction in single-shot holographic two-wavelength contouring. Opt. Eng. **56**, 124101 (2017)

[Ago17b] Agour, M., Klattenhoff, R., Falldorf, C., Bergmann, R.B.: Speckle noise reduction in single-shot holographic two-wavelength contouring. In: Proceedings of SPIE, vol. 10233, p. 102330R (2017)

[Ago18] Agour, M., Falldorf, C., Bergmann, R.B.: Fast inspection of micro-parts by utilizing spatial multiplexing and autofocus in holographic contouring. Opt. Express **26**, 28576–28588 (2018)

[Ber12] Bergmann, R.B., Huke, P.: Advanced methods for optical nondestructive testing. In: Osten, W., Reingand, N. (eds.) Optical Imaging and Metrology: Advanced Technologies, pp. 393–412. Wiley-VCH Verlag GmbH & Co., Weinheim, Germany (2012)

[Bou93] Bourdet, P., Lartigue, C., Leveaux, F.: Effects of data point distribution and mathematical model on finding the best-fit sphere to data. Precis. Eng. **15**, 150–157 (1993)

[Bra00] Bradski, G.: The OpenCV libarary. Dr. Dobb's Journal of Software Tools, vol. 25, pp. 120–126 (2000)

[Cha17] Chaurasia, A., Culurciello, E.: LinkNet: exploiting encoder representations for efficient semantic segmentation (2017). arXiv:1707.03718

[Cho16] Chollet, F.: Xception: deep learning with depthwise separable convolutions (2016). arXiv:1610.02357

[Cle15] Clevert, D.A., Unterthiner, T., Hochreiter, S.: Fast and accurate deep network learning by exponential linear units (elus) (2015). arXiv:1511.07289

[Fal12] Falldorf, C., Huferath-von Luepke, S., von Kopylow, C., Bergmann, R.B.: Reduction of speckle noise in multiwavelength contouring. Appl. Opt. **51**, 8211–8215 (2012)

[Fal12a] Falldorf, C., Agour, M., von Kopylow, C., Bergmann, R.B.: Phase retrieval for optical inspection of technical components. J. Opt. **14**, 065701 (2012)

[Fal15] Falldorf, C., Agour, M., Bergmann, R.B.: Digital holography and quantitative phase contrast imaging using computational shear interferometry. Opt. Eng. **54**, 24110 (2015)

[Fla01] Flack, D.: Measurement Good Practice Guide No. 41: CMM Measurement Strategies. National Physical Laboratory, London, UK (2001)

[Flo14] Flosky, H., Vollertsen, F.: Wear behaviour in a combined micro blanking and deep drawing process. CIRP Ann. Manuf. Technol. **63**(1), 281–284 (2014)

[Gan96] Ganesh Sundara Raman, S., Padmanabhan, K.A.: Effect of prior cold work on the room temperature low-cycle fatigue behaviour of AISI 304LN stainless steel. Int. J. Fatigue **18**(2), 71–79 (1996)

[Ger10] Gerhard, C., Stephen, A., Vollertsen, F.: Limits for interferometric measurements on rough surfaces in streaming inhomogeneous media. Prod. Eng. Res. Dev. **4**(2), 141–146 (2010)

[Goc91] Goch, G.: Algorithm for the combined approximation of continuously differentiable profiles composed of straight lines and circle segments. CIRP Ann. **40**(1), 499–502 (1991)

[Gro15] de Groot, P.: Principles of interference microscopy for the measurement of surface topography. Adv. Opt. Photon. **7**, 1–65 (2015)

[Gru69] Grubbs, F.E.: Procedures for detecting outlying observations in samples. Technometrics **11**, 1–21 (1969)

[Han06] Hansen, H., Carneiro, K., Haitjema, H., De Chiffre, L.: Dimensional micro and nano metrology. CIRP Ann. **55**(2), 721–743 (2006)

[Hon03] Hong, S., Hoffmann, H.: Study of scaling effect on mechanical properties for milli-forming of sheet metal–Tensile test of a very thin sheet. In: 1st Colloquium Process Scaling, pp. 145–151. BIAS Verlag, Bremen (2003)

[Hua17] Huang, G., Liu, Z., Weinberger, K.Q., van der Maaten, L.: Densely connected convolutional networks. In: Proceedings of IEEE Conference on Computer Vision and Pattern Recognition, vol. 1, p. 3 (2017)

[Hut63] Hutchison, M.M.: The temperature dependence of the yield stress of polycrystalline iron. Phil. Mag. **8**(85), 121–127 (1963)

[Iof15] Ioffe, S., Szegedy, C.: Batch normalization: accelerating deep network training by reducing internal covariate shift (2015). arXiv:1502.03167

Jan07] Janssen, P.J.M.: First-order size effects in the mechanics of miniaturised components. Dissertation, Eindhoven University of Technology (2007)

[Kal96] Kals, R., Vollertsen, F., Geiger, M.: Scaling effects in sheet metal forming. In: Kals, H.J.J., Shirvani, B., Singh, U.P., Geiger, M. (eds.) 4th International Conference on Sheet Metal, pp. 65–75. University of Twente Enschede (1996)

[Kay93] Kay, S.M.: Fundamentals of Statistical Signal Processing: Estimation Theory, vol. 1, p. 1. Prentice Hall (1993)

[Kim07] Kim, G.-Y., Koc, M., Ni, J.: Modeling of the size effects on the behavior of metals in microscale deformation processes. J. Manuf. Sci. Eng. **129**, 470–476 (2007)

[Kin14] Kingma, D., Ba, J.: Adam: a method for stochastic optimization (2014). arXiv:1412.6980

[Köh10a] Köhler, B., Bomas, H., Hunkel, M., Lütjens, J., Zoch, H.-W.: Yield strength behaviour of carbon steel microsheets after cold forming and after annealing. Scripta Mat. **62**, 548–551 (2010)

[Köh10b] Köhler, B., Bomas, H., Zoch, H.-W., Stalkopf, J.: Werkstoffprüfung an Mikroproben und –halbzeugen. MP Mater. Test. **52**(11–12), 759–764 (2010)

[Köh11a] Köhler, B., Bomas, H., Zoch, H.-W.: Bewertung von Zugversuchen an Mikroblechen des Stahls DC01. Tagungsband Koll. In: Kraft, O., Haug, A., Vollertsen, F., Büttgenbach, S. (eds.) Mikroproduktion und Abschlusskolloquium des SFB 499, Oct. 11–12, 2011, Karlsruhe, Deutschland, pp. 139–146. KIT Scientific Publication Karlsruhe (2011)

[Köh13] Köhler, B., Bomas, H.: Flow stress. In: Vollertsen, F. (ed.) Micro Metal Forming, pp. 69–89. Springer, Berlin (2013) ISBN: 978-3-642-30915-1

[Köh17] Köhler, B., Clausen, B., Zoch, H.-W.: Einfluss der Vorschubgeschwindigkeit beim Rundkneten des Stahls X5CrNi18-10 (1.4301) auf deren mechanische Eigenschaften sowie Gefügeeigenschaften. In: Vollertsen, F., Hopmann, C., Schulze, V., Wulfsberg, J. (eds.) Tagungsband 8. Kolloquium Mikroproduktion, 27–28 Nov. 17, pp. 91–98. Bremen, Deutschland (2017)

[Köh18] Köhler, B., Clausen, B., Zoch, H.-W.: Development and application of methods to characterize micro semi-finishes products and micro components. In: Proceedings of 5th ICNFT 2018, Sep. 19–21, 2018, Bremen, Germany. MATEC Web of Conferences, vol. 190, p. 15012. https://doi.org/10.1051/matecconf/ (2018)

[Kop13] von Kopylow, C., Bergmann, R.B.: Optical metrology. In: Vollertsen, F. (ed.) Micro Metal Forming, pp. 392–404. Springer, Berlin (2013)

[Kov17] Kovac, J., Mehner, A., Köhler, B., Clausen, B., Zoch, H.W.: Mechanical properties, microstructure and phase composition of thin magnetron sputtered TWIP steel foils. HTM J. Heat Treat. Mater. **72**(3), 168–174 (2017)

[Kri12] Krizhevsky, A., Sutskever, I., Hinton, G.E.: ImageNet classification with deep convolutional neural networks. In: Proceedings of 25th International Conference on Neural Information Processing Systems, vol. 1, pp. 1097–1105 (2012)

[Kuh15] Kuhfuss, B., Moumi, E., Clausen, B., Epp, J., Koehler, B.: Investigation of deformation induced martensitic transformation during incremental forming of 304 stainless steel wires. In: Proceedings of the 18th International ESAFORM Conference on Material Forming, ESAFORM 2015, Apr. 15–17, 2015, Graz, Austria. Key Engineering Materials, vol. 651–653, pp. 645–650 (2015)

[Liu09] Liu, J.-P., Poon, J.-P.: Two-step-only quadrature phase-shifting digital holography. Opt. Lett. **34**, 250–252 (2009)

[Liu16] Liu, J., Liu, C., Tan, J., Yang, B., Wilson, T.: Super-aperture metrology: overcoming a fundamental limit in imaging smooth highly curved surfaces. J. Microsc. **261**, 300–306 (2016) (Wiley Online Library)

[Lüb10] Lübke, K., Sun, Z., Goch, G.: Ganzheitliche Approximation eines Gerade-Kreis-Gerade-Profils mit automatischer Trennung in Einzelprofile. In: Scholl, G. (ed.) XXIV. Messtechnisches Symposium des Arbeitskreises der Hochschullehrer für Messtechnik e.V. (AHMT), pp. 77–90. Shaker Verlag, Aachen (2010)

[Lüb12] Lübke, K., Sun, Z., Goch, G.: Three-dimensional holistic approximation of measured points combined with an automatic separation algorithm. CIRP Ann. Manuf. Technol. **61**(1), 499–502 (2012)

[Mar05] Marquet, P., Rappaz, B., Magistretti, P.J., Cuche, E., Emery, Y., Colomb, T., Depeursinge, C.: Digital holographic microscopy: a noninvasive contrast imaging technique allowing quantitative visualization of living cells with subwavelength axial accuracy. Opt. Lett. **30**, 468–470 (2005)

[McC79] McCool, J.: Systematic and random errors in least squares estimation for circular contours. Precis. Eng. **1**, 215–220 (1979)

[Mic14] Michihata, M., Fukui, A., Hayashi, T., Takaya, Y.: Sensing a vertical surface by measuring a fluorescence signal using a confocal optical system. Meas. Sci. Technol. **25**, 064004 (2014)

[Mik17] Mikulewitsch, M., von Freyberg, A., Fischer, A.: Adaptive Qualitätsregelung für die laserchemische Fertigung von Mikroumformwerkzeugen. In: Vollertsen, F., Hopmann, C., Schulze, V., Wulfsberg, J. (eds.) Fachbeiträge 8. Kolloquium Mikroproduktion, Bremen, 27–28 Nov. 2017, pp. 21–26. BIAS Verlag (2017) (online)

[Mik18] Mikulewitsch, M., Auerswald, M., von Freyberg, A., Fischer, A.: Geometry measurement of submerged metallic micro-parts using confocal fluorescence microscopy. Nanomanufacturing Metrol. **1**(3), 171–179 (2018)

[Neo14] Neogi, N., Mohanta, D.K., Dutta, P.K.: Review of vision-based steel surface inspection systems. EURASIP J. Image Video Process. **2014**(1), 50 (2014)

[Noz15] Nozawa, J., Okamoto, A., Shibukawa, A., Takabayashi, M., Tomita, A.: Two-channel algorithm for singleshot high-resolution measurement of optical wavefronts using two image sensors. Appl. Opt. **54**, 8644–8652 (2015)

[Ron15] Ronneberger, O., Fischer, P., Brox, T.: U-net: convolutional networks for biomedical image segmentation. In: International Conference on Medical Image Computing and Computer-Assisted Intervention, vol. 9351, pp. 234–241. Springer (2015)

[Rüt08] Rüttinger, S., Buschmann, V., Krämner, B.: Comparison and accuracy of methods to determine the confocal volume for quantitative fluorescence correlation spectroscopy. J. Microsc. **232**, 343–352 (2008)

[Sav07] Savio, E., De Chiffre, L., Schmitt, R.: Metrology of freeform shaped parts. CIRP Ann. Manuf. Technol. **56**(2), 810–835 (2007)

[Sch94] Schnars, U., Jüptner, W.: Direct recording of holograms by a CCD target and numerical reconstruction. Appl. Opt. **33**, 179–181 (1994)

[Sim17] Simic, A., Freiheit, H., Agour, M., Falldorf, C., Bergmann, R.B.: In-line quality control of micro parts using digital holography. In: Holography: Advances and Modern Trends V, Proceedings of SPIE, vol. 10233, pp. 1023311–1023316 (2017)

[Sob90] Sobel, I.: An isotropic 3 × 3 image gradient operator. In: Machine Vision for Three-dimensional Scenes, pp. 376–379 (1990)

[Ste10] Stephen, A., Vollertsen, F.: Mechanisms and processing limits in laser thermochemical machining. CIRP Ann. Manuf. Technol. **59**(1), 251–254 (2010)

[Sto10] Stock, H.-R., Köhler, B., Bomas, H., Zoch, H.-W.: Characteristics of aluminum-scandium alloy thin sheets obtained by physical vapour deposition. Mater. Des. **31**, 76–81 (2010)

[Tay56] Taylor, G.I.: Strain in crystalline aggregates. In: Proceedings of the Colloquium on Deformation and Flow of Solids, Madrid, vol. 1955, pp. 3–12. Springer, Berlin (1956)

[Vol09] Vollertsen, F., Biermann, D., Hansen, H.N., Jawahir, I.S., Kuzman, K.: Size effects in manufacturing of metallic components. CIRP Ann. Manuf. Technol. **58**(2), 566–587 (2009)

[War94] Wardle, S.T., Lin, L.S., Cetel, A., Adams, B.L.: Orientation imaging microscopy: monitoring residual stress profiles in single crystals using an imaging quality parameter, IQ. In: Bailey, G.W., Garratt-Reed, A.J. (eds.) Proceedings of the 52nd Annual Meeting of the Microscopy Society of America, pp. 680–681 (1994)

[Wei16] Weimer, D., Scholz-Reiter, B., Shpitalni, M.: Design of deep convolutional neural network architectures for automated feature extraction in industrial inspection. CIRP Ann. Manuf. Technol. **65**(1), 417–420 (2016)

[Wes06] Westkämper, E., Stotz, M., Effenberger, I.: Automatic segmentation of measurement point clouds to geometric primitives. Technisches Messen **73**, 60–66 (2006)

[Wya98] Wyant, J.C.: White light interferometry. In: Proceedings of SPIE, vol. 4737, p. 98 (2002)

[Xie08] Xie, X.: A review of recent advances in surface defect detection using texture analysis techniques. Electron. Lett. Comput. Vis. Image Anal. **7**, 1–22 (2008)

[Yu15] Yu, F., Koltun, V.: Multi-scale context aggregation by dilated convolutions (2015). arXiv:1511.07122

[Zha13] Zhang, P., Mehrafsun, S., Goch, G., Vollertsen, F.: Automatisierung der laserchemischen Feinbearbeitung und Qualitätsprüfung mittels Interferometer. In 6. Kolloquium Mikroproduktion 10, B34 (2013)

[Zha11] Zhang, P., Mehrafsun, S., Lübke, K., Goch, G., Vollertsen, F.: Laserchemische Feinbearbeitung und Qualitätsprüfung von Mikrokaltumformwerkzeugen. In: Kraft, O., Haug, A., Vollertsen, F., Büttgenbach, S., 5. Kolloquium Mikroproduktion und Abschlusskolloquium SFB 499, Karlsruhe, vol. 7591, pp. 169—176. KIT Scientific Publishing (2011)

[Zha17] Zhang, P., von Freyberg, A., Fischer, A.: Closed-loop quality control system for laser chemical machining in metal micro-production. Int. J. Adv. Manuf. Technol. **93**, 3693 (2017)

Chapter 6
Materials for Micro Forming

Hans-Werner Zoch, Alwin Schulz, Chengsong Cui, Andreas Mehner, Julien Kovac, Anastasiya Toenjes and Axel von Hehl

H.-W. Zoch (✉) · A. Schulz (✉) · C. Cui · A. Mehner (✉) · J. Kovac · A. Toenjes (✉) · A. von Hehl
Leibniz Institute for Materials Engineering—IWT, University of Bremen, Bremen, Germany
e-mail: zoch@iwt-bremen.de

A. Schulz
e-mail: aschulz@iwt-bremen.de

A. Mehner
e-mail: mehner@iwt-bremen.de

A. Toenjes
e-mail: toenjes@iwt-bremen.de

F. Vollertsen et al. (eds.), *Cold Micro Metal Forming*, Lecture Notes in Production Engineering, https://doi.org/10.1007/978-3-030-11280-6_6

6.1 Introduction to Materials for Micro Forming

Hans-Werner Zoch

Thin metallic foils with a thickness below 50 μm are required for deep drawing of sub millimeter micro components. Commercial metallic foils for deep drawing are typically made of highly ductile materials such as pure aluminum, copper or stainless steel. Yet these materials are not suited for deep drawing of high strength micro components. Alloys with high mechanical strength are not available as thin foils below 50 μm due to significant strain hardening during the cold rolling process. Therefore, an alternative process based on physical vapor deposition (PVD) was developed, in order to produce thin metallic foils with improved mechanical properties. Thin foils of high strength aluminum alloys containing scandium and zirconium (Al–Sc–Zr), heat treatable martensitic chromium steels (X70Cr13) and high manganese austenitic steels (X5MnSiAl25-3-3) were produced by PVD magnetron sputtering onto thin substrate foils which were removed by selective etching after the PVD deposition process. Microstructure, mechanical properties and the effect of post annealing of these monometallic PVD foils were studied in order to find optimized processing parameters for deep drawing of micro components with advanced mechanical properties. Additionally, bimetallic PVD foils were produced and successfully tested for deep drawing of high strength micro components. Also, a prototype device for continuous physical vapor deposition onto thin metallic substrate tapes for an industrial scalable production of thin bimetallic PVD foils was developed, built and successfully tested.

Heat treatment processes within the manufacturing chain of micro components, either have the target to restore ductility or deformability after a cold rolling or deep drawing process, or are used to increase the final strength of the finished component. For single piece production and high volume production as well, short-term heat treatment processes have advantages compared to long-term heat treatments.

Because of the very small dimensions of the micro components and their distortion sensitivity, a new furnace design was developed—the drop-down tube furnace. Parts falling through an indirect heated tube of 6 m height are heated by radiation during falling down. As there is no mechanical contact, no distortion or damage of the parts occurs. Surface reactions are prevented by protective gas, e.g. when components see a short recrystallization annealing to recover deformability after cold working. Cooling in air or nitrogen gas at the furnace outlet allows to martensitic harden heat treatable steels. In the case of precipitation hardening of aluminum alloys, the heating times are too short for solution annealing, yet aging treatments during falling down of still supersaturated alloys as created by PVD sputtering processes are possible to increase strength.

In forming processes for micro components specific properties are required for the raw materials. If the principal dimensions of the workpiece and the corresponding tool are reduced to the micro scale these requirements are even more pronounced. Especially for the tools the microstructural phase distribution should be fine and very homogenous. When hard carbides are necessary for wear resistance powder metallurgical methods are suitable to achieve the desired properties. Among those technologies spray forming has the capability to produce free-standing near net shape products from the melt. The combination of two spray-forming assemblies, the so called co-spray forming, gives the opportunity to produce bulk materials with different local chemical compositions and a gradient in between. This allows the design of tools with different properties in specific areas. Using tool steels gives additional possibilities by adjusting hardness via specific heat treatments. In this chapter co-spray forming is shown to produce graded high-alloyed tool steels. The challenge of simultaneously hardening of different microstructures is solved by induction heat treatment, finally leading to materials that could be hard machined to swaging tools. These tools have been used for micro swaging of wires from austenitic stainless steels.

6.2 Tailored Graded Tool Materials for Micro Cold Forming via Spray Forming

Alwin Schulz* **and Chengsong Cui**

Abstract In micro cold forming, the tools are loaded differently in various functional areas, and tailored material properties are therefore required. To meet this requirement, different tool steels can be applied in the specific regions of the tools, with a gradual material transition in between to ensure a good bonding. These kinds of composite material can be produced via two newly developed material manufacturing processes: co-spray forming and successive spray forming. The spray-formed materials have been hot formed to eliminate porosity and break up the carbide network. Moreover, a selective heat treatment based on middle frequency induction heating has been developed to simultaneously austenitize the tool steels at different temperatures since the different steels may require different heat treatment conditions. Finally, micro rotary swaging tools made of graded tool steels have been precisely machined in hardened condition due to the fine and homogeneous microstructure in the steels. The tools have been successfully applied to form wires of AISI 304 stainless steel.

Keywords Manufacturing process · Tool steel · Spray forming

6.2.1 Introduction

In cold forming of micro metallic components, the loads on the forming tools vary in different functional regions, so tailored material properties in the specific regions are required. For example, the reduction zone of an infeed rotary swaging tool should be hard and wear-resistant to ensure high compact pressure and high friction, while the calibration zone of the tool should be strong but also micro-machinable, since the radius of the calibration zone is smaller than 1 mm. To meet these requirements, different tool steels may be applied in the specific regions of the tools. A gradual material transition in between is expected to reduce critical stresses at the interface and ensure good bonding of the different materials.

Components adapted to the specific loads in different areas by material modification or sequential construction have long been manufactured by means of plating, cladding, thermal spraying, electroplating, PVD and CVD. These methods have in common that a relatively sharp transition between the base material and functional layer is formed. The chemical bonding at the interfaces is weakened due to different crystal structures and lattice parameters [Wan96]. Process-related residual stresses at the interfaces further limit the loading capacity of the composites [Sun01]. In addition, different thermal expansions of the base material and the functional layer at elevated temperatures, e.g. as a result of unlubricated frictional contact, lead to additional thermal stresses [Kho00]. These limitations are counteracted by an intermediate layer between the functional layer and the base material, which serves to mediate differences in the crystal structure, lattice parameters and thermal stresses [Mus92]. If the functional layer is formed by the reaction of several components, as in the case of TiN layer via PVD or CVD, gradients can be built up within a layer. Graded layer structures can also be generated during thermal spraying by mixing different powders in the flame [Mus92]. These gradients are metal–matrix composites (NiCr–ZrO_2) with different mixing ratios, but the grading is limited by the particle size, since the fine powders are difficult to process [Peu05]. Moreover, the thermal cycles during thermal spraying are so short that homogenization of the material is hard to achieve by mixing of the melts or by diffusion over longer distances [Söl92].

Spray forming is an advanced material manufacturing process, in which a stream of alloy melt is atomized and the resultant small droplets are spray-deposited on a moving substrate, resulting in a near-net-shaped product with fine and homogeneous microstructure [Hen17]. As a new development of the process, two different alloys are melted simultaneously, and broken up into droplets in two spray cones via two free-fall atomizers, resulting in two-layered flat deposits as shown in Fig. 6.1a [Mey09]. If the two spray cones partially overlap, a gradient zone is generated in the two-layered flat deposits [Cui13]. An alternative spray forming process to produce graded materials is to spray two different alloys into a ring-shaped deposit (see Fig. 6.1b) [Cui16]. A gradual material transition can also be generated in the deposit if the two melts are mixed incrementally in the tundish. These two spray forming processes, which overcome the problems encountered by the conventional processes mentioned above, are expected to result in graded tool

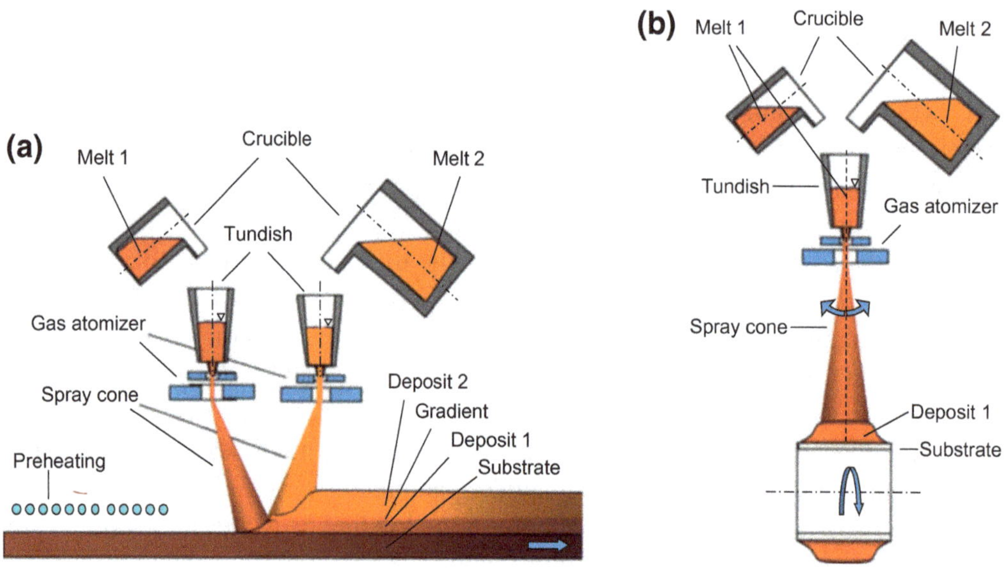

Fig. 6.1 Schematic drawings of **a** co-spray forming of a flat graded deposit, and **b** successive spray forming of a ring-shaped graded deposit [Cui16]

materials that meet the requirements of micro cold forming. In the following section, these two spray forming processes will be introduced, and the spray-formed graded tool materials will be characterized and evaluated.

Moreover, a selective heat treatment based on middle frequency induction heating and inductor oscillation has been developed and tested to simultaneously austenitize the tool steels at different temperatures since the different steels may require different heat treatment conditions to achieve optimal material properties.

Finally, the machinability of the graded tool materials will be assessed and the performance of the graded tools in micro cold forming will be demonstrated.

6.2.2 Production of Graded Tool Materials

6.2.2.1 Materials Selection

Three tool steels have been selected for the spray forming of graded tool materials for micro cold forming. The nominal chemical composition of the steels is listed in Table 6.1. The large amount of vanadium-rich carbides in the high-speed steel HS6-5-3C is responsible for its high hardness and excellent wear resistance. The HS6-5-2C is similar to the HS6-5-3C, and it shows relatively higher fracture toughness due to lower contents of V and C and hence a smaller amount of carbides. Compared with the high-speed steels, the cold work tool steel X110CrMoV8-2 has an even lower content of hard carbides and significantly higher fracture toughness, which leads to better micro-machinability and fracture toughness. In this study, two material combinations for the graded tool materials have been investigated: HS6-5-3C/HS6-5-2C, and HS6-5-3C/X110CrMoV12.

6.2.2.2 Spray Forming of Graded Tool Materials

Spray forming of graded tool materials is illustrated in Figs. 6.1 and 6.2: co-spray forming of flat deposits [Cui13], and successive spray forming of ring-shaped deposits [Cui16].

During the former process two tool steels are melted and gas-atomized separately and co-spray-deposited on a translating flat substrate, resulting in a flat

Table 6.1 Nominal chemical composition of tool steels

Steel grade		Chemical element, wt%				
Description	AISI type	C	Cr	W	Mo	V
HS6-5-3C	M3:2	1.3	4.2	6.3	5.0	3.0
HS6-5-2C	M2	0.9	4.0	6.4	5.0	1.9
X110CrMoV8-2	–	1.0	8.0	–	2.5	0.3

(a)
(b)
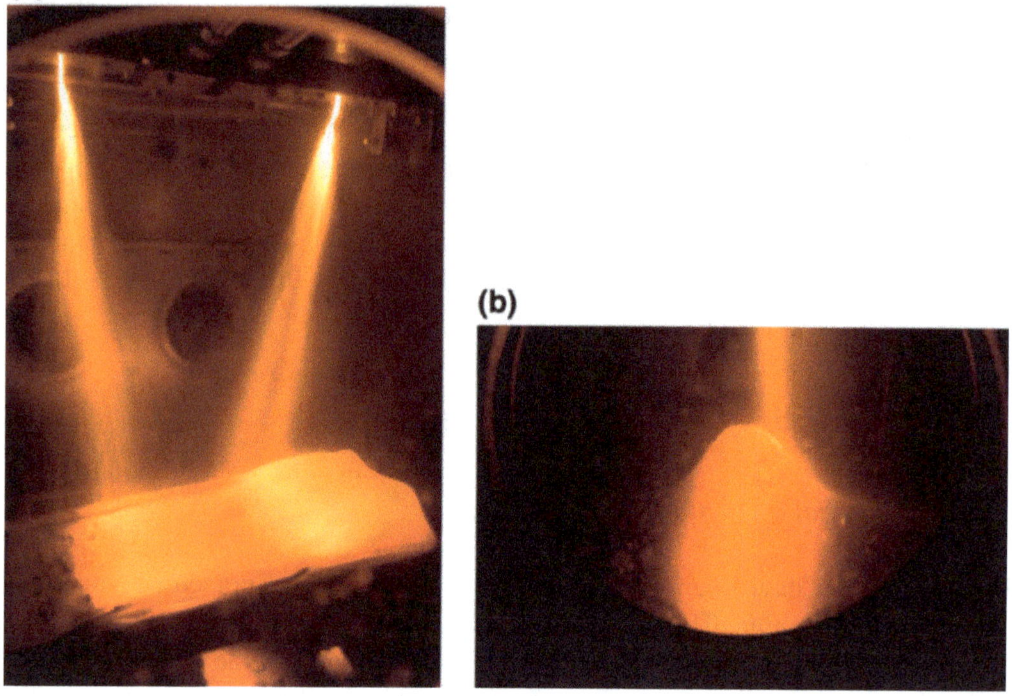

Fig. 6.2 Photographs of **a** co-spray forming of a flat graded deposit, and **b** successive spray forming of a ring-shaped graded deposit

composite deposit with a gradual transition of the chemical composition in between when the two spray cones partially overlap [Cui13a]. By this method, different microstructure and mechanical properties can be combined in a single deposit. In addition, fine and homogeneous microstructure can be achieved in the spray-formed materials due to rapid solidification and droplet fragmentation during the spray forming process.

During the latter process two different tool steels are also melted separately in two crucibles. One melt is poured first into a tundish, followed by gas atomization and spray deposition on a rotating tubular substrate, resulting in a ring-shaped deposit (inner ring). When the first crucible is empty, the second steel melt is poured into the same tundish, gas-atomized and spray-deposited on top of the inner ring to form an outer ring. If there is still a quantity of the first melt in the tundish when the second melt is poured into it, they mix in the tundish and the proportion of the second steel increases gradually as it is continuously added. Consequently, a gradient is also generated inside the deposit, depending on the gradational mixture of the two melts in the tundish. In comparison with the co-spray forming process, the successive spray forming process is more suitable for the manufacture of thick deposits since less heat is accumulated in the deposit during spray forming due to rotation of the deposit, and therefore lower thermal stresses and lower distortion are generated in the deposit.

Co-spray forming of the graded tool steels was carried out in the facility SK1+ at the University of Bremen. Firstly, flat deposits of HS6-5-3C/HS6-5-2C and HS6-5-3C/X110CrMoV8-2 with a thickness of approx. 30 mm were spray-formed (Fig. 6.3a). The tilting angles of the spray cones were varied at 6°, 9° and 12° to achieve different thicknesses of the gradient zone. In addition, the atomization gas pressure was adjusted to achieve different thermal conditions, which influence the porosity and the microstructure of the deposits. Secondly, flat deposits with a thickness of 55–60 mm (Fig. 6.3b) were produced by reducing the translational speed of the substrate; therefore, after hot deformation, the graded materials are still thick enough for making micro forming tools. In order to provide thicker semi-finished materials for fabricating the micro forming tools and to further increase the degree of deformation of the deposits, the co-spray forming process was modified to produce flat graded deposits with a thickness of about 100 mm (Fig. 6.3c). This modified spray forming process consisted of two phases. In the first phase, co-spray forming was conducted, similar to that for the deposits with a thickness of 55–60 mm. In the second phase, the substrate moved backward when the melt for the lower layer of the deposit was ended, and the melt for the upper layer of the deposit was continuously sprayed over the graded deposit. Consequently, thicker deposits were produced.

Successive spray forming of graded tool steels was also carried out in the facility SK1+ [Cui16]. During this process, the tool steels (HS6-5-3C and X110CrMoV8-2) were melted separately and spray-deposited successively on a rotating tubular substrate, resulting in a ring-shaped graded deposit with an outer diameter of about 335 mm and an inner diameter of 114 mm.

To reduce the hardness and remove stresses, the spray-formed deposits were soft annealed by heating to 840 °C at the rate of 50 K/h, holding for 4 h, followed by cooling to 500 °C at the rate of 20 K/h and uncontrolled cooling in the furnace to room temperature.

6.2.2.3 Densification of Graded Tool Materials

Porosity is an unavoidable characteristic of spray-formed materials [Hen17]. Four approaches have been applied for the densification of the spray-formed graded tool steels:

Fig. 6.3 Transverse sections of flat deposits of different thickness around **a** 30 mm, **b** 60 mm, and **c** 100 mm

(1) Samples from the 30 mm thick flat deposits were HIPped at 1140 °C under the pressure of 100 MPa for 3 h. Generally, the porosity is eliminated by HIPping. However, for deposits with open porosity, the HIPping should be conducted in capsule. It was also found that the carbide networks in the spray-formed deposits remained unchanged after HIPping, which would impair the fracture toughness of the materials.
(2) Samples from the 30 mm thick flat deposits were hot rolled on a lab rolling mill to eliminate porosity and break up the carbide network [Cui14]. The samples (thickness 24–28 mm after milling) were preheated in a furnace to 1100 °C, held at that temperature for 10–30 min, and rolled in three passes to a final thickness of approximately 10 mm. For each rolling pass, the true strain and the true strain rate of the rolled samples were in the range of 0.2–0.4 and 2–3 s^{-1}, respectively. The total strain of the samples was about 0.9. Samples from the 60 mm thick flat deposits were also hot rolled on this lab rolling mill. The samples (thickness 44–49 mm after milling) were preheated to 1100 °C and rolled in three passes to a final thickness of approximately 24 mm. For each rolling pass, the true strain and true strain rate of the sample were 0.2–0.3 and 1–2 s^{-1}, respectively. The total strain of the sample was around 0.7.
(3) The 100 mm thick flat deposits were hot forged to eliminate porosity and break up the carbide network. The deposits were forged in a temperature range from 1150 °C to 950 °C to a thickness of approximately 50 mm and cooled slowly in dry sand.
(4) The spray-formed ring-shaped deposit was machined to a ring preform (318 mm OD × 125 mm ID × 50 mm), and forwarded to a rolling mill for direct ring production. The deposit was preheated in a furnace to 1100 °C and rolled to the final dimensions (543 mm OD × 431 mm ID × 39 mm). The area reduction of the ring was 2.23. After ring rolling, the sample was cooled slowly in the furnace from 860 °C to room temperature.

6.2.2.4 Heat Treatment

The different materials in the graded deposits usually require different heat treatment conditions. For example, the recommended austenitization temperatures for high-speed steels are much higher than those for cold work tool steels. If the graded deposits are austenitized at high temperatures, too much austenite may be retained in the cold work tool steels and the grain structure would be coarsened. If the graded deposits are austenitized at low temperatures, insufficient dissolution of carbides may hinder the secondary hardening of the high-speed steels. Due to that, for the graded deposits made of different steels, any compromise for heat-treating in a furnace should lead to insufficient hardening results.

To demonstrate this, samples of the graded deposits were austenitized in a vacuum furnace at 1080 °C for 20 min or at 1180 °C for 10 min, and quenched with nitrogen at 0.6 MPa, followed by triple tempering at 550 °C for 2 h. It was found that the austenitization temperature of 1180 °C was suitable for the HS6-5-3C and HS6–5-2C, but too high for the X110CrMoV8-2 (see Fig. 6.11). The appropriate austenitization temperature for X110CrMoV8-2 must not exceed 1080 °C. Therefore, the traditional austenitization in a furnace is not suitable for hardening of graded materials like HS6-5-3C/X110CrMoV8-2.

To overcome this problem, a selective heat treatment based on induction heating with an oscillating inductor has been developed, as shown in Fig. 6.4. An oscillating ring-shaped inductor is used to heat the cylindrical sample. The upper region of the sample is HS6-5-3C and the lower region is X110CrMoV8-2. Between the two alloys is the gradient zone. The desired austenitization temperatures for HS6-5-3C and X110CrMoV8-2 are T1 and T2, respectively. Middle frequency induction is applied for a thorough heating from the surface to the core of the sample. The induction heating process is divided into three phases:

Phase 1: The complete sample is heated up to the temperature T2 (desired for X110CrMoV8-2) by oscillating the inductor over the complete sample.

Phase 2: The inductor continues to oscillate over the complete sample but the inductor current is reduced, so that the sample temperature is held at T2.

Phase 3: The inductor oscillates over the upper region (HS6-5-3C) and the local temperature is further increased to T1 (desired for HS6-5-3C). In order to

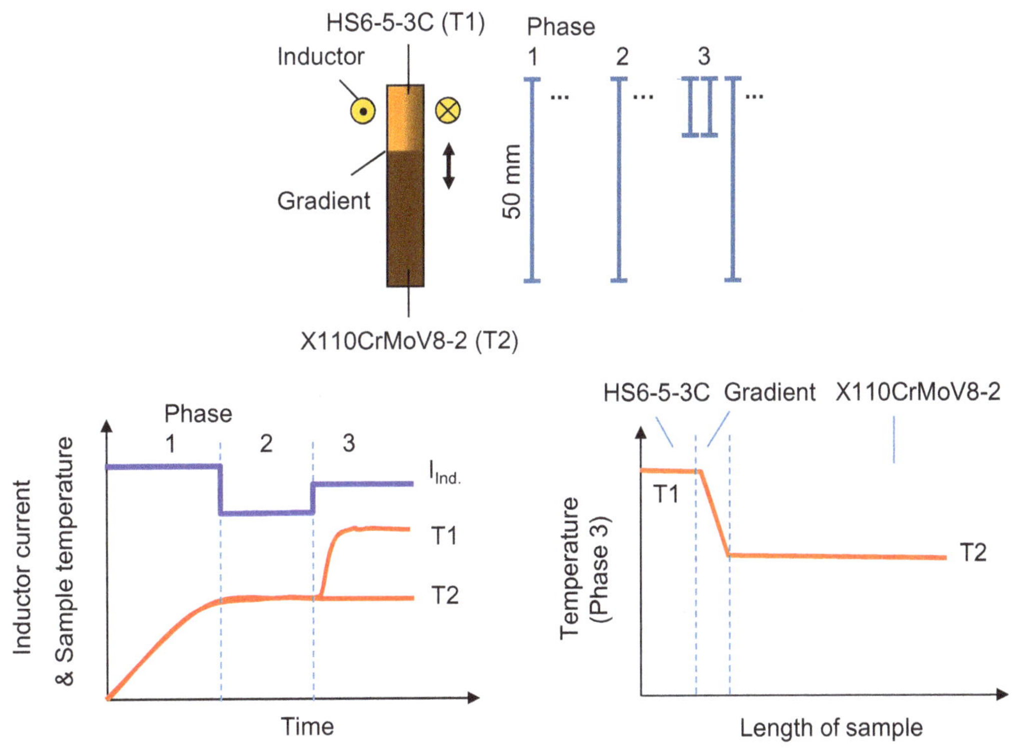

Fig. 6.4 Schematic of the selective heat treatment of a cylindrical sample of the graded material HS6-5-3C/X110CrMoV8-2 based on middle frequency induction heating and oscillation of the inductor [Cui17a] Copyright 2017 by MS&T17. Used with permission

hold the temperature of the lower region at T2, the inductor must also oscillate over the complete sample after several oscillations over the upper region. This combined oscillation is repeated and a tailored austenitization of the composite material is achieved.

In this study, cylindrical samples (∅20 mm × 50 mm) were machined along the thickness of the hot rolled ring for the selective heat treatment. The HS6-5-3C region was approximately 10 mm long, the gradient zone 10 mm, and the X110CrMoV8-2 region 30 mm. For temperature measurement, initial tests were done with samples with ∅1.1 mm holes of different depths for placing sheath thermocouples (∅1 mm, type K).

The induction heat treatment facility used in this study was a VL1000 SINAC 200/300 S MFC from EFD Induction GmbH (see Fig. 6.5). The inductor was a water-cooled single turn copper coil with an inner diameter of 22 mm, a cross-section of 6 × 6 mm^2, and a wall thickness of 1 mm. Thermal insulation (glass fiber) was applied at both ends of the sample to reduce heat loss. In all experiments, an infrared pyrometer (KTR 1075 from Maurer) and a 2-color (ratio) pyrometer in combination with a video camera with a short wavelength infrared filter (ISR 6-TI Advanced, LumaSense) were used to measure the surface temperature of the sample during the heat treatment.

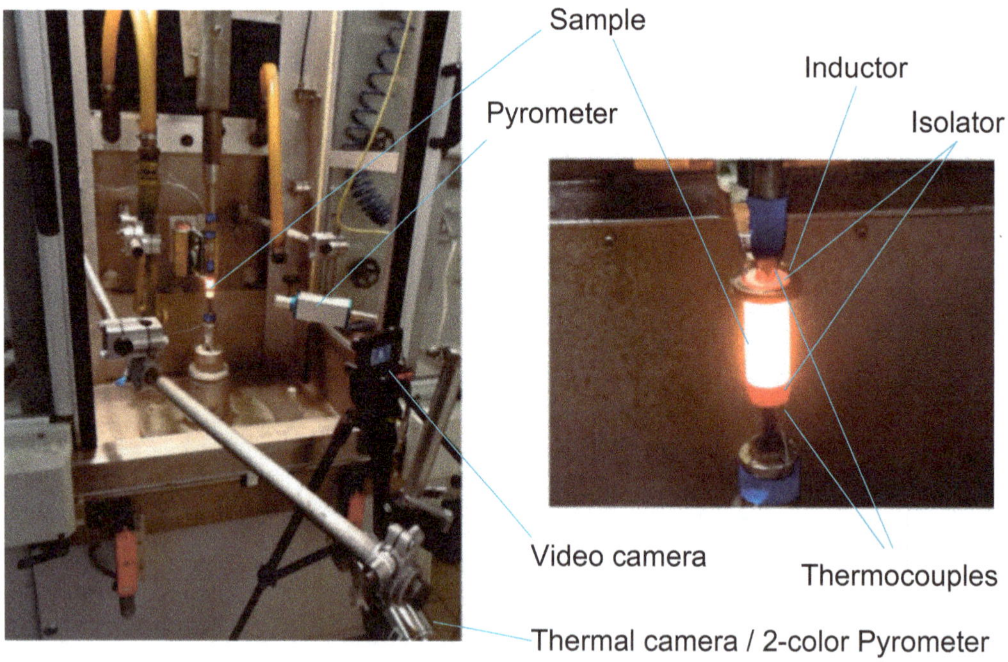

Fig. 6.5 Setup of the selective heat treatment of a cylindrical sample of the graded material HS6-5-3C/X110CrMoV8-2 based on middle frequency (14 kHz) induction heating and oscillation of the inductor

After induction heating, the samples were cooled in air since the tool steels are air hardenable. Finally, the composite samples were tempered three times in a protective atmosphere at 550 °C for 2 h.

The influence of the inductive process parameters (inductor current, motion of inductor, etc.) on the temperature distribution and the result of hardening of the graded materials have been investigated. The microstructure and hardness of the graded materials have been correlated with the austenitizing conditions. More details of the experiments can be found in [Cui17a].

6.2.3 Evaluation of the Graded Tool Materials

6.2.3.1 Co-spray-Formed Material

The distributions of the chemical elements in the co-spray-formed graded materials are most clearly represented by the element vanadium. The measured contents of vanadium over the deposit thickness are given in Fig. 6.6. For both material combinations, the concentration profiles are similar. The effect of the tilting angle of the atomizers on the gradient zone is significant. Tilting the atomizers leads to overlapping of the spray cones and a mixing of the droplets of the two different steels.

Representative microstructure of the graded deposits after hardening and tempering is presented in Figs. 6.7 and 6.8 [Cui13]. The micrographs exhibit equiaxed grain structures with fine carbides, as well as a carbide network at the primary austenite grain boundaries. For both material combinations, there are more MC type carbides in the HS6-5-3C than in the HS6-5-2C or X110CrMoV8-2. The gradient zone shows some intermediate microstructure.

Porosity and carbide networks, which impair the fracture toughness of the materials, have been frequently observed in the spray-formed tool steels. After hot

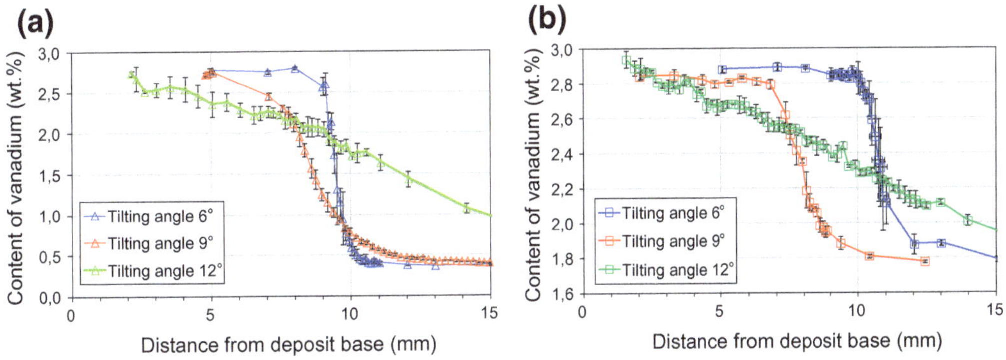

Fig. 6.6 Distribution of the element vanadium in the spray-formed graded deposits of 30 mm thickness depending on the tilting angles of the spray cones: **a** HS6-5-3C/110CrMoV8-2, and **b** HS6-5-3C/HS6-5-2C [Cui13]

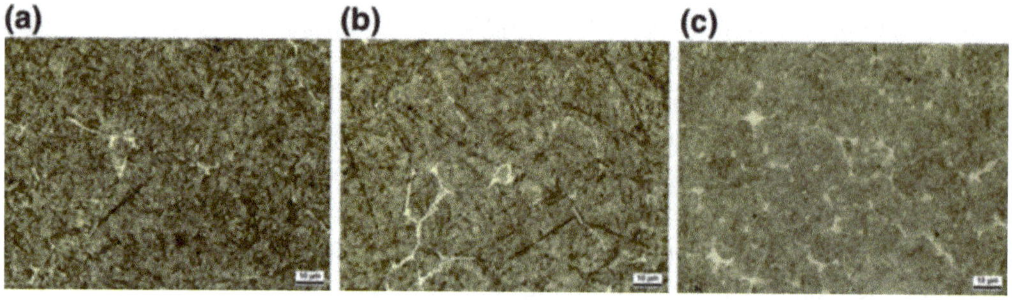

Fig. 6.7 Microstructure of a graded deposit of 30 mm thickness, hardened at 1080 °C and triple tempered at 550 °C: **a** X110CrMoV8-2, **b** X110CrMoV8-2+HS6-5-3C, **c** HS6-5-3C [Cui13]

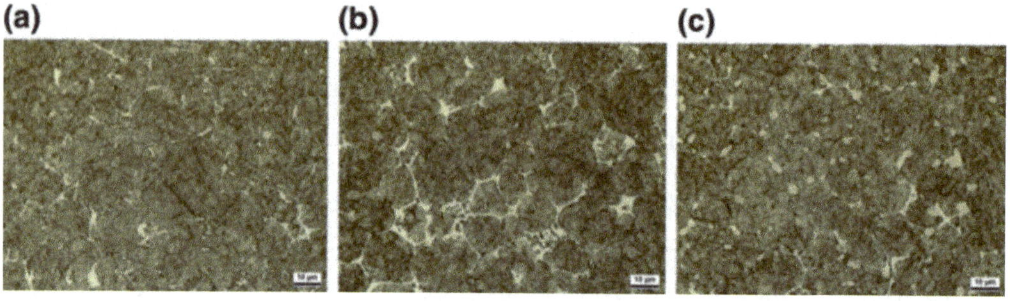

Fig. 6.8 Microstructure of a graded deposit of 30 mm thickness, hardened at 1180 °C and triple tempered at 550 °C: **a** HS6-5-2C, **b** HS6-5-2C+HS6-5-3C, **c** HS6-5-3C [Cui13]

deformation, the porosity has been essentially eliminated, and the coarse carbides have been broken into small pieces. During hot deformation the grain structure might experience a dynamic recrystallization, too. This also resulted in a fine and homogeneous microstructure, as shown in Figs. 6.9 and 6.10.

The hardness values of the graded deposits after hardening as well as after hardening plus tempering are shown in Fig. 6.11:

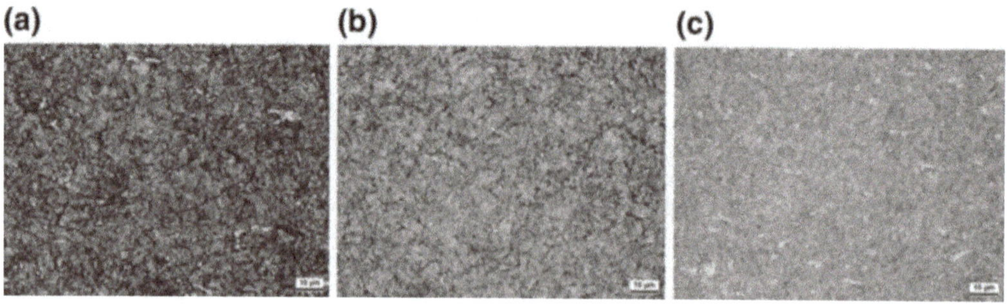

Fig. 6.9 Microstructure of a graded deposit of 30 mm thickness, hot rolled to 10 mm, hardened at 1080 °C and triple tempered at 550 °C: **a** X110CrMoV8-2, **b** X110CrMoV8-2+HS6-5-3C, **c** HS6-5-3C

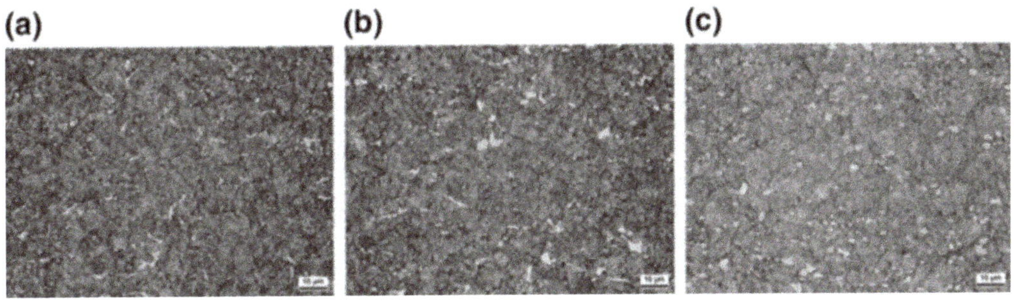

Fig. 6.10 Microstructure of a graded deposit of 30 mm thickness, hot rolled to 10 mm, hardened at 1180 °C and triple tempered at 550 °C: **a** HS6-5-2C, **b** HS6-5-2C+HS6-5-3C, **c** HS6-5-3C

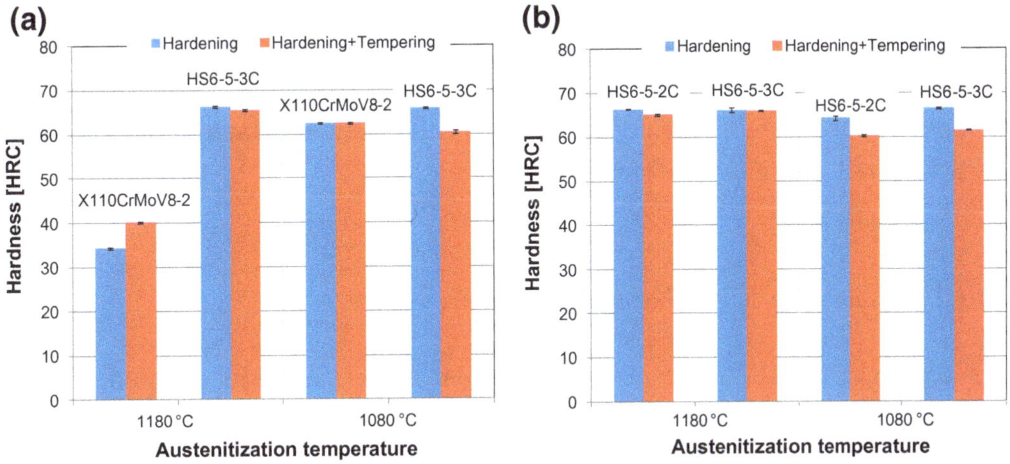

Fig. 6.11 Hardness of graded deposits after hardening and tempering: **a** X110CrMoV8-2/HS6-5-3C, and **b** HS6-5-2C/HS6-5-3C [Cui13]

(1) The hardness of the as-hardened X110CrMoV8-2 with the austenitization temperature of 1180 °C is very low (around 34 HRC). Its hardness increases to 62 HRC with the austenitization temperature of 1080 °C. This indicates that carbide dissolution in the matrix of X110CrMoV8-2 is too high during austenitization at 1180 °C. It results in a very low martensitic transformation temperature and therefore too much retained austenite after quenching. The hardness of the as-hardened HS6-5-3C with the austenitization temperature of 1180 °C is about 66 HRC. It is nearly the same for the austenitization temperature of 1080 °C due to the high carbon content and high alloying element content of this steel. After three times of tempering at 550 °C, the hardness of the X110CrMoV8-2 austenitized at 1180 °C increases to 40 HRC, indicating further transformation of retained austenite. For the X110CrMoV8-2 austenitized at 1080 °C, its hardness remains nearly the same after tempering. For the HS6-5-3C austenitized at 1180 °C, its hardness decreases slightly after tempering. However, for the HS6-5-3C austenitized at 1080 °C, its hardness decreases significantly to about 61 HRC. The relatively low austenitization

temperature leads to a higher degree of softening during tempering of the high-alloyed steel, and this could not be compensated by secondary hardening because fewer carbides were dissolved during austenitization. It is therefore concluded that a tailored austenitization for this material combination is needed to achieve the optimal material properties for both steels.

(2) For the graded steels HS6-5-3C/HS6-5-2C, the hardness of both steels is higher than 60 HRC in the hardened condition as well as in the tempered condition. When austenitized at 1180 °C, the hardness of both the as-hardened steels is around 66 HRC. After tempering, it is about 65–66 HRC because secondary hardening occurs. When austenitized at 1080 °C, the hardness values of the as-hardened HS6-5-2C and HS6-5-3C are about 62 HRC and 66 HRC, respectively. Fewer carbides dissolve in the austenite matrix at this temperature, and particularly the highly stable carbides like tungsten and molybdenum carbides do not dissolve. After tempering, their hardness reduces to approximately 60 HRC due to decomposition of martensite and less secondary hardening. To achieve high hardness, the austenitization temperature of 1180 °C is preferred for this material combination.

6.2.3.2 Successive Spray-Formed Material

The distributions of the main alloying elements C, Cr, Mo, W and V in a ring-shaped graded material (HS6-5-3C/X110CrMoV12) via successive spray forming and ring rolling are presented in Fig. 6.12. The gradient zone between the inner HS6-5-3C and the outer X110CrMoV8-2 is approximately 10 mm thick, starting at the position about 12 mm from the inner surface of the ring. The slope of the gradient of the elements depends on the level of mixture of the two steel melts in the tundish. If there is less melt of the first steel in the tundish when the second steel melt is added, or the second melt is added more quickly, the gradient zone in the spray-formed ring would be narrower.

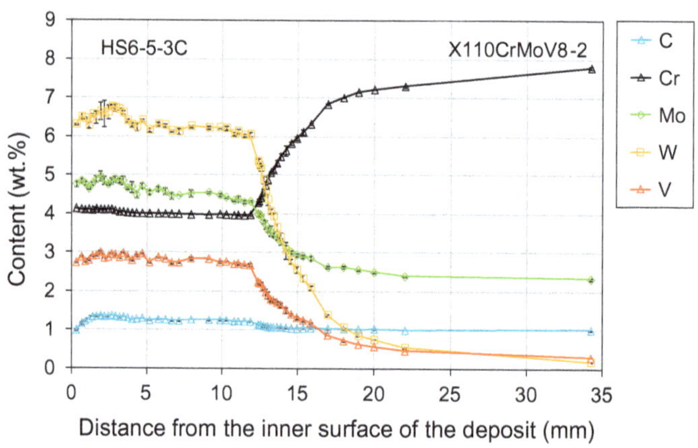

Fig. 6.12 Distributions of main chemical elements in the as-rolled ring-shaped graded deposit HS6-5-3C/X110CrMoV12 [Cui16]

The microstructure of the graded materials after the selective induction hardening and tempering, as described before, is presented in Fig. 6.13 [Cui17a]. The MC-type carbides were precipitated in the HS6-5-3C region, and mainly eutectic carbides were observed in the X110CrMoV8-2 region. In the gradient zone, a mixed structure of the two steels was constituted. With high austenitization temperatures (1220 °C/1120 °C), the X110CrMoV8-2 region showed coarse martensite and a large amount of retained austenite, while the gradient zone showed very coarse austenite grains and very little martensite. The microstructure of the graded materials austenitized at 1180 °C/1080 °C and 1150 °C/1050 °C looked similar, except that the microstructure in the gradient zone was somewhat coarser under the austenitization condition at 1180 °C/1080 °C. It was hard to find retained austenite in the graded materials processed under these conditions.

The hardness profiles of the graded materials processed under various induction heating conditions are shown in Fig. 6.14. The maximum austenitization temperatures reached in the various regions of the samples are also plotted in the diagrams. It can be summarized as follows:

(1) Graded temperatures are achieved in all the samples. The relatively broad temperature gradients are caused by the heat conduction from the high temperature region to the low temperature region. To reduce the temperature gradient, the heating time for the samples should be shortened.

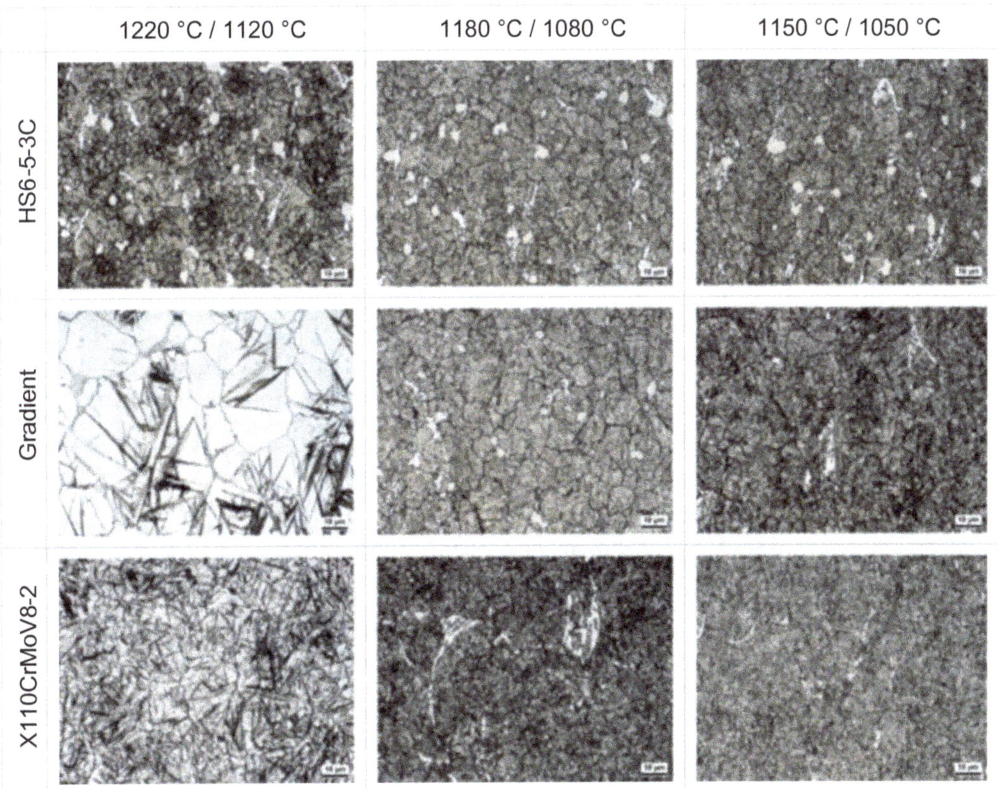

Fig. 6.13 Microstructure of the graded tool steels after selective induction hardening and tempering [Cui17a] Copyright 2017 by MS&T17. Used with permission

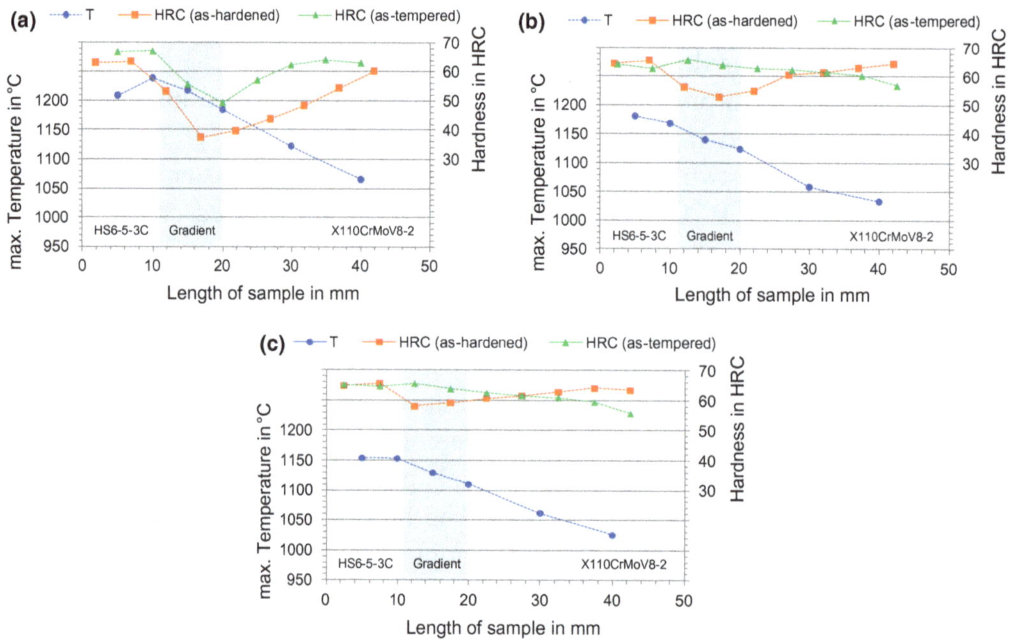

Fig. 6.14 Hardness and maximum austenitization temperature of the graded tool steels processed under various induction heating conditions (target temperatures): **a** 1220 °C/1120 °C, **b** 1180 °C/1080 °C, and **c** 1150 °C/1050 °C [Cui17a] Copyright 2017 by MS&T17. Used with permission

(2) The hardness of the gradient zone in the as-hardened condition is low. It increases as the maximum austenitization temperature decreases due to reduced retained austenite. If the austenitization temperature is too high, for example above 1150 °C, excessive dissolution of carbides in the matrix lowers the martensite transformation temperature and results in excessive retained austenite.

(3) The hardness of the gradient zone increases after triple tempering at 550 °C. This is due to the precipitation of fine carbides and further transformation of retained austenite to martensite during the tempering at high temperature. However, if the austenitization temperature is too high, carbide precipitation and further transformation of retained austenite to martensite would be insufficient due to stabilization of the austenite, resulting in a soft gradient zone (see Fig. 6.14a).

(4) In Fig. 6.14b the maximum austenitization temperature is approximately 1180 °C in the HS6-5-3C region and 1060 °C in the X110CrMoV8-2 region. After induction hardening the hardness of the HS6-5-3C steel and the X110CrMoV8-2 steel can reach 65 HRC, while the hardness of the gradient zone is low (52 HRC at the length position 17 mm). After tempering, the hardness of the gradient zone increases to 62–65 HRC. At lower austenitization temperatures (target temperatures: 1150 °C/1050 °C), the hardness of the gradient zone is further increased (Fig. 6.14c). After tempering, the hardness profiles are similar to that shown in Fig. 6.14b.

(5) If the austenitization temperature for X110CrMoV8-2 is very low (for example below 1050 °C), secondary hardening is insufficient and its hardness decreases after tempering (position around 40 mm in Fig. 6.14b, c). A high austenitization temperature is preferred for HS6-5-3C since it contributes to carbide dissolution and secondary hardening.

6.2.4 Fabrication of Micro Cold Forming Tools

Micro forming tools have been fabricated from the co-spray-formed graded steels. For plunge rotary swaging tools the tool surface consisted of HS6-5-3C (hard and wear-resistant), the tool body of X110CrMoV8-2 or HS6-5-2C (fracture-resistant and tough), with the gradient zone in between. For infeed rotary swaging tools, the reduction zone consisted of HS6-5-3C (wear-resistant, high friction) and the calibration zone of X110CrMoV8-2 or HS6-5-2C (fracture-resistant and highly micro-machinable), and the gradient zone was located between (see examples in Figs. 6.15 and 6.16).

Micro rotary swaging tools were manufactured in the hardened state by means of micro milling. The machining was carried out on a DMG Sauer Ultrasonic 20 linear machine tool. The machining operation comprised multiple consecutive roughing and finishing steps. CAD/CAM programming was used for tool path generation. In micro machining, the procedure of tool path generation is essential to achieve the desired shape accuracy and surface finish [Bri13]. Conventional path planning with a projected line pitch is not sufficient, as this would result in an inhomogeneous contour of the micro rotary swaging tools. In this study, the tool path generation was carried out with the line pitch arranged tangential to the targeted contour of the rotary swaging tools to meet the required machining quality. A detailed overview of all machining operations for the generation of the rotary swaging tool's micro contour can be found in [Cui17].

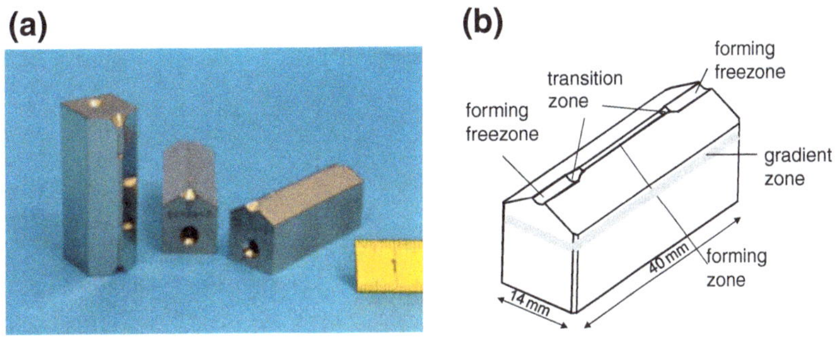

Fig. 6.15 Micro plunge rotary swaging tools of graded tool steel (X110CrMoV8-2/HS6-5-3C) processed by co-spray forming and hot deformation: **a** tool segments; and **b** different functional zones of the tools [Cui15]

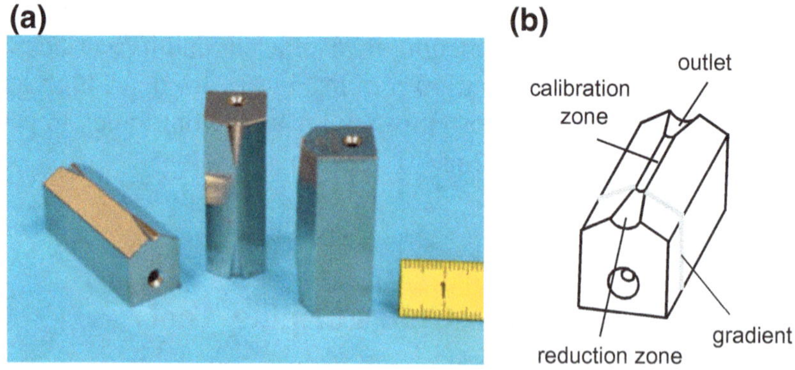

Fig. 6.16 Micro infeed rotary swaging tools of the graded tool steel (HS6-5-2C/HS6-5-3C) processed by co-spray forming and hot deformation: **a** tool segments; and **b** different functional zones of the tools

The functional surfaces of the forming tools after micro milling were investigated by means of scanning electron microscopy. Stereoscopic images were used to retrieve 3D information on the tool surfaces (using MeX software from Alicona), as seen in Fig. 6.17. It shows that the tool surfaces were smooth and the fine geometrical structures of the tools were precisely machined. Only slight traces of micro milling were observed at the tool surfaces. The excellent micro-machinability of the graded steels was guaranteed due to their fine and homogeneous microstructure.

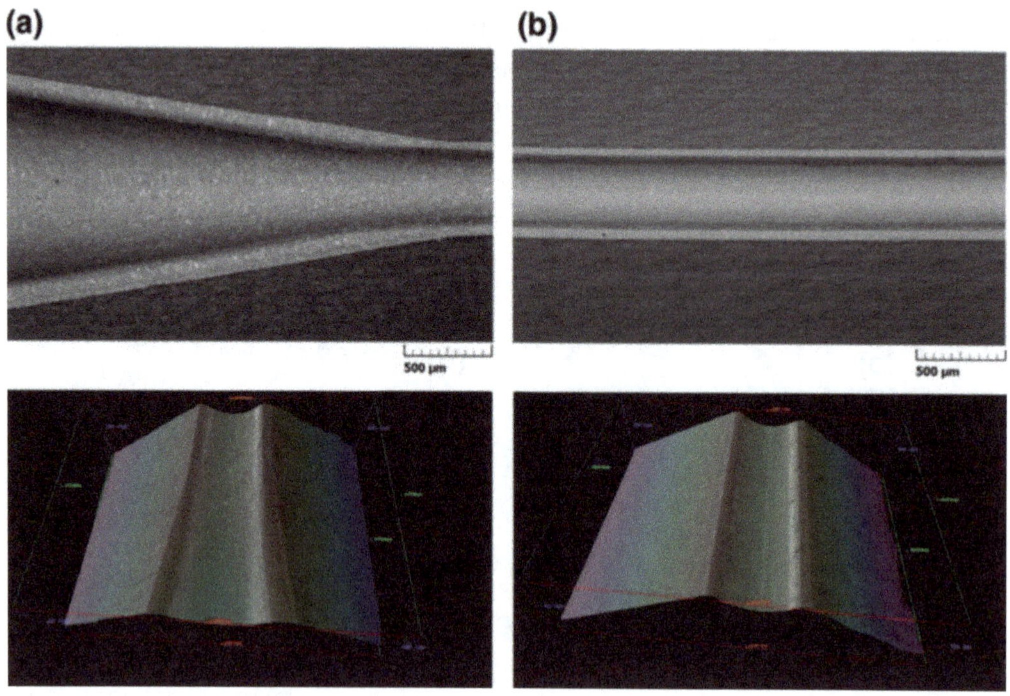

Fig. 6.17 Functional zones of a set of micro infeed rotary swaging tools (X110CrMoV8-2/HS6-5-3C) after micro milling: **a** reduction zone; **b** calibration zone

Refinement of the hard carbides by faster cooling during spray forming and more fracture of the carbides during heavier hot deformation would further improve the micro-machinability of the tool materials.

6.2.5 Performance of Micro Forming Tools

Rotary swaging is an incremental forming process for tubes, bars and wires. The workpieces are deformed in the swaging head by the forming tools (at least two segments), which rotate around the workpieces and perform radial oscillating movements at high frequency. There are two main variations of the process: plunge rotary swaging and infeed rotary swaging [Kuh08]. The diameter of the workpiece is reduced over the complete fed length during infeed rotary swaging, and it allows only local diameter reductions during plunge rotary swaging.

Sets of micro forming tools, which were fabricated from the graded tool steels, were tested in the micro rotary swaging machine (Type Felss HE 3/DE). Wires of stainless steel 1.4301 (AISI 304) in annealed state (surface hardness 300 HV0.1) were used for the test. The functional surfaces of the forming tools before and after the rotary swaging were investigated by means of scanning electron microscopy.

In plunge rotary swaging, the diameter of the stainless steel wires was reduced from 1.0 mm to the final diameter of 0.8 mm. The tool oscillated with a frequency of 100 Hz and the radial displacement was set to 35 μm/s. The total strain of the wires was 0.45. The gradient zone in the tool segments did not show signs of fracture or cracking after the test (up to 2500 pieces).

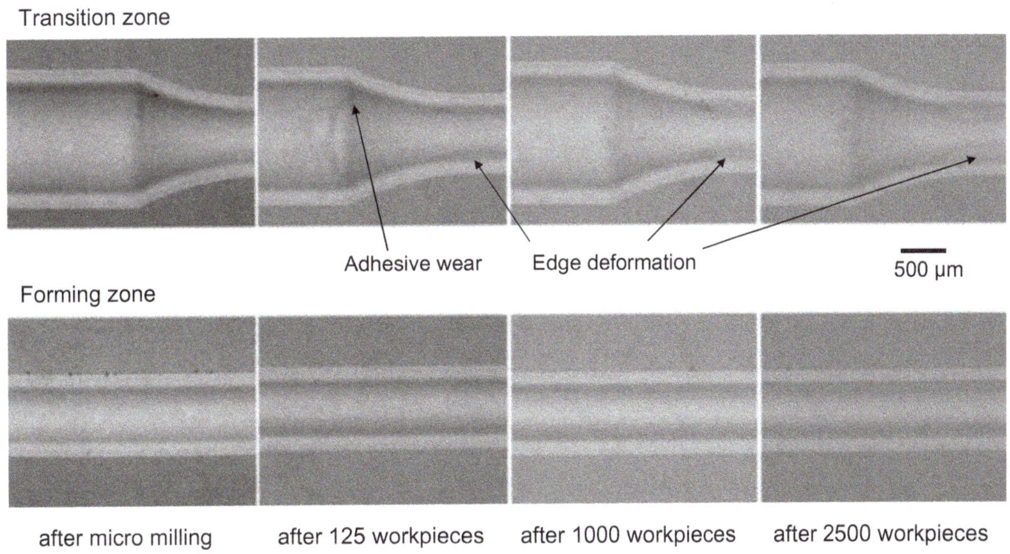

Fig. 6.18 Functional zones of the micro plunge rotary swaging tools before and after deforming workpieces of austenitic steel 1.4301 (forming zone Ø = 0.8 mm) [Cui15]

The functional surfaces of a forming tool segment (X110CrMoV8-2/HS6-5-3C) before and after micro rotary swaging are shown in Fig. 6.18. Slight adhesion of the workpiece on the tool surfaces and slight deformation/abrasive wear in the transition zone appeared. The forming zone of the tools was almost unchanged after processing 2500 workpieces. The shape, dimensional accuracy, and surface quality of the processed wires were satisfactory.

In infeed rotary swaging, the diameter of the stainless steel wires was reduced from 1.0 to 0.5 mm in one step, which corresponds to a total strain of 1.38. The tool oscillation took place with a frequency of 100 Hz and the workpiece axial feed velocity was 3 mm/s, which also means a displacement of 0.03 mm per stroke. The tools were investigated after forming wires of 5, 15, 35 and 85 m, respectively. The graded tool segments survived the test, and no fracture or cracking in the gradient zone was observed.

The functional surfaces of a forming tool segment (HS6-5-2C/HS6-5-3C) after micro infeed rotary swaging of AISI 304 wire of different lengths are shown in Fig. 6.19. The geometrical structures and the surface quality of the forming tools were not significantly changed by the swaging process. Slight abrasion in the reduction zone occurred after cold forming 5 m of wire. Slight adhesion of the wire on the tool surface was also observed both in the reduction zone and in the calibration zone. The abrasion and adhesion in the reduction zone are more pronounced than in the calibration zone since the loads (impact forces during swaging) and the friction between the wire and the tool were much higher in the reduction zone.

The degree of deformation and wear of the forming tools increased with the increasing number/length of workpieces. This might lead to poor surface roughness, low size and shape accuracy of the workpieces and even complete failure of the deformation (fracture of the workpieces). In order to reduce the deformation and

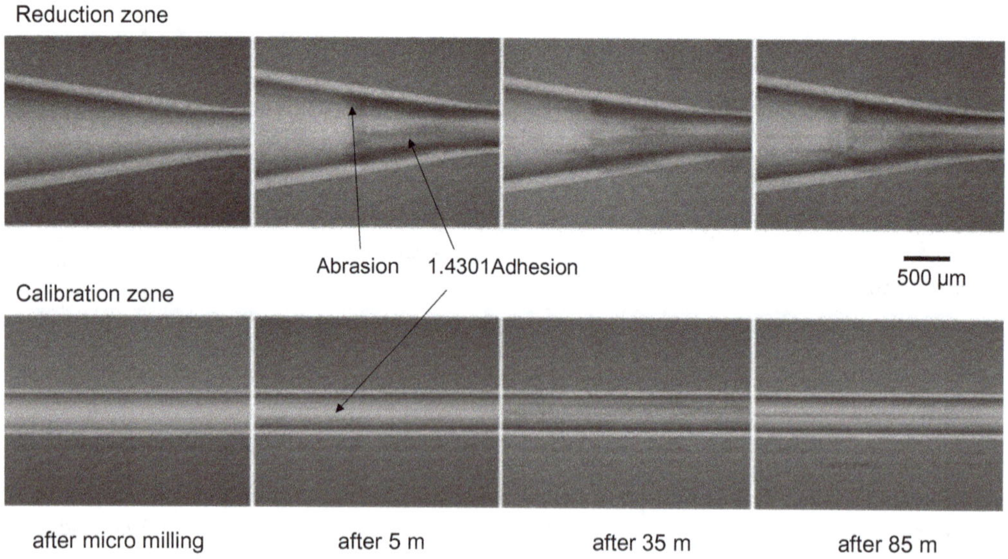

Fig. 6.19 Functional zones of the micro infeed rotary swaging tools before and after deforming workpieces of austenitic steel 1.4301 (forming zone Ø = 0.5 mm) [Cui16]

wear of the swaging tools, several strategies have been proposed to further improve the behavior of the swaging tools: (1) applying stronger tool materials in the tool surfaces, (2) reducing the deformation ratio of the workpieces, (3) reducing the temperature of the tools by coolant, and (4) improving the lubrication of the tools to reduce the friction between the workpieces and the tools.

6.3 Production of Thin Sheets by Physical Vapor Deposition

Andreas Mehner*, **Julien Kovac and Hans-Werner Zoch**

Abstract Thin metallic foils with thicknesses below 50 µm are required for the production of hollow micro components by deep drawing. The production of such foils by conventional metallurgical methods and cold rolling requires highly ductile materials such as pure aluminum, copper or stainless steel. Hardenable high-strength alloys are not available as thin foils due to the limited manufacturing process for such materials. Therefore, an alternative process based on physical vapor deposition (PVD) was developed in order to produce thin metallic foils with improved mechanical strength. The PVD-based production process and the micro structure and mechanical properties of the resulting thin metallic foils will be presented for several hardenable high strength aluminum alloys and steels. Selected foils were successfully tested for deep drawing of high-strength sub-millimeter micro cups.

Keywords Sheet metal · Composite · Physical vapor deposition (PVD)

6.3.1 Introduction

There is an increasing industrial demand for micro parts with enhanced mechanical properties and reliability. Micro deep drawing of submillimeter parts with complex shapes requires thin metallic foils with a thickness below 50 µm. Common metallic foils for deep drawing are produced by cold rolling of ductile materials such as pure aluminum, copper or stainless steel [Beh16]. These materials are characterized by high ductility and low mechanical strength. Although a few authors have investigated the drawability of some special high-strength materials [Had13], these materials are generally difficult to obtain as thin foils with a thickness below 50 µm because cold rolling of high-strength materials is very limited due to strain hardening, which requires a series of multiple cold rolling and annealing processes in order to restore the ductility of these materials after each cold rolling process.

Physical vapor deposition (PVD) and particularly magnetron sputtering have been proposed as an alternative process to produce thin foils of high-strength alloys [Sto10]. Such PVD thin foils were obtained by deposition of the considered materials onto substrate foils which were removed by chemical etching after the deposition process, in order to obtain freestanding thin foils. Research was necessary to find appropriate process parameters as well as a suitable method to separate the deposited films from the substrate foils. There are several methods to separate the deposited films from a substrate. Firstly, there is the use of non-adhesive substrates such as oxidized steel or the deposition of soluble interlayers such as NaCl that easily dissolve in water after the deposition process [Mat02]. A second method is removing the substrate by etching. The etchant has to be selected carefully so that the coating material remains undamaged. Muggleton has provided a variety of appropriate substrates and etchants to produce thin freestanding PVD films for different materials [Mug87]. Instead of removing the metallic substrate foils, it is also possible to use PVD-coated substrate foils as bimetallic foils. The mechanical strength of the substrate foils made of conventional ductile materials such as pure aluminum, copper or austenitic stainless steel could be increased by the deposition of thin layers of high-strength aluminum alloys or martensitic steels, resulting in high-strength bimetallic foils. It was shown that some mechanical properties of bimetallic sheets or foils are deduced from rules of mixtures. Hence, for aluminum/stainless steel bimetallic sheets, the apparent elastic modulus, the tensile strength and the yield strength coefficient (according to Hollomon's law) follow rules of mixtures, whereas the uniform elongation and the strain hardening exponents are deduced by force weighted laws of mixtures [Lee88]. In each case, a significant increase of strength and ductility compared to pure aluminum was observed. The deep drawing behavior of aluminum/stainless steel bimetallic sheets was investigated by various authors. Parsa et al. investigated the evolution of the limit drawing ratio in single drawing and redrawing for bi-layered sheets with various layer ratios of aluminum and stainless steel through simulation and experiments [Par01]. They found that with increasing steel layer thickness (i.e. decreasing aluminum layer thickness) the limit drawing ratio increased. The limit drawing ratio also depends on the orientation of the sheets in the experimental setup. It was lower if the aluminum layer faced toward the punch.

Promising alloys for micro cold forming are high-strength aluminum alloys such as scandium- or zirconium-containing aluminum alloys. Aluminum–scandium alloys are used for several high tech applications, in particular in the construction of aircraft parts, where the combination of light weight and high strength is particularly important. Among the different alloying elements for aluminum alloys, scandium is the one that results in the best increase of hardness per fraction of element [Roy05]. Alloying aluminum with scandium leads to the formation of nanometer-scale spheroidal Al_3Sc precipitates which efficiently impede the motion of dislocations due to their high coherency with the aluminum matrix and their high dispersivity. The Al_3Sc precipitates increase the mechanical strength, hardness and also the recrystallization temperature of these alloys. It was also shown that the addition of scandium improves the corrosion resistance and weldability of these alloys. A possible way to reduce the price of these alloys is partial substitution of the rare and expensive scandium by zirconium, which is much less expensive. In thermodynamic equilibrium, aluminum–zirconium alloys contain Al_3Zr precipitates with a tetragonal lattice structure which are less coherent with the aluminum matrix compared to Al_3Sc, resulting in limited reinforcement [Mur92]. However, during heat treatment, metastable cubic Al_3Zr precipitates form primarily and the tetragonal equilibrium phase only appears after long annealing times [Kni08]. The metastable cubic Al_3Zr precipitates have similar properties to those of the Al_3Sc precipitates and thus also increase the mechanical strength, though a little less than scandium. Another issue of aluminum–scandium alloys is that the size of the precipitates rapidly increases. Beyond a critical radius, they become incoherent, resulting in a decrease of hardness. In order to overcome this issue, aluminum alloys containing both scandium and zirconium were developed. As a result, small amounts of zirconium are incorporated into Al_3Sc precipitates. $Al_3(Sc_{1-x}Zr_x)$ precipitates are more stable and coarsen more slowly due to the reduced diffusivity of Zr in aluminum [Ful05]. Moreover, a substantial increase of strength has been reported for Al–Sc–Zr alloys compared to pure Al–Sc alloys [Son11].

6.3.2 Methods

6.3.2.1 The Magnetron Sputtering Process

In the following, the PVD magnetron sputtering process to produce thin Al–Sc–Zr foils will be presented. Magnetron sputtering is a physical vapor deposition (PVD) process for the deposition of thin films or coatings. The most common applications are the deposition of hard coatings for cutting and forming tools, hard coatings for wear and corrosion protection, low friction coatings, conducting and semi-conducting thin films for microelectronic devices, and photovoltaic cells or special coatings for medical applications. Sputtering is based on the evaporation of target materials by low-pressure plasmas. The vaporized target materials condense onto the substrates and also on the walls of the vacuum chamber. The sputtering process is performed in a vacuum chamber. After evacuation to about 10^{-6} mbar residual pressure, an inert gas (usually argon, neon or xenon) is introduced into the chamber until a working pressure of about 10^{-3} mbar is achieved. The metallic targets are connected to the cathode of a generator with a negative voltage of several hundreds of volts, whereas the rest of the chamber and the substrate are electrically grounded. A low-pressure plasma is created in front of the target due to collision ionization of the inert gas atoms by accelerated electrons emitted from the cathode (target). The positively charged inert gas ions created, such as Ar^+, are accelerated into the direction of the negatively charged target and hit with high kinetic energy onto the target where the surface atoms are sputtered (evaporated). The sputtered atoms move in the opposite direction to the substrate and progressively form a coating. Magnetron sputtering is an advanced sputtering process using permanent magnets behind the targets in order to concentrate the low-pressure plasma in front of the target, which increases the sputter and deposition rate [Tho86]. A more detailed description of the process can be found in reference books such as those of Matox [Mat02] or Ohring [Ohr02].

For conventional direct current magnetron sputtering, the maximum plasma power density is limited due to the formation of arcs if the plasma power density exceeds approx. 25 Wcm^{-2}. The risk of arc formation is reduced if the plasma power is pulsed. Pulsed DC magnetron sputtering (DCMS) allows an increase of the maximum plasma power density up to approx. 50 Wcm^{-2}. For pulsed DCMS, the duration of the pulse and off-cycle are comparable. The pulse frequency is in the range from 20 to 70 kHz. High-power impulse magnetron sputtering (HiPIMS) was developed for a further increase of the plasma power density [Sar10]. For HiPIMS, the duration of the power pulse is much shorter than the duration of the off-cycle. The duty cycle is defined as the ratio of the duration of the pulse to the duration of the whole cycle. For HiPIMS, the duty cycle is typically between 1% and 20%. For low-duty cycles, the power density could be increased up to 1000 Wcm^{-2}. For example, in a HiPIMS process with a defined frequency of 1000 Hz, a duty cycle of 0.2 ms (20%) and an average power of 2 kW, the maximum peak power could be increased up to 10 kW. The principal advantage of HiPIMS is the higher kinetic energy of the sputtered atomic species, resulting in a higher film density and better adhesion to the substrate [Mül87].

6.3.2.2 Separation of the Foils from the Substrate

For the production of freestanding metallic foils, it is necessary to separate the coated material from the substrate foil. This separation step is not required for the production of bimetallic PVD foils, since the substrate foils remain as a component of the bimetallic foils. For the production of freestanding monometallic aluminum foils by chemical etching, however, an appropriate choice of the substrate material is required. For the production of PVD aluminum alloy foils, low alloyed ferritic steel substrate foils like DC01 are adequate since these steels easily dissolve in concentrated nitric acid, whereas the aluminum alloys remain undamaged during the etching process due to their high corrosion resistance. For the production of freestanding PVD-steel foils, the use of thin copper substrate foils is appropriate because copper can be removed using an etching solution of ammonia and trichloroacetic acid, which does not attack the resulting steel foil [Mug87].

6.3.2.3 Continuous PVD Coating Process for Thin Substrate Foils

For mass production of bimetallic PVD foils, a continuous PVD magnetron sputtering process is required. A prototype device for continuous physical vapor deposition of thin metallic substrate tapes in an industrial PVD magnetron sputtering unit (CemCon 800/CC) was developed, constructed and tested. Figure 6.20 shows the scheme for continuous PVD magnetron sputtering in a PVD vacuum chamber with six targets. During the coating process, the metallic substrate tape from a donor coil is transported in front of the six targets. The coated tape is wound onto a receiver coil. The substrate tape is led by pulley rolls. For the practical realization, a substrate tower device was designed, as shown in Fig. 6.20b. A synchronic wind-off and wind-up of the substrate tape from the donor coil to the receiver coil is achieved by a central gear wheel which drives the gear wheels of both coilers. The pulley rolls were equipped with tension springs in order to control the tension of the substrate tape.

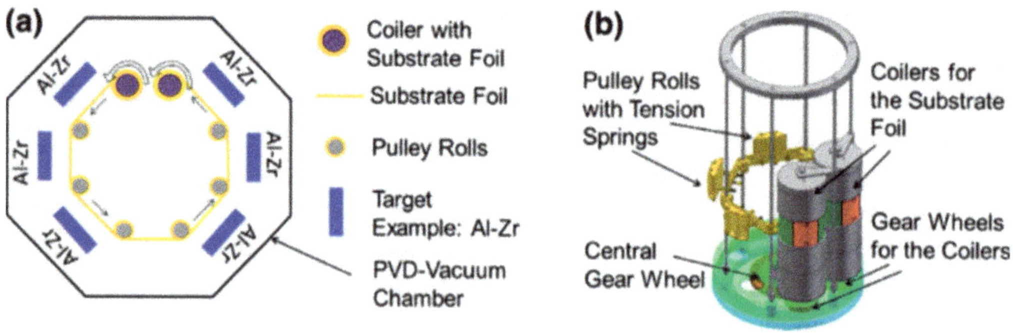

Fig. 6.20 Scheme of continuous magnetron sputtering in a PVD vacuum chamber (**a**). Design drawing of the constructed substrate tower for continuous PVD coating of long substrate tapes (**b**)

The deposited foil thickness is inversely proportional to the band uncoiling speed and it is directly proportional to the target number, the plasma power and the exposition length of the band between the two coils. As a result, in practice a higher band speed will lead to smaller thickness. Producing thicker foils requires a slowdown of the band speed or an increase of the deposition rate. With the current unit, the maximum exposition length is limited to about one meter. Producing Al–Sc–Zr/steel bimetallic foils with a layer thickness of 5 μm, for instance, would require a band speed of 36 mm/min. As a result, the production of 100 m of this foil would take about 2 days.

6.3.3 Results for PVD Al–Sc, Al–Zr and Al–Sc–Zr Thin Foils

Scandium and zirconium alloyed aluminum thin foils were produced in different types of magnetron sputtering units, which also influenced their structural and mechanical properties. Firstly, a laboratory-scale unit equipped with a single target and a non-moving substrate holder was used. Here, the substrate temperature was kept constant with a heating/cooling system. Secondly, an industrial PVD unit equipped with four or six targets and a rotating substrate holder was used for the production of larger foils.

6.3.3.1 Results for Al–Sc, al–Zr and Al–Sc–Zr Thin Foils Produced in a PVD Unit with Fixed Substrate Holder and Controlled Substrate Temperature

In the PVD unit with a non-moving substrate holder and a single target, the angular flow of atoms impinging onto the substrate follows a cosine law [Ohr02], resulting in an inhomogeneous distribution of the film thickness, as shown in Fig. 6.21a. Nonetheless, the flow of atoms is particularly strong and results in a very close-packed columnar structure as shown in Fig. 6.21b and c. The deposition process results in non-equilibrium thermodynamic conditions due to the quick condensation of energetic gaseous species onto a cold substrate (below 40 °C). The result of these conditions is an oversaturated aluminum solid solution, nearly free of precipitates [Kov13].

The mechanical properties of PVD–Al–Sc, Al–Zr, and Al–Sc–Zr thin foils were studied by tensile tests. The chemical composition of the target materials and of the produced PVD foils was measured by spark optical emission spectroscopy (S-OES) and by glow discharge optical emission spectroscopy (GDOES). The PVD–Al–Sc foils contained 2.0 mass% scandium, the PVD–Al–Zr foils contained 4.0 mass% zirconium and the PVD–Al–Sc–Zr foils had 1 mass% scandium and 0.5 mass% zirconium.

Several rectangular tensile test specimens were cut from each foil variant. In Fig. 6.22, selected representative tensile curves for each variant show the impact of the target power, the substrate temperature and a post-deposition heat treatment on the mechanical properties of the resulting Al–Sc, Al–Zr and Al–Sc–Zr foils. The foils have different tensile characteristics. Al–Sc foils show tensile strengths up to 400 MPa and high elongations at fracture up to approx. 10%. In comparison, the tensile strength of Al–Zr foils is higher but the elongation at fracture does not exceed 2%. Al–Sc–Zr foils achieved the highest strength but are more ductile than Al–Zr foils.

The experimental results show that the target plasma power, the substrate temperature as well as post annealing have a significant influence on the mechanical strength of the different foils. When the target plasma power is increased, the

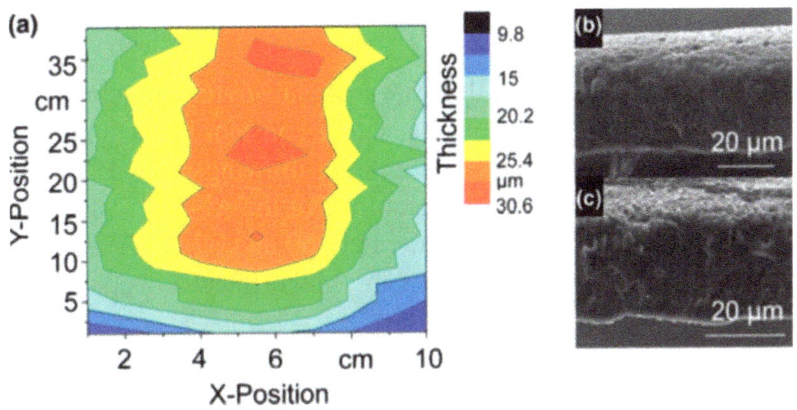

Fig. 6.21 Spatial distribution of the film thickness for an Al–Sc–Zr foil produced with a non-moving substrate holder and a single Al–Sc–Zr target (plasma power 1 kW) after 3.5 h of deposition (**a**). SEM pictures of the cross-sections of a PVD Al–Sc foil (**b**) and a PVD Al–Zr foil (**c**) produced with a non-moving substrate holder and a single target (plasma power 2 kW)

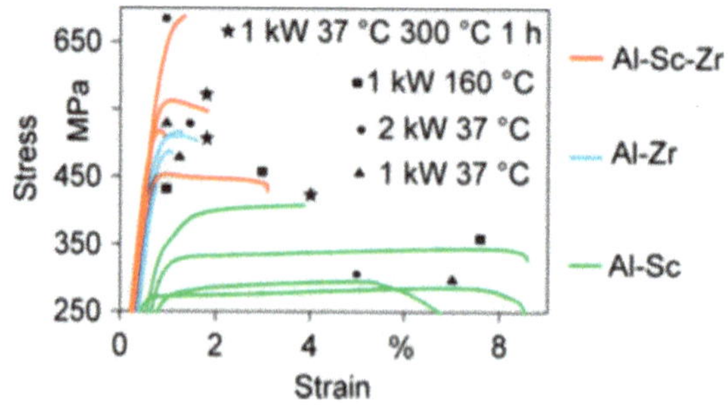

Fig. 6.22 Selected tensile curves of PVD Al–Sc (2 mass-% Sc), Al–Zr (4 mass-% Zr) and Al–Sc–Zr (1 mass-% Sc + 0.5 mass-% Zr) foils produced with target powers of 1 and 2 kW and substrate temperatures of 37 and 160 °C with and without post-deposition heat treatment at 300 °C for 1 h in normal air atmosphere

sputtered atoms hitting onto the substrate have higher kinetic energy. Due to momentum transfer, the surface atoms on the coating gain a higher mobility, which induces less porosity [Mül87]. The high kinetic energy of the sputtered atoms also increases the density of crystal defects as dislocations [Ohr02], which increase the mechanical strength of the foils. With increasing substrate temperature, different effects occur. A higher temperature may activate precipitation and generates thermal stress, which increases the strength. But also dynamic defects recovery is activated, which decreases the strength. The mechanical strength may thus increase or decrease, depending on the competition of these effects. During the post heat treatment at 300 °C, diffusion of the alloying elements occurs, resulting in the formation of precipitates and thus increasing the strength. However, recovery may also occur, which compensates the rise of the strength as observed for PVD Al–Zr foils produced at respectively 2 and 3 kW and aged at different temperatures for various times [Ego16].

PVD Al–Sc and Al–Zr foils deposited at substrate temperatures of 37 °C and 160 °C were tested by deep drawing with a constant blank holder pressure of 1 N/mm^2 [Beh12]. For Al–Sc foils, the limiting drawing ratio (LDR) was always 1.6, independently of the substrate temperature. For the Al–Zr foils, higher LDRs of 1.7 and 1.8 were reached for foils deposited at 160 and 37 °C. The higher LDR of the Al–Zr foils may be a consequence of the higher mechanical strength of these foils (Fig. 6.22).

6.3.3.2 Results for Al–Sc–Zr Foils Produced in a PVD Unit with Multiple Targets and Rotating Substrate Holder

The main disadvantage of a single target and non-moving substrate configuration is the limited size (110 × 500 mm^2) of the foils and an inhomogeneous thickness distribution, as shown in Fig. 6.21. Larger foils with a homogeneous thickness distribution were obtained using a magnetron sputtering unit with a rotating substrate holder and multiple targets, as shown in Fig. 6.23a. The consequence of this configuration is a lower average flow of sputtered atoms with a lower kinetic energy. If the rotating substrate holder and multiple targets are used, the flow of sputtered atoms and their kinetic energy varies periodically in time. The time-averaged value of the flow of sputtered atoms and their average kinetic energy was lower for experiments in the PVD unit with rotating substrate holders, which leads to a notable drop of the deposition rate. An averaged deposition rate of approx. 5.7 µm/h was measured for the experiments using the PVD unit with a single DC target with 1 kW plasma power and a non-moving substrate holder. For the PVD unit with four targets (each 1 kW plasma power) and a rotating substrate holder, the measured deposition rate was approx. 3.0 µm/h, with much better homogeneity of the thickness distribution, as shown in Fig. 6.23b.

Al–Sc–Zr foils were produced using a PVD unit with a rotating substrate holder and with three DC Al–Sc targets (2 mass% Sc) and one DC Al–Zr target (2 mass-% Zr) or with two HiPIMS Al–Sc–Zr targets (1 mass-% Sc, 0.5 mass-% Zr) [Kov18]. The DC target plasma power was 2 or 4 kW. The morphology and the mechanical properties of these foils were compared to foils produced by HiPIMS using an averaged target power of 2 kW (duty cycle 20%) at a frequency of 1000 Hz. The scandium and zirconium content of the foils was measured by glow discharge optical emission spectroscopy (GDOES). Both DC Al–Sc–Zr foils contained 1.8 mass-% Sc and 0.3 mass-% Zr. The HiPIMS Al–Sc–Zr foil contained 1.8 mass-% Sc and 0.6 mass-% Zr. Figure 6.24 shows SEM and optical microscopy pictures of the cross-sections of all three foils. All foils showed surface porosities resulting from shadowing effects during the PVD process. These shadowing effects increase with the rising film thickness. The porosity increased with decreasing kinetic energy of the deposited atoms. This is because the mobility of the surface atoms on the coating is reduced and pores are less likely to be filled [Mül87]. Thus, an increase of the DC power or using HiPIMS results in fewer and smaller pores, as shown in Fig. 6.24.

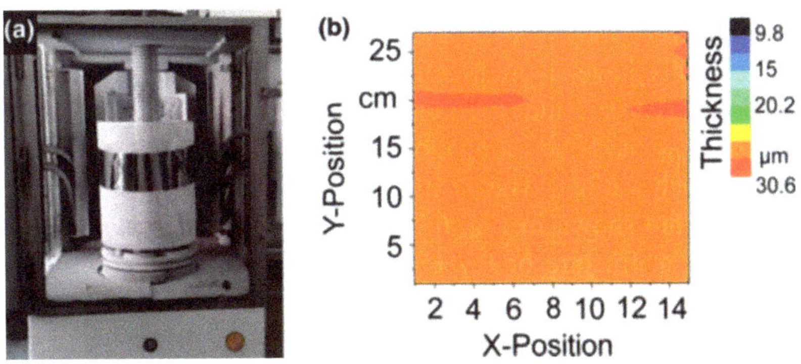

Fig. 6.23 Macro image of a polished steel foil substrate in the PVD unit with rotating substrate holder (**a**) and a plot of the resulting thickness distribution for a PVD Al–Sc–Zr foil produced in this PVD unit (**b**)

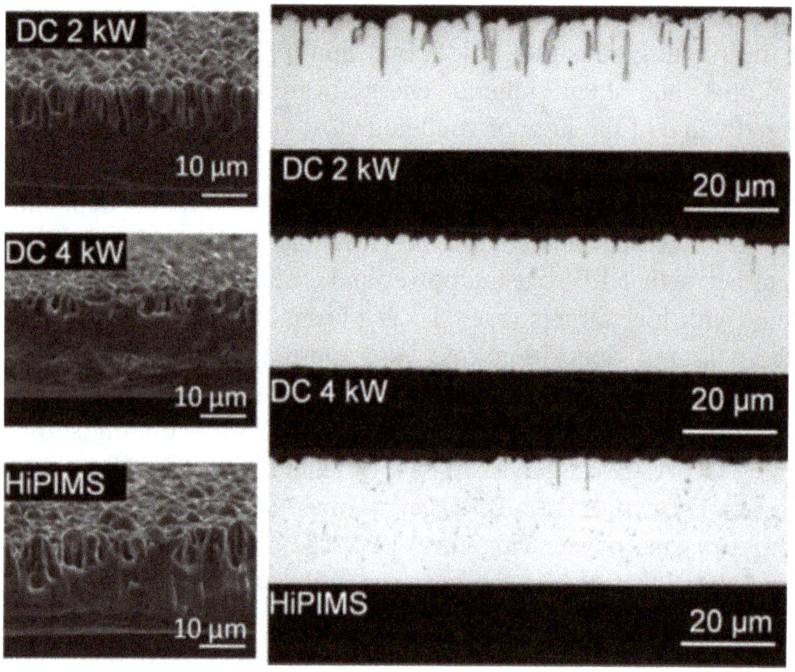

Fig. 6.24 SEM pictures of the free surface of freestanding Al–Sc–Zr foils and optical microscopy pictures of the cross-sections of these foils

Table 6.2 Average Young's modulus, yield strength ($Rp_{0.01}$), tensile strength (Rm) and strain at fracture (A) of Al–Sc–Zr monometallic foils

Al–Sc–Zr foil	E (GPa)	$Rp_{0.01}$ (MPa)	Rm (MPa)	A (%)
2 kW DC	58 ± 4	269 ± 42	292 ± 63	0.6 ± 0.2
4 kW DC	69 ± 2	314 ± 29	400 ± 38	1.5 ± 1.3
2 kW HiPIMS	65 ± 1	230 ± 9	260 ± 12	5.7 ± 3.9

The results of tensile tests with Al–Sc–Zr foils deposited using DCMS with different target powers and HiPIMS showed that the microstructure and the morphology of the foils have a great impact on their mechanical properties. The Young's modulus (E), the ultimate tensile strength (Rm), the yield strength at 0.01% strain ($Rp_{0.01}$) and the elongation at fracture (A) were determined by 10 tensile tests for each variant. The results are summarized in Table 6.2. Selected tensile test curves are shown in Fig. 6.25. The low Young's modulus of 59 MPa and the low strain at fracture of 0.7% for the DC Al–Sc–Zr foil at 2 kW target power compared to those of the 4 kW DC and the 2 kW HiPIMS foils are related to the high surface porosity of this foil, as shown in Fig. 6.24. The yield strength and the tensile strength of the Al–Sc–Zr foils produced in the PVD unit with the moving substrate holder were lower than for the foils produced in the PVD unit with the static substrate holder shown in Fig. 6.22. This effect may be the result of a higher average kinetic energy of the hitting atoms for the latter foils. Similar to the foils

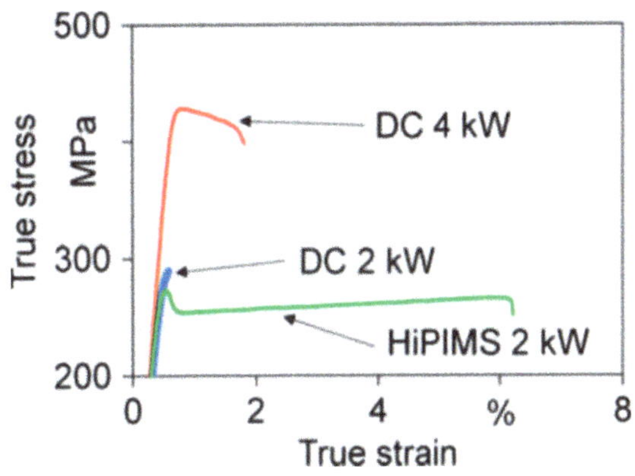

Fig. 6.25 Selected tensile test curves of the different Al–Sc–Zr foils

produced with the static substrate holder, an increase of the DC plasma target power from 2 to 4 kW results in an increase of the yield and tensile strength, whereas the 2 kW HiPIMS deposited foil showed comparable strength to that of the 2 kW DC foil.

The results of micro deep drawing with freestanding PVD Al-Sc-Zr foils depend on the orientation of the foil facing the die. This orientation effect is due to the different surface morphologies of the top side compared to the bottom side which was in contact with the substrate foil, as shown in Fig. 6.24. On the top side (upper side or free surface), the columnar grains grow freely, resulting in high roughness and porosity. If the free surface of the foils faced towards the die, the LDR was only 1.7, whereas a LDR of 1.8 was reached if the bottom side faced toward the die [Kov18]. This is because strong tensile stress always develops on the outer shell of the cup. Due to pores (Fig. 6.26b), the free surface is expected to sustain less stress, resulting in crack formation for lower drawing ratios compared to the denser bottom surface (Fig. 6.26a). Thus, higher LDRs are achieved if the bottom surface faces towards the die.

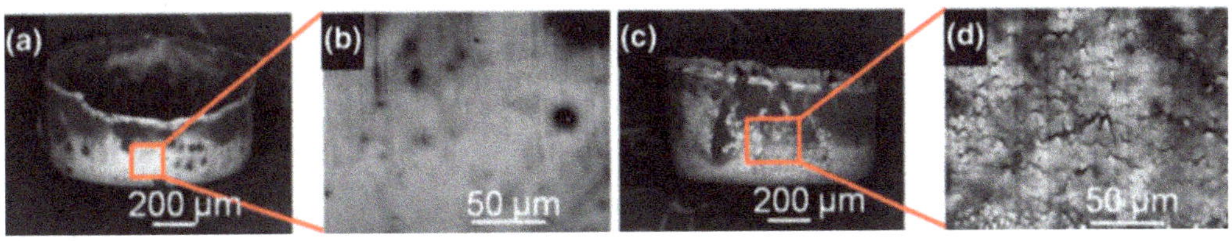

Fig. 6.26 SEM pictures of Al–Sc–Zr cup obtained by deep drawing with drawing ratio of 1.7 with the free surface facing toward the punch (inside) (**a**, **b**) and with the free surface facing toward the die (**c**, **d**)

6.3.4 Results for PVD Al-Sc-Zr/Stainless Steel Bimetallic Foils

The mechanical properties and the drawability of high-strength PVD Al–Sc–Zr foils were further improved by combination with more ductile but lower strength austenitic stainless X5CrNi18-10 steel foils. The mechanical properties of bimetallic foils or complex laminates are typically described by rules of mixtures, stating that the mechanical properties of laminates are averaged values of the properties of the constituting compounds weighted by their volume fraction. Hence, bimetallic Al–Sc–Zr/X5CrNi18-10 foils are expected to show strength and ductility intermediate between that of pure Al–Sc–Zr foils and X5CrNi18-10, depending on the thickness of each layer. The rule of mixtures was studied by depositing 5, 10 and 15 μm thick layers of Al–Sc–Zr onto thin commercially available X5CrNi18-10 foils with thicknesses of 25, 20 and 15 μm so that the total thickness of the resulting bimetallic foil was always 30 μm. The X5CrNi18-10 substrate foils (width 50 mm) were fixed to the rotating substrate holder as described in Sect. 6.3.3.2. A DC target power of 4 kW was used. Figure 6.27 shows typical tensile tests curves of the different bimetallic foils and of a freestanding Al–Sc–Zr foil and a commercial X5CrNi18-10 foil. The average values and standard deviations of the measured tensile properties are given in Table 6.3.

The measured strength and ductility of the bimetallic foils was between those of the monolayer Al–Sc–Zr and X5CrNi18-10 foils. All the bimetallic foils showed higher yield and tensile strength compared to the monometallic Al–Sc–Zr foils. The elongation at fracture was comparable to monometallic stainless steel foils. These properties also increased with the layer thickness (volume fraction). The yield and tensile strength as well as the apparent Young's modulus of the bimetallic foils are well predicted by rules of mixtures [Kov18]. Despite being submitted to higher strains than the elongation at fracture of monometallic Al–Sc–Zr foils, the Al–Sc–Zr layer in the bimetallic foils did not crack prematurely in the tensile tests. Thus, the bimetallic foils behaved similarly to non-composite materials. According to Choi et al. [Cho97], the impressive augmentation of strength and ductility is related

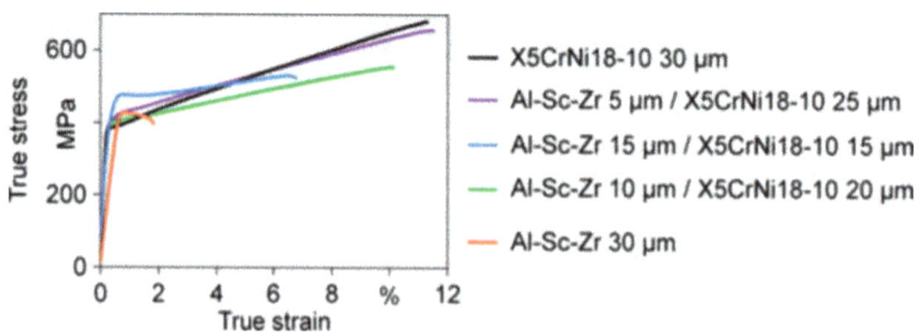

Fig. 6.27 Selected tensile test curves of bimetallic Al–Sc–Zr/X5CrNi18-10 foils with different Al–Sc–Zr layer thickness, of 30 μm thick monometallic Al–Sc–Zr and X5CrNi18-10 foils

Table 6.3 Average Young's modulus, yield strength (Rp0.01), tensile strength (Rm) and elongation at fracture of Al–Sc–Zr/stainless steel bimetallic foils and of Al–Sc–Zr and X5CrNi18-10 monometallic foils

Foil type	Thickness (μm) Al–Sc–Zr/ X5CrNi18-10	E (GPa)	$Rp_{0.01}$ (MPa)	Rm (MPa)	A (%)
Al–Sc–Zr	30/0	69 ± 2	314 ± 29	400 ± 38	1.5 ± 1.3
Bimetallic	15/15	148 ± 25	329 ± 43	526 ± 21	11.1 ± 8
Bimetallic	10/20	177 ± 14	327 ± 9	548 ± 10	12.1 ± 3
Bimetallic	5/25	192 ± 8	339 ± 12	662 ± 22	12.6 ± 1.1
X5CrNi18-10	0/30	197 ± 25	355 ± 13	684 ± 41	11.5 ± 1.8

to the wrapping of the bimetallic foils during tensile tests, due to the different lateral contraction of steel and aluminum.

The micro deep drawing experiments showed that the increase of ductility and strength observed in the tensile tests also resulted in better drawability. Indeed, whereas monometallic PVD Al–Sc–Zr foils achieved a limiting drawing ratio (LDR) of either 1.7 or 1.8, depending on the side facing towards the matrix, bimetallic foils of 25 μm of stainless steel and 5 μm of Al–Sc–Zr achieved a LDR of 1.9, independently of the orientation of the bimetallic foils in the drawing die [Kov18].

6.3.5 Production of Monometallic and Bimetallic Steel Foils

The production of thin steel foils by magnetron sputtering is more challenging than the production of thin aluminum alloy foils. Firstly, ferromagnetic steel targets such as ferritic chromium steels (X40Cr13) shield the magnetic field of the magnetron targets, resulting in a less concentrated plasma in front of the steel target, which reduces the sputter and the deposition rate. Secondly, the melting point of steels is higher than that of aluminum alloys, resulting in reduced surface diffusion, which supports the formation of pores in the foils [Kov13a]. Furthermore, austenitic steels may present an unexpected metastable ferrite structure after magnetron sputtering if the substrate temperature is low [Dah70], and the steel deposit may also be over-saturated in crystal defects such as dislocations or point defects, resulting in particularly brittle foils. Therefore, heat treatment of the foils may be necessary to reduce the density of crystal defects and improve the ductility of the PVD steel foils [Kov16]. Despite these constraints, PVD steel foils with promising properties were successfully produced by magnetron sputtering, as shown in the following.

The first example are high manganese austenitic X5MnSiAl25-3-3 steel foils. This steel belongs to the group of TWIP steels (TWIP: twinning-induced plasticity) which are characterized by very high ductility due to the TWIP effect. The austenite lattice structure of this steel is characterized by a relatively low stacking fault energy, which enables twinning as the dominant plastic deformation mechanism. Because of the TWIP effect, the tensile strength and elongation at fracture of these alloys are much higher than those of conventional stainless steels. For the X5MnSiAl25-3-3, a tensile strength of up to 800 MPa and an elongation at fracture of 90% were reported [Grä00]. Therefore, this steel is interesting for the deep drawing of micro components with a high limit drawing ratio and strength. But the deposition of X5MnSiAl25-3-3 foils by magnetron sputtering resulted in foils with relatively high porosity and brittleness. Therefore, it is important to reduce the residual porosity by increasing the plasma target power and by biasing the substrate [Kov13a]. Also, using mirror polished substrate foils contributed to lowering the shadowing effects responsible for porous structures [Kov17]. Following these adjustments, denser and smoother deposits were obtained. Nevertheless, after the deposition, the foils still showed high brittleness due to a fine-grained structure, a high density of crystal defects and a ferritic instead of an austenitic lattice structure. Therefore, an annealing process was required in order to obtain a fully austenitic recovered and recrystallized structure [Kov16]. Prior to the annealing process, the copper substrate foils were removed by etching as described in Sect. 6.3.2.2. The microstructure and the mechanical properties of magnetron sputtered X5MnSiAl25-3-3 foils after annealing at different temperatures are presented in Fig. 6.28a and b. Foils treated at low temperatures (500–600 °C) presented a finely grained phase mixture of austenite and ferrite with high tensile strength (>1100 MPa) but low ductility (1%). Only after annealing at high temperature (800 °C), the foils achieved a coarse-grained austenitic structure. The strength decreased to approx. 750 MPa, which is higher than that of X5CrNi18-10 foils, but their ductility was quite comparable to X5CrNi18-10 foils of the same thickness. But the maximum elongations at fracture of up to 90% as reported for the bulk materials [Grä00] were not achieved.

Another example of bimetallic foils which could be used for the production of high-strength micro components, are bimetallic foils of martensitic hardenable high carbon chromium steel (X75Cr17) onto ductile austenitic steel (X5CrNi18-10). The combination of these materials allows a good compromise between high strength and ductility. However, it was observed that the deposited X75Cr17 layers fail at low strains (<1%) during tensile tests and that the strength of the bimetallic foils was below the strength of monometallic X5CrNi18-10 foils. This behavior can be explained by the more similar contraction of both steel layers, preventing wrapping during deformation [Cho97], but also by the high brittleness of the X75Cr17 resulting from the fine grain structure and porosity. Small improvements of the tensile strength, ductility and critical strain (i.e. the strain at which the X75Cr17 layer begins to crack) of the bimetallic foils were reached by decreasing the X75Xr17 layer thickness and by application of a substrate bias voltage [Kov17a]. Nonetheless, the only way to obtain a higher tensile strength than pure stainless steel foils is annealing of the bimetallic foils. As shown in Fig. 6.28c and d, after

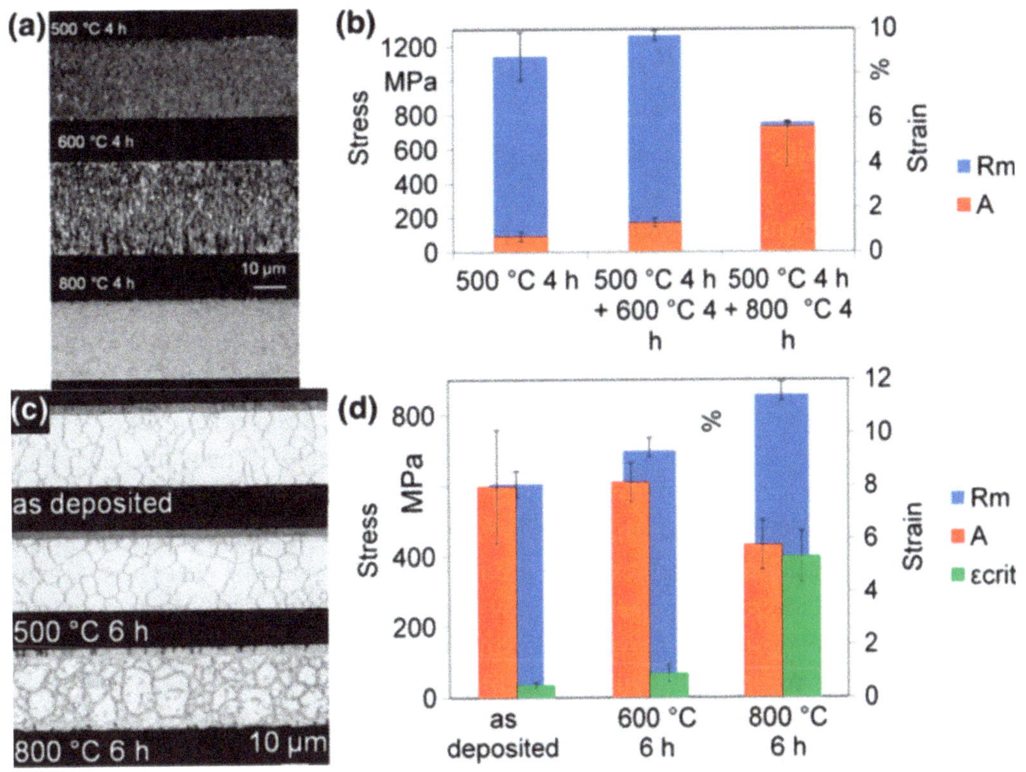

Fig. 6.28 Light microscopic pictures of the cross-sections and average tensile properties of X5MnSiAl 25-3-3 TWIP steel foils (**a**, **b**) and X75Cr17/X5CrNi18-10 bimetallic foils (**c**, **d**) after annealing at different temperatures

annealing at 500 °C the X75Cr17 foils failed at a critical strain of 0.9% and the strength of the bimetallic foil reached 700 MPa, which was slightly above that of the X5CrNi18-10 substrate foils. After annealing at 800 °C, no fracture of the X5Cr17 foil was observed during tensile tests and the strength reached by the bimetallic foil was above 850 MPa, whereas an appreciable strain at fracture of 5% was measured. This increase of tensile strength may, however, be partially due to carbon diffusion from the X75Cr17 coating to the X5CrNi18-10 substrate foil, resulting in a phase mixture of ferrite and austenite similarly to duplex steels. The drawability of these bimetallic foils was tested by micro deep drawing tests. Different limiting drawing ratios were obtained, depending on the orientation of the bimetallic foils to the drawing die. If the smooth X5CrNi18-10 substrate foil faced towards the die, a LDR of 1.8 was reached. If the brittle X75Cr17 layer faced towards the die, the micro cups failed more easily and the maximum LDR was only 1.7.

6.4 Heat Treatment of Micro Semi-finished Products and Micro Cold Formed Components

Anastasiya Toenjes*, Axel von Hehl and Hans-Werner Zoch

Abstract Heat treatment is an elementary process step in the production of metallic components. It serves on the one hand to adjust the formability of the semi-finished components by annealing processes and on the other hand to optimize the functional properties, for example, in terms of strength and hardness. An exactly adjusted heat treatment is particularly crucial for parts with thin wall thicknesses of less than 100 μm, since here the number of grains in the transverse section is low and thus weak points or defects have a considerably bigger impact on the failure behavior of components on the micro than on the macro scale [Köh10]. Besides, aspects such as handling and charging have to be considered particularly in the transfer of conventional processes from the macro to the micro scale. Thus, for the heat treatment of micro components, a new facility to carry out the heat treatment during falling was designed to avoid deformations during charging and processing of the sensitive components. This 5.5 m high drop-down tube furnace enables a variety of heat treatment processes for micro components made of both steel and aluminum alloys. In this regard, a wide range of macro processes could be scaled down to micro processes tailored to the drop-down tube furnace. Thus, recrystallization annealing, hardening, tempering, nitriding as well as age hardening can be performed during falling.

Keywords Heat treatment

6.4.1 Annealing Processes

The formability of a metal decreases with the increasing degree of deformation, which can be traced back to a rising dislocation density. Dislocations that occur during cold forming are carriers of plastic deformation. Due to their increasing number and interaction, however, they simultaneously impede further plastic deformation. Local stress fields of the dislocations superimpose during their interaction, leading to an increase of the macroscopic stress that must be overcome in order to uphold further plastic deformation [Eck69].

In order to recover the formability for further cold forming operations, heat treatment processes such as recrystallization annealing can be carried out for both steel and aluminum alloys. The aim of the process is to undo work hardening associated with cold forming and to restore plastic formability. In the recrystallization mechanisms, a distinction is made between primary recrystallization, secondary (/ternary) recrystallization, strain-induced grain boundary movement, and general grain growth [Eck69].

The primary recrystallization of a metallic structure requires a sufficient amount of cold forming (critical forming ratio) before heating above the recrystallization temperature, which corresponds to about 0.4 of the melting temperature of the material in Kelvin. Both the recrystallization temperature and the critical deformation degree are material-dependent factors. During recrystallization, high-angle grain boundaries begin to move through the material and eliminate dislocations that were formed during plastic deformation. The primary recrystallization is a rapid process compared to the other softening mechanisms, and generates a fully new microstructure with minimized dislocation density and renewed grain size. The resulting grain size is mainly dependent on the degree of deformation. The higher the degree of deformation, the smaller the grain size after recrystallization annealing [Eck69].

Recrystallization annealing can also be carried out during very short time cycles, benefiting the heat treatment of cold formed micro components. This can be seen from the example of the rotary swaged wire sections made of X5CrNi18-10. After a short-time heat treatment in the drop-down tube furnace at 1300 °C, the hardness can be reduced to less than 4 s by about 4250 H_{iT} 0.01/10/10/10 (Fig. 6.29). In addition, relationships between the degree of deformation, the temperature, the heat treatment time and the hardness are clearly visible [Bar12].

At the drop-down tube furnace temperature of 1300 °C, the recrystallization of the samples with a deformation degree of 3.22 is completed in just a few seconds. Figure 6.30a shows the microstructure of the sample after rotary swaging. After 1.3 s a new grain formation can be seen, but the old elongated structures have not completely disappeared (Fig. 6.30b). After 2.3 s, a completely new formation of

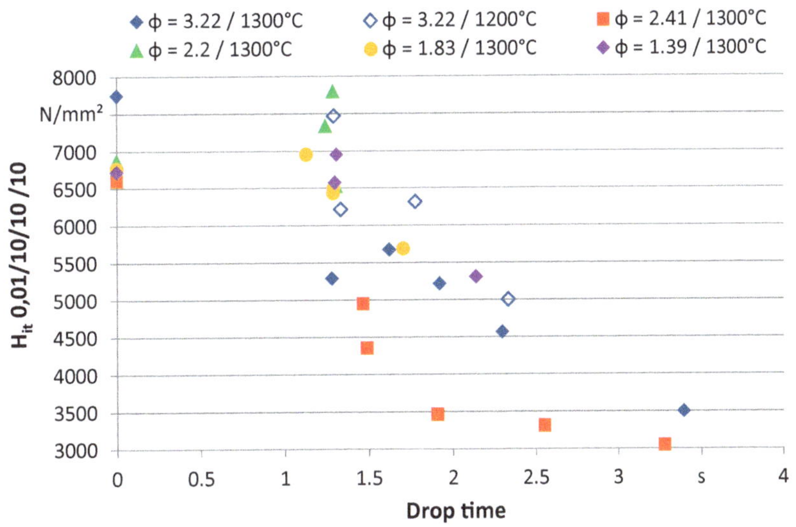

Fig. 6.29 Results of the universal micro hardness measurements indentation hardness versus drop time depending on the drop-down tube furnace temperature and the degree of deformation [Bar12]

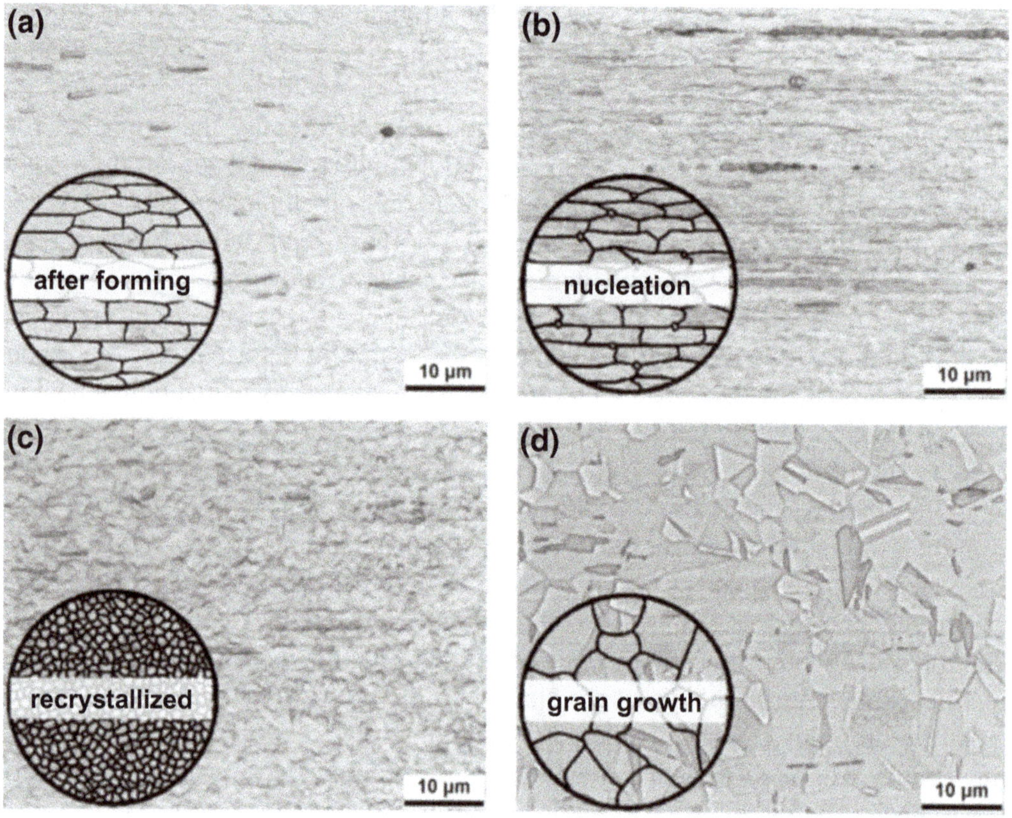

Fig. 6.30 Microstructure of a specimen after rotary swaging with $\varphi = 3.22$ **a** after forming and after subsequent short-time recrystallization annealing at 1300 °C for **b** 1.3 s, **c** 2.3 s and **d** 3.4 s; etched with V2A etchant for 120 s at 40 °C. Material: X5CrNi18-10 [Bar12, Sei14]

the grain structure with a grain size in the range of a few micrometers can be seen (Fig. 6.30c). Even after 3.4 s, a homogeneous grain growth occurs with the resulting grain sizes between 5 and 10 µm (Fig. 6.30d) [Bar12].

6.4.2 Martensitic Hardening of Steels

The aim of hardening is a considerable increase in hardness through the formation of martensite in hardenable steels. This is achieved by a phase transformation of ferrite/pearlite to austenite during heating from room temperature to austenitizing temperature, followed by a transformation of the austenite to martensite during rapid cooling. Depending on the alloying elements, the austenitizing temperature of steel is about 800 to 950 °C. The holding time at austenitizing temperature depends on the homogeneity of the material, the alloy composition, the dimensions of the part and the temperature. However, the time must be long enough to dissolve all the carbides and to achieve a homogeneous distribution of the alloying elements [Eck69]. The austenitizing time can be taken from time–temperature austenitization diagrams (TTA). The resulting microstructure and hardness can be estimated with continuous-cooling transformation diagrams (CCT). If the steel is quenched (cooled very rapidly) after austenitizing, a diffusionless martensitic transformation takes place. According to the Bain model, a lattice-deforming transformation of the face-centered cubic austenite into the tetragonal distorted body-centered martensite occurs, as shown in Fig. 6.31(right) [Eck69]. Simultaneously, a lattice- maintaining deformation of the austenite must take place. This deformation is accompanied with shearing (gliding) and dilatation, parallel and perpendicular to the invariant habit plane, which has to be regarded as the phase boundary between the austenite and martensite phase.

In order to increase the production rate by reducing the production times, the heat treatment duration should be as short as possible. In this case of the micro components, the base material has to be as homogeneous as possible, because the higher the homogeneity, the shorter the time until a homogeneous austenite is reached [Eck69]. Due to the low volume-to-surface ratio and thus generally high cooling rate, the quench medium has no significant effect on the resulting microstructure and its mechanical properties. The martensitic microstructure depends in this case on the austenite grain size [Mac11]. The martensite morphology is coarser the larger the austenite grain size is. In order to simplify the charging and to increase the productivity, particular furnace concepts, such as drop-down tube furnaces (Sect. 6.4.4) can be used for heat treatment of micro components.

In a short-time heat treatment, the relationship of temperature and time is highly significant. For example, heat treatment of soft annealed micro components made of C100 with a wall thickness of approximately 100 μm at temperatures below 1100 °C, times between 1.5 to 3.5 s, and water quenching, does not lead to any changes of the microstructure compared to the initial state. From 1100 °C and above, however, the change to a martensitic hardening structure can be recognized, and this becomes

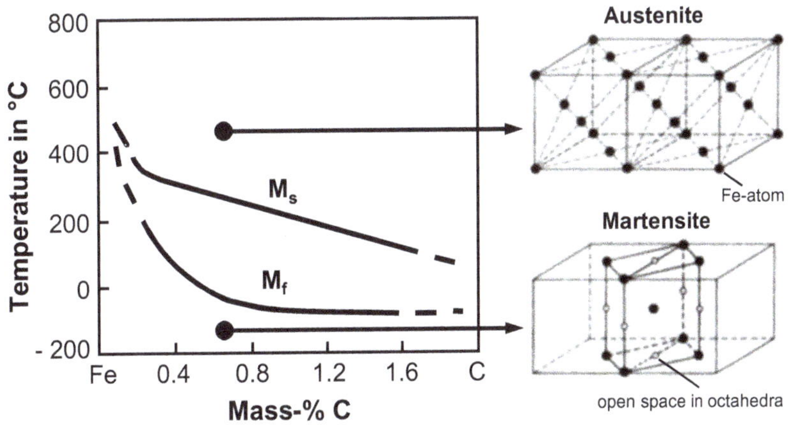

Fig. 6.31 Left: start M_s and finish temperature M_f of the transformation of austenite into martensite depending on the carbon content. Right: Relation of the lattice orientation of austenite and martensite according to E. C. Bain [Mac11]

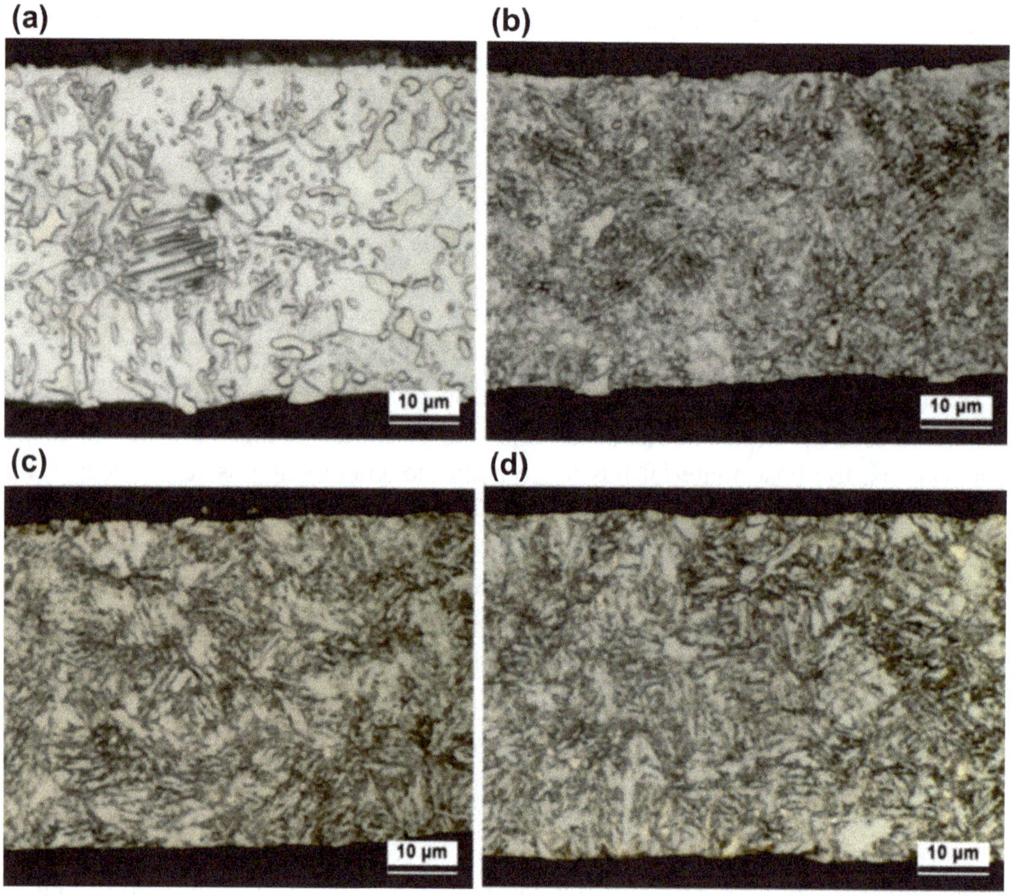

Fig. 6.32 Microstructure of C100 deep drawn micro components: **a** soft annealed and subsequently austenitized in the drop-down tube furnace at **b** 1100 °C, **c** 1200 °C, **d** 1300 °C, water quenched, etched with 3% alc. HNO_3 for 25 s [Bar10]

more homogeneous with the increasing furnace temperature (Fig. 6.32). Since it is a hypereutectoid steel and the austenitizing period is very short, not all the carbides are dissolved, so there are still very coarse carbide precipitations. However, the amount and the size decrease with increasing temperature. For quenching in air, similar results with regard to the structure formation have been determined [Bar10].

6.4.3 Precipitation Hardening of Aluminum Alloys

Aluminum alloys have to be distinguished between hardenable and non-hardenable. For non-hardenable aluminum alloys, the strength can be increased by solid solution and work hardening, while the strengthening effects of precipitations can be neglected. However, non-hardenable aluminum alloys do not achieve the high strengths of hardenable alloys. Hardenable aluminum alloys exploit the effect of nano-scale precipitations, which greatly increase the strength values, providing sufficient elongation at rupture at the same time [Ost07].

The significantly higher strengths compared to the non-hardenable alloys are achieved by a particular heat treatment, known as precipitation hardening or age hardening, to be adapted to the respective alloy system.

The basic steps of precipitation hardening are shown in Fig. 6.33. In the first step, the solution annealing, the workpiece is heated and held at temperatures below the solidus temperature to dissolve the alloying elements atoms as far as possible in the aluminum α matrix lattice. Depending on the alloy, the solution annealing temperatures range between 470 and 560 °C. Subsequently, the workpiece is quenched, for example, in water to room temperature, whereby a supersaturated solid solution results. During the following aging step, the alloying elements sequentially build finely distributed precipitations as intermetallic phases from the solid solution. The aging treatment can be carried out either at room temperature (natural aging) or at elevated temperature (artificial aging). These precipitates are an obstacle to dislocation movement and therefore increase the yield strength and tensile strength [Ost07].

However, the drop-down tube furnace described in Sect. 6.4.4 is not suitable for solution annealing, since the solution annealing time of the hardenable aluminum alloys is several minutes and the falling time in the drop-down tube furnace is just a few seconds. Therefore, alloys that are suitable for precipitation hardening in the drop-down tube furnace should already be in the supersaturated solid solution condition before heat treatment. This can be achieved with pre-processes with cooling rates higher than e.g. conventional casting processes in order to bring the

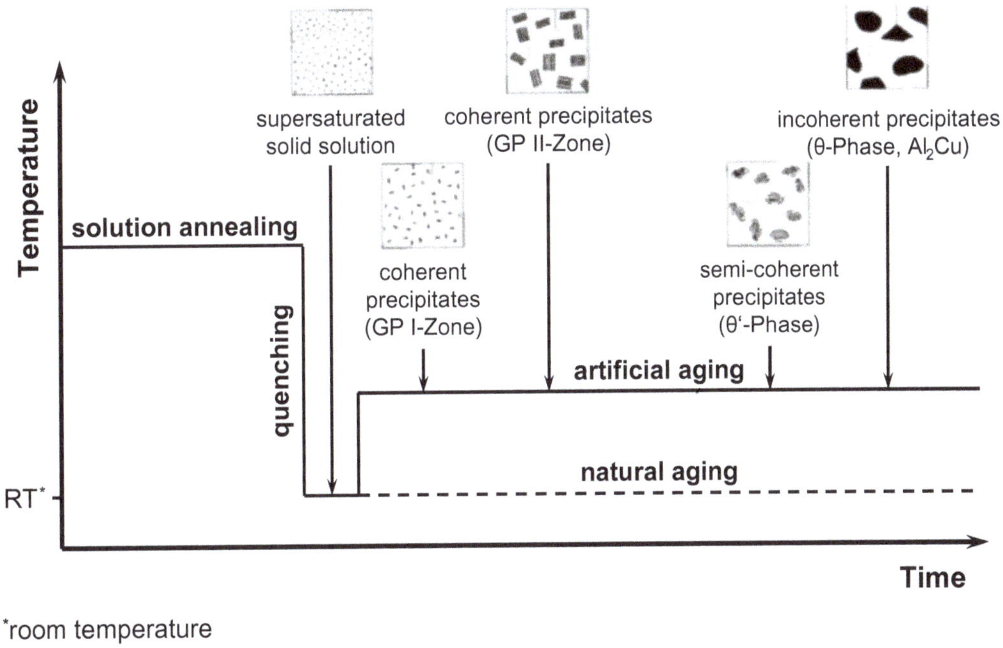

Fig. 6.33 The basic steps of precipitation hardening of Al–Cu alloy according to [Mac11]

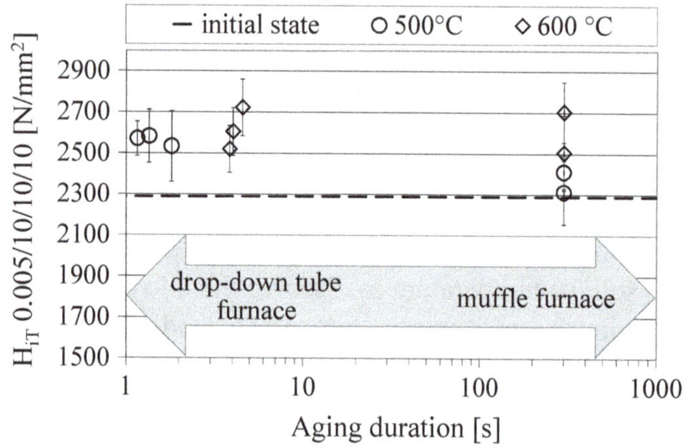

Fig. 6.34 Micro hardness measurements on deep drawn cups of Al–Zr before and after aging treatments at 500 and 600 °C in drop-down tube furnace and standard air atmosphere muffle furnace [Toe18]

alloying elements as far as possible into supersaturated solution. Processes that enable a maximum supersaturation for precipitation hardening are, for example, spray forming, also known as melt atomization, and laser additive manufacturing, which both exploit the extremely rapid solidification of the small melt volumina. As a process that enables a supersaturation without cooling from high temperatures by implementing the alloy elements via sputtering at relatively low temperature, PVD has been established into the micro production chain (see Sect. 6.1). For example, in binary Al–Zr alloys the maximum solubility of zirconium in aluminum solid solution is 0.28 mass-% maximum at 650 °C (melting point of the aluminum). By generating a supersaturated solution of 4 mass-%, it was proven that PVD makes it possible to overcome the solubility limit. However, the hardening potential of such Al–Zr alloy is dependent on the setting of the plasma power [Ego16]. The precipitation hardening results of deep drawn cups, which were made from PVD-generated Al–Zr foils, after both short-time heat treatment in the drop-down tube furnace and standard heat treatment in an atmosphere muffle furnace, are shown in Fig. 6.34. The short-time heat treatment increased the hardness of the parts from $H_{iT} \approx 2290$ MPa to $H_{iT} \approx 2720$ MPa. Thus, after about 4 s, the same hardness could be achieved in the drop-down tube furnace as after 5 min in a muffle furnace [Toe18].

6.4.4 Drop-Down Tube Furnace for Heat Treatment of Micro Components

The handling of small and damage-sensitive components is one of the challenges in the heat treatment of micro components. Conventional ways of charging such as bulk material on racks or in boxes are not practical with micro components since they can lead to damage of the parts. The contact between the components at elevated temperatures would increase the probability of welding effects and of plastic deformation and lead to a high scrap rate. To avoid these effects, a new heat treatment technology has been developed in which micro components are dropped one after another through a vertical tube furnace. With this technology, for example, the austenitization and final quench-hardening of large series of steel components is feasible. However, the requirements for such a furnace are high in order to have a sufficiently long drop duration that matches, for example, the austenitizing time needed. Using excess temperatures to maximize the austenitizing temperature enables an acceleration of the mechanisms activated on the micro-structural level, such as diffusion or recrystallization. An example of a drop-down tube furnace that meets these requirements consists of a tube with a total heated length of 5 m, inner tube diameter of 54 mm, and electric heating to a maximum temperature of 1300 °C. A schematic of the drop-down tube furnace as well as a manual feed device for the single sample operation and the opposing gas flow device can be seen in Fig. 6.35.

The drop-down tube furnace can be used for quench-hardening and tempering, as well as for age-hardening and recrystallization annealing, whereby the drop duration can be influenced by means of an opposing gas flow from the bottom. In order to avoid damage to the components due to oxidation, it is necessary to use inert gas. In addition, by using nitrogen it is even possible to nitride the components. Processes that can be performed in the drop-down tube furnace are summarized in Table 6.4.

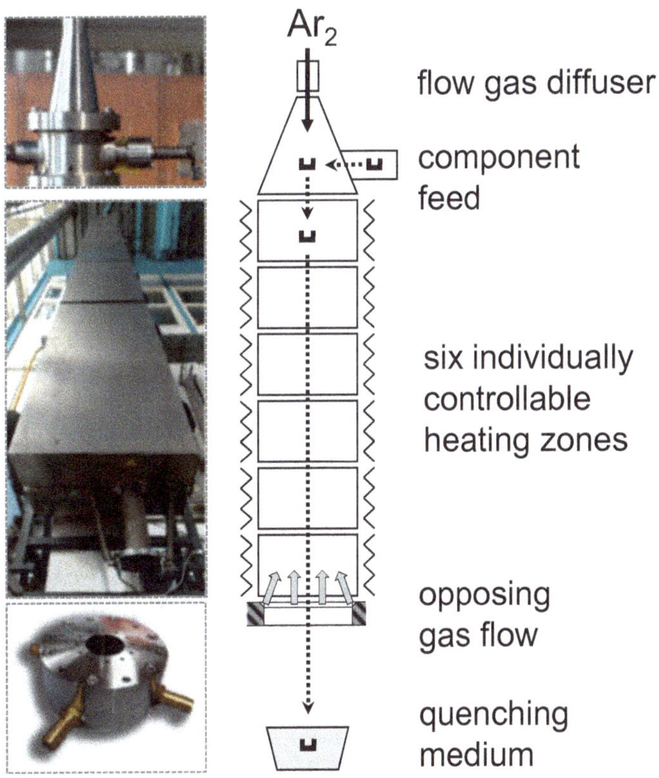

Fig. 6.35 Drop-down tube furnace for heat treatment of micro components during falling

Table 6.4 Processes that can be performed in the drop-down tube furnace

Initial state	Process	End state
Cold-formed micro components	Recrystallization annealing	Recrystallized micro components
Cold-formed steel components	Hardening	Hardened steel components
Hardened steel components	Tempering	Tempered steel components
Cold-formed steel components + N2	Nitriding	Nitrided steel components
Cold-formed AlSc2-components	Age hardening	Precipitation hardened AlSc2 components

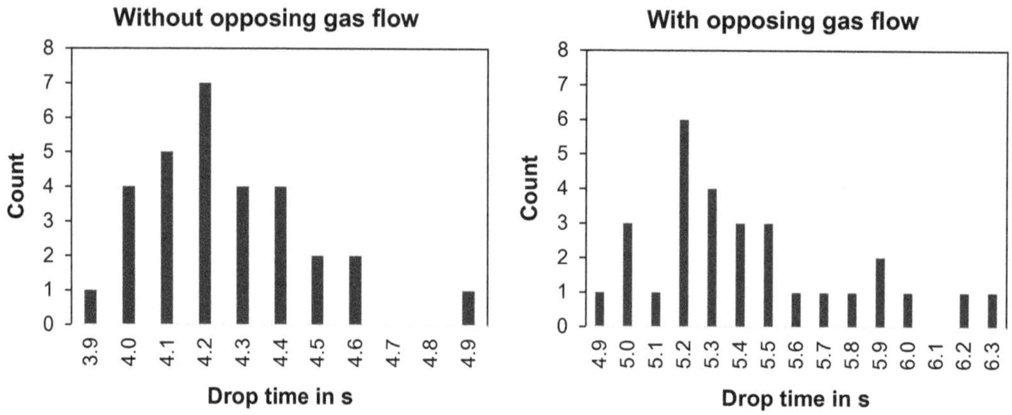

Fig. 6.36 Drop duration of the blanks with diameter of 4 mm and thickness of 20 μm of C100 steel at 20 °C furnace temperature

Since heat treatment is defined by temperature and time, the scattering of the drop time from the micro components is of particular importance. In order to achieve repeatable heat treatment results, it is necessary to measure the process times and to keep them constant. However, the variance of the drop duration also depends on the geometry of the components. If the components wobble during the fall, the variance of the fall duration can be considerably increased. For example, the effect of different possible orientations during falling influences the falling duration of steel cups with a diameter of 1 mm by ±14%. When using the opposing gas flow from the bottom of the drop-down tube furnace, both the drop time and the variance of the drop times are increased (Fig. 6.36). This is also due to the orientation of the components as they enter the gas stream.

References

[Bar12] von Bargen, R., von Hehl, A., Zoch, H.-W.: Kurzzeit-Rekristallisationsglühen von Mikrobauteilen aus X5CrNi18-10 im Fallrohrofen. HTM J. Heat Treat. Mater. **67**(6), 386–392 (2012)

[Bar10] von Bargen, R., von Hehl, A., Zoch, H.-W.: Härten von Mikrobauteilen aus Stahl im Fall. J. Heat Treat. Mater. **65**(2), 55–62 (2010)

[Beh16] Behrens, G., Trier, O., Tetzel, H., Vollertsen, F.: Influence of tool geometry variations on the limiting drawing ratio in micro deep drawing. Int. J. Mater. Form. **9**(2), 253–258 (2016)

[Beh12] Behrens, G., Kovac, J., Köhler, B., Vollertsen, F., Stock, H.-R.: Drawability of thin magnetron sputtered Al-Zr foils in micro deep drawing. Trans. Nonferrous Soc. China **22** (Supplement 2), 268–274 (2012)

[Bri13]. Brinksmeier, E., Vollertsen, F., Riemer, O., Flosky, H., Behrens, G., Böhmermann, F.: Mikrofräsbearbeitung zur Herstellung leistungsfähiger Mikroumformwerkzeuge. In: Proceedings of 6th Kolloquium Mikroproduktion, Brunswick, Germany **6**, A23-1–A23-8, Oct. 8 2013

[Cui17] Cui, C., Schulz, A., Moumi, E., Kuhfuß, B., Böhmermann, F., Riemer, O.: Co-spray-formed graded steel for micro rotary swaging tools. Int. J. Mechatron. Manuf. Syst. **10**(3), 185–205 (2017)

[Cui17a] Cui, C., Schulz, A., Nadolski, D.: Selective heat treatment of spray-formed composite tool steel. In: Contributed Papers from Materials Science and Technology 2017 (MS&T17), Oct. 8–12, 2017, pp. 1164–1171. David L. Lawrence Convention Center, Pittsburgh, PA, USA (2017)

[Cui16] Cui, C., Schulz, A., Moumi, E., Kuhfuß, B., Böhmermann, F., Riemer, O.: Grade tool material for micro cold forming via a novel spray forming process. In: Proceedings of 10th Tool Conference, Bratislava, Oct. 4–7, pp. 365–374 (CD) (2016). ISBN: 978-3-200-04786-0

[Cui15] Cui, C., Schulz, A., Steinbacher, M., Moumi, E., Kuhfuß, B., Böhmermann, F., Riemer, O.: Development of micro rotary swaging tools of graded tool steel via co-spray forming. Manuf. Rev. **2**, 22 (2015). Published by EDP Sciences (2015) (online) https://doi.org/10.1051/mfreview/2015024

[Cui14] Cui, C., Schulz, A., Uhlenwinkel, V.: Materials characterization and mechanical properties of graded tool steels processed by a new co-spray forming technique. Mat.-wiss. u. Werkstofftech. **45**(8), 652–665 (2014)

[Cui13] Cui, C., Schulz, A., Uhlenwinkel, V.: Co-spray forming of gradient deposits from two sprays of different tool steels using scanning gas atomizers. Steel Res. Int. **84**(11), 1075–1084 (2013)

[Cui13a] Cui, C., Schulz, A.: Modeling and simulation of spray forming of clad deposits with graded interface using two scanning gas atomizers. Metall. Mater. Trans. B **44**, 1030–1040 (2013)

[Cho97] Choi, S.H., Kim, K.H., Oh, K.H., Lee, D.N.: Tensile deformation behavior of stainless steel clad aluminum bilayer sheet. Mater. Sci. Eng. **A222**(2), 158–165 (1997)

[Dah70] Dahlgren, S.D.: Equilibrium phases in 304L stainless steel obtained by sputter deposition. Metall. Trans. **1**, 3095–3099 (1970)

[Ego16] Egorova, A., Kovac, J., von Hehl, A., Mehner, A., Zoch, H.-W.: On the relation between plasma power and aging treatment in the production of thin aluminum zirconium foils by magnetron sputtering. Materialwissenschaft und Werkstofftechnik **47**(11), 989–996 (2016)

[Eck69] Eckstein, H.-J.: Wärmebehandlung von Stahl—Metallkundliche Grundlagen. VEB Deutscher Verlag für Grundstoffindustrie, Leipzig (1969)

[Ful05] Fuller, C.B., Murray J.L., Seidman, D.N.: Temporal evolution of the nanostructure of Al (Sc, Zr) alloys: Part I: Chemical compositions of $Al_3(Sc_{1-X}Zr_X)$ precipitates. Acta Materialia **53** (200), 5401–5413 (2005)

[Grä00] Grässel, O., Kruger, L., Frommeyer, G., Meyer, L.W.: High strength Fe-Mn-(Al, Si) TRIP/TWIP steels development–properties–application. Int. J. Plast. **16**(10), 1391–1409 (2000)

[Had13] Hadi, S., Kiet Tieu, A., Lu, C., Zhu, H.: A micro deep drawing of ARB processed aluminum foil AA1235. Int. J. Mater. Product Technol. **47**(1–4), 175–187 (2013)

[Hen17] Henein, H., Uhlenwinkel, V., Fritsching, U.: Metal Sprays and Spray Deposition. Springer International Publishing AG (2017)

[Kho00] Khor, K.A., Dong, Z.L., Gu, Y.W.: Influence of oxide mixtures on mechanical properties of plasma sprayed functionally graded coating. Thin Solid Films **368**, 86–92 (2000)

[Kuh08] Kuhfuß, B., Moumi, E., Piwek, V.: Micro rotary swaging: process limitations and attempts to their extension. Microsyst. Technol. **1412**, 1995–2000 (2008)

[Kni08] Knipling, K.E., Dunand, D.C., Seidman, D.N.: Precipitation evolution in Al-Zr and Al-Zr-Ti alloys during aging at 450–600 °C. Acta Materialia **56**(6), 1182–1185 (2008)

[Kov18] Kovac, J., Heinrich, L., Köhler, B., Mehner, A., Clausen, B., Zoch, H.W.: Tensile properties and drawability of thin bimetallic aluminum-scandium-zirconium/stainless steel foils and monometallic Al–Sc–Zr fabricated by magnetron sputtering. In: Vollertsen, F., Dean, T.A., Qin, Y., Yuan, S.J. (Eds.), Proceedings of the 5th International Conference on New Forming Technology (ICNFT 2018), Vol. 190, p. 15001. MATEC Web Conference (2018)

[Kov17] Kovac, J., Mehner, A., Köhler, B., Clausen, B., Zoch, H.W.: Mechanical properties, microstructure and phase composition of thin magnetron sputtered TWIP steel foils. HTM J. Heat Treat. Mater. **72**(3), 168–174 (2017)

[Kov17a] Kovac, J., Mehner, A., Köhler, B., Clausen, B., Zoch, H.-W.: Ermittlung der mechanischen Eigenschaften dünner PVD-X75Cr17/X5CrNi18-10 Bimetallfolien. Tagungsband 8. In: Vollertsen, F., Hopman, C., Schulze, V., Wulfsberg, J. (Eds.), Kolloquium Mikroproduktion, pp. 113–122. BIAS-Verlag, Bremen (2017)

[Kov16] Kovac, J., Epp, J., Mehner, A., Köhler, B., Clausen, B., Zoch, H.-W.: On the potential of magnetron sputtering for manufacturing of thin high Mn TWIP steel foils. Surf. Coat. Technol. **308**, 136–146 (2016)

[Kov13] Kovac, J., Stock, H-.R., Köhler, B., Bomas, H., Zoch, H.-W.: Tensile properties of magnetron sputtered aluminum-scandium and aluminum-zirconium thin films: a comparative study. Surf. Coat. Technol. **215**, 369–375 (2013)

[Kov13a] Kovac, J., Köhler, B., Mehner, A., Clausen, B., Zoch, H.-W.: Neue Konzepte zur Herstellung dünner TWIP Stahlfolien mittels physikalischer Gasphasenabscheidung. Tagungsband 6. In: Tutsch, R. (ed.) Kolloquium Mikroproduktion. Shaker-Verlag, Braunschweig (2013)

[Köh10] Köhler, B., Bomas, H., Hunkel, M., Lütjens, J., Zoch, H.-W.: Yield-strength behaviour of carbon steel microsheets after cold forming and after annealing. Scripta Materialia **62**(8), 548–551 (2010)

[Lee88] Lee, D.N., Kim, Y.K.: Tensile properties of stainless steel-clad aluminum sandwich sheet metals. J. Mater. Sci. **23**, 1436–1442 (1988)

[Mat02] Mattox, D.M.: Handbook of Physical Vapor Deposition (PVD) Processing. Elsevier, Albuquerque, NM, USA (2002)

[Mug87] Muggleton, H.A.F.: Deposition techniques for the preparation of thin films nuclear targets. Vacuum **37**(11), 785–817 (1987)

[Mül87] Müller, K.-H.: Stress and microstructure of sputter-deposited thin films: molecular dynamics investigations. J. Appl. Phys. **62**(5), 1796–1799 (1987)

[Mur92] Murray, J., Peruzzi, A., Abriata, J.P.: The Al-Zr (Aluminum-Zirconium) system. J. Phase Equilib. **13**(3), 277–291 (1992)

[Mey09] Meyer, C., Uhlenwinkel, V., Ristau, R., Jahn, P., Müller, H.R., Krug, P., Trojahn, W.: Thermal simulation of multilayer materials generated by spray forming. In: International Conference on Spray Deposition and Melt Atomization, Bremen, Germany, September, 2009 (CD Version) ISBN: 9783887227104

[Mac11] Macherauch, E., Zoch, H.-W.: Praktikum in Werkstoffkunde. Vieweg+Teubner, Wiesbaden (2011)

[Mus92] Musil, J., Fiala, J.: Plasma spray deposition of graded metal-ceramic coatings. Surf. Coat. Technol. **52**, 211–220 (1992)

[Ohr02] Ohring, M.: Material Science of Thin Films (second edition). Academic Press, San Diego, San Francisco, New York, Boston, London, Sydney, Tokyo (2002)

[Ost07] Ostermann, F.: Anwendungstechnologie Aluminium. Springer, Berlin, Heidelberg (2007)

[Peu05] Peukert, W., Schwarzer, H.-C., Stenger, F.: Control of aggregation in production and handling of nanoparticles. Chem. Eng. Proc. **44**, 245–252 (2005)

[Par01] Parsa, M.H., Yamaguchi, K., Takakura, N.: Redrawing analysis of aluminum-stainless-steel laminated sheet using FEM simulations and experiments. Int. J. Mech. Sci. **43**, 2331–2447 (2001)

[Roy05] Royset, J., Ryum; N.: Scandium in aluminum alloys. Int. Mater. Rev. **50**(1), 19–44 (2005)

[Sar10] Sarakinos, K., Alami, J., Konstantidinis, S.: High power pulsed magnetron sputtering: a review on scientific and engineering state of the art. Surf. Coat. Technol. **204**(11), 1661–1684 (2010)

[Son11] Song, M., He, Y., Fang, S.: Effects of Zr content on the yield strength of an Al–Sc alloy. J. Mater. Eng. Perform. **20**(3), 377–381 (2011)

[Sei14] Seidel, W., Hahn, F., Thoden, B.: Werkstofftechnik: Werkstoffe—Eigenschaften—Prüfung—Anwendung. Carl Hanser Verlag München (2014)

[Sto10] Stock, H.R., Köhler, B., Bomas, H., Zoch, H.-W.: Characteristics of aluminum-scandium alloy thin sheets obtained by physical vapour deposition. Mater. Des. **31**(Supplement 1), 76–86 (2010)

[Söl92] Sölter, H.-J., Müller, U., Lugscheider, E.: High-speed temperature measurement for on-line process control and quality assurance during plasma spraying: 1. Identification of important process parameters—measurement principle and potential as an industrial process control instrument. Powder Metall. Int. **24**(3), 169–174 (1992)

[Sun01] Sun, S., Pugh, M.: Interfacial properties in steel-steel composite materials. Mater. Sci. Eng. A **318**, 320–327 (2001)

[Tho86] Thornton, J.A.: The microstructure of sputter-deposited coating. J. Vac. Sci. Technol. **A4** (6), 3059–3065 (1986)

[Toe18] Toenjes, A., Kovac, J., Koehler, B., von Hehl, A., Mehner, A., Clausen, B., Zoch, H.-W.: Process chain for the fabrication of hardenable aluminium-zirconium micro-components by deep drawing. In: 5th International Conference on New Forming Technology (ICNFT2018), MATEC Web Conf. 190 15013 (2018)

[Wan96] Wang, R., Fichtborn, K.A.: Computer simulation of metal thin-film epitaxy. Thin Solid Films **272**, 223–228 (1996)

www.ingramcontent.com/pod-product-compliance
Lightning Source LLC
LaVergne TN
LVHW082114200926
848484LV00008B/1948
* 9 7 8 1 0 1 3 2 7 2 6 5 3 *